Contents

Preface

Thematic Semester on Probabilistic Methods in Geometry, Topology and Spectral Theory was held at Centre de Recherches Mathématiques in Montreal from August until December 2016. Probabilistic methods have played an increasingly important role in many areas of mathematics, from the study of random groups and random simplicial complexes in topology, to the theory of random Schrödinger operators in mathematical physics. The Thematic Semester at CRM (organized by the CRM Mathematical Analysis Laboratory and the CRM Probability Laboratory) included five intensive one-week workshops: *Frontiers in Mathematical Physics* workshop in Honour of Barry Simon's 70th Birthday; *Probabilistic Methods in Spectral Geometry and PDE* held on August 22–26, 2016; *Random Growth Problems and Random Matrices*; *Probabilistic Methods in Dynamical Systems and Applications*; and *Probabilistic Methods in Topology,* held on November 14-18. Workshops on Probabilistic Methods in Spectral Geometry and PDE, and on Probabilistic Methods in Topology are described below in more detail. The thematic semester featured an emphasis on interconnections and cross-fertilization of ideas between these topics, leading to new investigations and fruitful collaborations between participants. There were three Aisenstadt chairs during the semester: Yuval Peres, Scott Sheffield and Nalini Anantharaman. Nalini Anantharaman gave her lectures during the workshop on Probabilistic Methods in Spectral Geometry and PDE, whose participants contributed many papers to this volume. Other papers in this volume are related to the workshop on Probabilistic Methods in Topology, which included a series of introductory lectures by Matthew Kahle, and a lecture by Misha Gromov.

The workshop on Probabilistic Methods in Spectral Geometry and PDE was organized by Yaiza Canzani (Harvard), Linan Chen (McGill), Dmitry Jakobson (McGill), Armen Shirikyan (Cergy-Pontoise), Lior Silberman (UBC) and John A. Toth (McGill). The workshop brought together some of the leading researchers in quantum chaos, semi-classical theory, ergodic theory and dynamical systems, partial differential equations, probability, random matrix theory, mathematical physics, conformal field theory, and random graph theory. Its emphasis was on the use of ideas and methods from probability in different areas, such as quantum chaos (study of spectra and eigenstates of chaotic systems at high energy); geometry of random metrics and related problems in quantum gravity; solutions of partial differential equations with random initial conditions. The workshop was part of a series of related workshops held at the CRM and elsewhere, including a workshop on infinite-dimensional geometry (MSRI, December 2013), a workshop on the geometry of eigenvalues and eigenfunctions, and a workshop on manifolds of metrics and probabilistic methods in geometry and analysis (both held at the CRM in 2012).

Many of the talks concerned delocalization (equidistribution) of eigenfunctions at high energy, or the so-called quantum ergodicity. In particular this was the main topic of Nalini Anantharaman's first two Aisenstadt lectures. Quantum ergodicity is also the subject of three papers in the current volume.

In the paper *Quantum Limits of Eisenstein Series in* $\mathbb{H}^3$, Niko Laaksonen studies the quantum limits of Eisenstein series off the critical line for $\mathrm{PSL}_2(\mathcal{O}_K)$ where K is an imaginary quadratic field of class number one. He generalizes previous results of Petridis, Raulf and Risager on $\mathrm{PSL}_2(\mathbb{Z})\backslash\mathbb{H}^2$.

In the paper *Observability and Quantum Limits for the Schrödinger Equation on* $\mathbb{S}^d$, Fabricio Macià and Gabriel Rivière describe their recent results on semiclassical measures for the Schrödinger evolution on Zoll manifolds. The authors focus on the particular case of eigenmodes of the Schrödinger operator on the sphere endowed with its canonical metric. They also explore the relation of this problem with the observability question from control theory.

In the paper *Entropy Bounds and Quantum Unique Ergodicity for Hecke Eigenfunctions on Division Algebras,* Lior Silberman and Akshay Venkatesh prove the Arithmetic Quantum Unique Ergodicity conjecture for non-degenerate sequences of Hecke eigenfunctions on quotients $\Gamma\backslash G/K$, where $G \simeq PGL_d(\mathbb{R})$, K is a maximal compact subgroup of G and $\Gamma < G$ is a lattice associated to a division algebra over $\mathbb{Q}$ of prime degree d. More generally, they introduce a new method of proving positive entropy of quantum limits, which applies to higher-rank groups.

Several talks at the workshop were devoted to the study of "random" eigenfunctions; that subject was considered in three papers in this volume.

Linan Chen and Na Shu contributed the paper *A Geometric Treatment of Log-correlated Gaussian Fields.* One way to regularize a log-correlated Gaussian free field (GFF) is to consider (functionals of) its spherical averages. In even dimensions, this regularization approach has been adopted in the construction of the Liouville Quantum Gravity (LQG) measure and the proof of the Knizhnik-Polyakov-Zamolodchikov (KPZ) formula. In this article, the authors combine the Fourier-Bessel expansion with the spherical averages of the GFF to extend such a regularization approach to treat log-correlated GFFs in odd dimensions. In particular, the authors established the existence of the LQG measure and the KPZ formula under this setting.

In the paper *Tangent Nodal Sets for Random Spherical Harmonics,* Suresh Eswarathasan considers a fixed vector field V on $\mathbb{S}^2$ and studies the distribution of points which lie on the nodal set (of a random spherical harmonic) where V is also tangent. He shows that the expected value of the corresponding counting function is asymptotic to the eigenvalue with a leading coefficient that is independent of the vector field V. This demonstrates, in some form, a universality for vector fields up to lower order terms.

In the survey paper *Random Nodal Lengths and Wiener Chaos,* Maurizia Rossi collects some of the recent results on the "nodal geometry" of random eigenfunctions on Riemannian surfaces. She focuses on the asymptotic behaviour, for high energy levels, of the nodal length of Gaussian Laplace eigenfunctions on the torus (arithmetic random waves) and on the sphere (random spherical harmonics). She gives some insight on both Berry's cancellation phenomenon and the nature of nodal length second order fluctuations (non-Gaussian on the torus and Gaussian on the sphere) in terms of chaotic components. Finally she considers the general case of monochromatic random waves, i.e. Gaussian random linear combination of eigenfunctions of the Laplacian on a compact Riemannian surface with frequencies from a short interval, whose scaling limit is Berry's Random Wave Model.

Another subject discussed at the workshop was the study of random metrics and random maps. This is related to two papers in the current volume: the paper of Linan Chen and Na Shu described earlier; and the paper contributed by V. Konakov, S. Menozzi and S. Molchanov, titled *The Brownian Motion on $Aff(\mathbb{R})$ and Quasi-Local Theorems.* This paper is concerned with Random walk approximations of the Brownian motion on the Affine group $Aff(\mathbb{R})$. The authors are in particular interested in the case where the innovations are discrete. In this framework, the return probabilities of the walk have fractional exponential decay in large time, as opposed to the polynomial one of the continuous object. The authors prove that integrating those return probabilities on a suitable neighbourhood of the origin, the expected polynomial decay is restored.

The workshop on Probabilistic Methods in Topology was the last workshop held during the thematic semester on Probabilistic Methods in Geometry, Topology and Spectral Theory at CRM. It was organized by D. Wise (McGill), M. Pichot (McGill), L. Silberman (UBC), P. Przytycki (McGill), I. Rivin (Temple), A. Nabutovsky (Toronto), M. Kahle (Ohio State). It brought together researchers working on random simplicial complexes and geometry of spaces of triangulations (with connections to manifold learning); topological statistics, and geometric probability; theory of random groups and their properties; random knots; and other problems.

Many talks at the workshop concerned the study of random graphs, random simplicial complexes and their properties, including the talks by Kahle, Bubenik, Addario-Berry, Abu Fraiha, Bobrowski, Behrstock, Lishak, Farber, Peled and Luczak. That was also the subject of the last Aisenstadt lecture by Nalini Anantharaman, who discussed asymptotic equidistribution of Laplace eigenvectors on large graphs. Two of the papers in the current volume are devoted to the study of random graphs.

In the paper *Formal Zeta Function Expansion and the Frequency of Ramanujan Graphs,* Joel Friedman shows that logarithmic derivative of the Zeta function of any regular graph is given by a power series about infinity whose coefficients are given in terms of the traces of powers of the graph's Hashimoto matrix. He then makes a formal calculation that suggests that for fixed d there is an $f(d) > 1/2$ such that a d-regular graph on n vertices is Ramanujan with probability at least $f(d)$ for n sufficiently large. A related argument for random covering graphs of degree n over a fixed, regular "base graph" suggests that for n large, a strict majority of random covering graphs are relatively Ramanujan.

The talks by Rivin and Even-Zohar concerned the study of random knots. An important tool in the study of knots are knot polynomials. The study of those polynomials is related to the study of Tutte (dichromatic) polynomials of graphs; and to the study of Bollobás-Riordan polynomials of *ribbon graphs* (graphs that are cellularly embedded into surfaces). In their paper *Rank and Bollobás-Riordan Polynomials: Coefficient measures and zeros*, D. Jakobson, T. Langsetmo, I. Rivin and L. Turner discuss some (numerical and theoretical) results about the coefficients and zeros of Tutte (dichromatic) polynomial of graphs of bounded degree whose size increases. In particular, certain natural coefficient measures converge to a delta-function, provided that the corresponding graphs converge in the sense of Benjamini-Schramm. The authors also establish related results for Bollobás-Riordan polynomials for large ribbon graphs.

Yaiza Canzani
Linan Chen
Dmitry Jakobson

Contemporary Mathematics
Volume **739**, 2019
https://doi.org/10.1090/conm/739/14891

A geometric treatment of log-correlated Gaussian free fields

Linan Chen and Na Shu

ABSTRACT. One way to regularize a log-correlated Gaussian free field (GFF) is to consider (functionals of) its spherical averages. In even dimensions, this regularization approach has been adopted in the construction of the Liouville Quantum Gravity (LQG) measure and the proof of the Knizhnik-Polyakov-Zamolodchikov (KPZ) formula (e.g. Chen and Jakobson, Ann. Henri Poincaré 15 (2014), pp 1245-1283 and Duplantier, Rhodes, Sheffield, and Vargas, Invent. Math. 185 (2011) pp 333-393). In this article, we combine the Fourier-Bessel expansion with the spherical averages of the GFF to extend such a regularization approach to treat log-correlated GFFs in odd dimensions. We also outline the proofs of the existence of the LQG measure and the KPZ formula under this setting.

1. Introduction

Many recent developments in statistical physics and probability theory have seen *Gaussian Free Field* (GFF) as an indispensable tool. Heuristically speaking, GFFs are analogues of the Brownian motion with multidimensional time parameters. Just as the Brownian motion is thought of as a natural interpretation of "random curve", GFFs are considered as promising candidates for modeling "random surfaces" or "random manifolds", which ultimately lead to the study of random geometry. Motivated by their importance, GFFs have been widely studied both in discrete and continuum settings. For example, a rich literature has been established on the geometry of the discrete log-correlated GFF on two-dimensional lattice, such as the distribution of the extrema and the size of the level sets of the GFF (e.g., see [**4**, **9**, **10**] and references therein). However, in continuum settings it is more challenging to obtain analytical results on the geometry of GFFs due to the *singularity* of the GFFs. More specifically, the log-correlated GFF on $\mathbb{R}^n$ (or a subset of $\mathbb{R}^n$), for $n \geq 2$, consists of tempered distributions, which means that if Θ is such a field and $\theta \in \Theta$ is a generic element in Θ, then "$\theta(x)$" is not defined for every $x \in \mathbb{R}^n$. In fact, as we will make rigorous in the next section, when viewed formally as a centered Gaussian family, the two-point covariance function of the values of θ has a logarithmic singularity along the diagonal. This is why we refer to such GFFs as *log-correlated* GFFs. To overcome the challenge posed by the singularity, we need to apply to θ a procedure known as *regularization* in physics literature.

2010 *Mathematics Subject Classification.* Primary 28C20, 42C99, 46E35, 60G57, 60G60.

Key words and phrases. Gaussian free field, Fourier-Bessel expansion, regularization, random measure, the KPZ formula.

The authors are partially supported by the NSERC Discovery Grant (No. 241023).

Various regularization procedures have been considered in the study of problems related to the geometry of log-correlated GFFs. One commonly used regularization procedure is to average θ over some sufficiently "nice" Borel sets near a given point x. This is the main idea that will be adopted in this article. Although it is a tempered distribution, θ can be integrated over sufficiently regular submanifolds, such as spheres. Since averaging is the natural way to "tame" the singularity of a tempered distribution, such a geometric regularization procedure is a natural choice in the study of the geometry of singular GFFs. For example, for Θ being the 2D log-correlated GFF, via averaging θ over circles centered at x, Duplantier and Sheffield (**[12]**) constructed a random measure formally represented as "$e^{\gamma\theta(x)}dx$" where dx is the 2D Lebesgue measure and $\gamma > 0$ is a physics parameter. This random measure has its origin in statistical physics and quantum field theory, and is referred to as the *Liouville Quantum Gravity* (LQG) measure. Further, the authors of **[12]** provided the first mathematically rigorous proof of the *Knizhnik-Polyakov-Zamolodchikov* (KPZ) formula which is a well celebrated conjecture proposed by the three named physicists in 1988 (**[17]**). Many other works on GFFs have been established based on the circle or sphere averaging regularization. Again for Θ being the 2D log-correlated GFF, Hu, Miller and Peres (**[15]**) adopted the circular averages of θ to introduce the notion of "thick points", which, analogous to the notion of extrema for discrete GFFs, are locations where θ achieves "unusually" large values. The authors of **[15]** further determined the Hausdorff dimensions of the sets consisting of thick points. There are similar results in higher dimensions. Chen and Jakobson (**[6]**) treated the log-correlated GFFs on $\mathbb{R}^{2n}$ for $n \geq 2$, where they devised a regularization procedure based on a functional of the spherical averages of θ, and extended the construction of the LQG measure and the proof of the KPZ formula to $\mathbb{R}^{2n}$. Cipriani and Hazra (**[7]**) used the same regularization as that in **[6]** to extend the study of thick point sets of the log-correlated GFF from 2D to 4D. In the next section we will review the properties of the circular/spherical averages of the log-correlated GFFs in even dimensions.

Another regularization procedure that has led to fruitful results on GFFs is the theory of Gaussian Multiplicative Chaos (GMC). The theory of GMC was first introduced by Kahane in 1985 (**[16]**), where he considered centered Gaussian families whose covariance functions are $\sigma-$positive definite. In fact, it was already clear in **[16]** that the GMC model leads to a random measure which is essentially equivalent to the LQG measure mentioned above, and such a measure is supported on a Borel set where the GMC instance achieves "unusually" large values. More recently, the theory of GMC was revived by a series of works (e.g., an incomplete list includes **[1, 4, 13, 19–21]**). In particular, under the assumption that the covariance function is $\sigma-$positive definite, Rhodes and Vargas (**[20]**) interpreted the log-correlated GFF in any dimension as a GMC model and conducted the multi-fractal analysis of the LQG measure, which also lead to a proof of the KPZ formula. Again in the GMC setting, Rhodes and Vargas (**[21]**) further determined the Hausdorff dimension of the support of the LQG measure, which can be identified as a thick point set. Since the GMC theory is not the focus of this article, we will not delve into the details here and refer interested readers to the works cited above.

Another approach that has been adopted to treat general Gaussian random functions on a manifold is to express the random function in the basis of eigenfunctions corresponding to a certain differential operator on the manifold, with the

CONTEMPORARY MATHEMATICS

739

Centre de Recherches Mathématiques Proceedings

Probabilistic Methods in Geometry, Topology and Spectral Theory

CRM Workshops

Probabilistic Methods in Spectral Geometry PDE
August 22–26, 2016

Probabilistic Methods in Topology
November 14–18, 2016
Centre de Recherches Mathématiques,
Université de Montréal, Québec, Canada

Yaiza Canzani
Linan Chen
Dmitry Jakobson
Editors

AMERICAN MATHEMATICAL SOCIETY

2010 *Mathematics Subject Classification.* Primary 05C80, 11F72, 33C55, 35P20, 58J51, 58J65, 60C05, 60G15, 60G60, 81Q50.

Library of Congress Cataloging-in-Publication Data

Names: Canzani, Yaiza, 1987- editor. | Chen, Linan, 1984- editor. | Jakobson, Dmitry, 1970- editor.

Title: Probabilistic methods in geometry, topology, and spectral theory / Yaiza Canzani, Linan Chen, Dmitry Jakobson, editors.

Description: Providence, Rhode Island : American Mathematical Society ; Montreal, Quebec, Canada : Centre de Recherches Mathematiques, 2019. | Series: Contemporary mathematics, 0271-4132 ; volume 739 | "CRM workshop on probabilistic methods in spectral geometry and PDE, August 22-26, 2016 [and] probabilistic methods in topology, November 22-26, 2016." | Includes bibliographical references.

Identifiers: LCCN 2019023076 | ISBN 9781470441456 (paperback)

Subjects: LCSH: Mathematical physics–Congresses. | Probabilities–Congresses. | Geometric analysis–Congresses. | Topology–Congresses. | Spectral theory (Mathematics)–Congresses. | AMS: Combinatorics {For finite fields, see 11Txx} – Graph theory {For applications of graphs, see 68R10, 81Q30, 81T15, 82B20, 82C20, 90C35, 92E10, 94C15} – Random graphs [See also 60B20]. | Special functions (33-XX deals with the properties of functions as functions) {For orthogonal functions, see 42Cxx; for aspects of combinatorics see 05Axx; for number-theoretic aspects see 11-XX; for} | Partial differential equations – Spectral theory and eigenvalue problems [See also 47Axx, 47Bxx, 47F05] – Asymptotic distribution of eigenvalues and eigenfunctions. | Global analysis, analysis on manifolds [See also 32Cxx, 32Fxx, 32Wxx, 46-XX, 47Hxx, 53Cxx] {For geometric integration theory, see 49Q15} – Partial differential equations on manifolds; differential op | Probability theory and stochastic processes {For additional applications, see 11Kxx, 62-XX, 90-XX, 91-XX, 92-XX, 93-XX, 94-XX} – Combinatorial probability – Combinatorial probability. | Probability theory and stochastic processes {For additional applications, see 11Kxx, 62-XX, 90-XX, 91-XX, 92-XX, 93-XX, 94-XX} – Stochastic processes – Gaussian processes. | Probability theory and stochastic processes {For additional applications, see 11Kxx, 62-XX, 90-XX, 91-XX, 92-XX, 93-XX, 94-XX} – Stochastic processes – Random fields. | Quantum theory – General mathematical topics and methods in quantum theory – Quantum chaos [See also 37Dxx].

Classification: LCC QC20 .P7564 2019 | DDC 519.2–dc23

LC record available at https://lccn.loc.gov/2019023076

DOI: https://doi.org/10.1090/conm/739

∞ The paper used in this book is acid-free and falls within the guidelines
established to ensure permanence and durability.
Visit the AMS home page at `https://www.ams.org/`

10 9 8 7 6 5 4 3 2 1 24 23 22 21 20 19

coefficients being independent copies of standard Gaussian random variables. Such an approach sees natural applications in the construction of random geometry in the manifold setting. In particular, it has generated various studies on geometric properties, such as the volume, the curvature, etc., when the underlying Riemannian metric is perturbed by a random series of eigenfunctions (see, e.g, [**2**, **3**, **8**]).

In this work, we study the regularization of the log-correlated GFF in arbitrary dimensions based on the spherical averages of the GFF. In particular, we hope to extend the circle/sphere averaging approach in $\mathbb{R}^{2n}$, as adopted in [**6**, **7**, **12**, **15**], to $\mathbb{R}^{2n+1}$ for $n \geq 1$. In §2 we will briefly review the theory of abstract Wiener space, which allows us to give mathematically rigorous interpretations of log-correlated GFFs in all dimensions. In §3 we will discuss the properties of the sphere averaging regularization. In particular, we will see that the spherical averages of the log-correlated GFF on $\mathbb{R}^{2n}$ have favorable properties as approximations of pointwise values of the GFF. However, such properties no longer hold in $\mathbb{R}^{2n+1}$ because, since the (pseudo-)differential operator associated with the log-correlated GFF on $\mathbb{R}^{2n+1}$ is non-local, only averaging locally is not sufficient to "capture" the behaviors of the GFF. To resolve this issue, we take advantage of the connection between the sphere averaging action and the Bessel functions, invoke tools of the special functions such as the *Fourier-Bessel expansion*, and devise a regularization based on "non-local" spherical averages of the GFF. By doing so, we obtain a regularization which recovers all the favorable properties as those in the even-dimensional setting. To further demonstrate that such a regularization makes a good candidate in studying the geometry of GFFs, we plan to adopt it to reproduce the construction of the LQG measure and the proof of the KPZ formula in odd dimensions. In §4 we outline the strategy and the main steps to achieve this goal in the three-dimensional setting.

The motivations of developing such a regularization approach are multi-fold: first, as we mentioned earlier, averaging a generic element θ of a singular GFF over the spheres centered at x is a natural approach to approximate "$\theta(x)$" and further to study the behaviors of θ near x; secondly, the action of taking the spherical averages can be carried out for general models of GFFs, and in particular we do not have to rely on the GMC theory and impose on the covariance function the strong constraint of being $\sigma-$positive definite; thirdly, although in general it remains open as whether the establishment of those results on the geometry of GFF is intrinsic or dependent on the choice of regularization, the method we propose offers a potential way of linking various regularizations to the one based on the sphere averages, and in turn we may be able to compare and connect different regularizations.

The article is a brief report on our ongoing work. Some claims are presented without proofs either for the proofs being straightforward but lengthy computations (such as Lemma 3.1 and Theorem 3.2), or because the claims are still being investigated (such as Theorem 4.2). We also plan to study further questions related to the regularization introduced in this work, e.g., "what is the relation between this regularization and that adopted in [**6**] for the log-correlated GFFs in even dimensions?", or "can this regularization be adapted to GFFs in non-Euclidean settings, such as GFFs on a manifold?" We will present a more complete exposition of our work, including the full proofs of all the results, in a separate article that will appear soon.

2. Abstract Wiener Space and Gaussian Free Fields

We begin with a brief review of the theory of abstract Wiener space, introduced by Gross in 1965 ([**14**]) as a mathematical construction of infinite dimensional Gaussian measures. Let Θ be a separable Banach space, H be a separable Hilbert space which is continuously embedded in Θ as a dense subspace, and $\mathcal{W}$ be a probability measure on Θ. The triple $(H, \Theta, \mathcal{W})$ is said to be an *Abstract Wiener Space* (*AWS*) if $\mathcal{W}$ has the characteristic function

$$\widehat{\mathcal{W}}(\lambda^*) := \mathbb{E}^{\mathcal{W}}\left[e^{i\langle \cdot,\lambda^*\rangle}\right] = \exp\left(-\frac{\|h_{\lambda^*}\|_H^2}{2}\right) \text{ for all } \lambda^* \in \Theta^*,$$

where Θ^* is the dual space of Θ, i.e., the space of continuous linear functionals on Θ, $\langle\ ,\ \rangle$ indicates the action between the element in Θ and the element in Θ^*, and h_{λ^*} is the unique element in H such that $(g, h_{\lambda^*})_H = \langle g, \lambda^*\rangle$ for every g in H. This probability measure $\mathcal{W}$, known as the *Wiener measure*, is centered, non-degenerate and locally finite with $\mathcal{W}(H) = 0$ if Θ is infinite dimensional. H is referred to as the *Cameron-Martin space*, whose inner product determines the covariance structure of $\mathcal{W}$. The theory of AWS guarantees that given any separable infinite dimensional Hilbert space H, there exists a separable Banach space Θ and a Wiener measure $\mathcal{W}$ on Θ such that $(H, \Theta, \mathcal{W})$ forms an AWS. In addition, there exists a unique linear isometry $\mathcal{I} : H \to L^2(\mathcal{W})$ such that $\mathcal{I}(h_{\lambda^*}) = \langle \cdot, \lambda^*\rangle$ for every $\lambda^* \in \Theta^*$, and $\{\mathcal{I}(h) : h \in H\}$ forms a centered Gaussian family in $L^2(\mathcal{W})$ with the covariance function

$$\mathbb{E}^{\mathcal{W}}[\mathcal{I}(h_1)\mathcal{I}(h_2)] = (h_1, h_2)_H \text{ for every } h_1, h_2 \in H.$$

We call $\mathcal{I}$ the *Payley-Wiener map* and its image $\mathcal{I}(h)$ a *Paley Wiener integral.* In particular, if $\{h_m : m \geq 1\}$ is an orthonormal basis of H, then $\{\mathcal{I}(h_m) : m \geq 1\}$ is a family of independent standard Gaussian random variables. Moreover, the series $\sum_{m=0}^{\infty} \mathcal{I}(h_m) h_m$ converges $\mathcal{W}$-almost surely in Θ as well as in $L^p(\mathcal{W})$ for every $p \in [1, \infty)$.

The framework of AWS provides a rigorous mathematical foundation for the construction of Gaussian random fields in arbitrary dimensions. One random field that will be explored in this article is the *Gaussian free field* (GFF), which can be seen as a natural analogue of the Brownian motion with multidimensional time parameters. There are various ways of introducing or interpreting GFFs. In this article, we adopt the perspective that a GFF can be seen as an AWS whose Cameron-Martin space is a certain Sobolev space. To be specific, consider the Sobolev space $H^s := H^s(\mathbb{R}^n)$ with $s \geq 1$ and $n \in \mathbb{N}$, which is the completion of the Schwartz function space $\mathcal{S}(\mathbb{R}^n)$ under the inner product

$$(f, g)_{H^s} := ((I - \Delta)^s f, g)_{L^2} = \frac{1}{(2\pi)^n} \int_{\mathbb{R}} \left(1 + |\xi|^2\right)^s \hat{f}(\xi)\overline{\hat{g}(\xi)} d\xi \tag{2.1}$$

for every $f, g \in \mathcal{S}(\mathbb{R}^n)$. Then, as mentioned before, there exists a separable Banach space $\Theta^s := \Theta^s(\mathbb{R}^n)$ and a probability measure $\mathcal{W}^s := \mathcal{W}^s(\mathbb{R}^n)$ on Θ^s such that $(H^s, \Theta^s, \mathcal{W}^s)$ forms an AWS. Let $\mathcal{B}_{\Theta^s}$ be the Borel σ-algebra of Θ^s. We refer to the probability space $(\Theta^s, \mathcal{B}_{\Theta^s}, \mathcal{W}^s)$ as the *dim-n order-s GFF*[1]. It is easy to see

[1]In physics literature, the notion "Gaussian free field" only refers to the case when $s = 1$. Here we will extend the terminology and continue using it when $s > 1$.

that, if $\lambda \in H^{-s}$, then $h_\lambda := (I-\Delta)^{-s}\lambda \in H^s$ and the Paley-Wiener integrals $\{\mathcal{I}(h_\lambda) : \lambda \in H^{-s}\}$ form a centered Gaussian family with the covariance function

$$\mathbb{E}^{\mathcal{W}^s}\left[\mathcal{I}(h_{\lambda_1})\mathcal{I}(h_{\lambda_2})\right] = (h_{\lambda_1}, h_{\lambda_2})_{H^s} = (\lambda_1, \lambda_2)_{H^{-s}} \text{ for every } \lambda_1, \lambda_2 \in H^{-s}. \tag{2.2}$$

Clearly the covariance structure of the Gaussian family above is determined by the integral kernel or the Green's function $G(x,y)$ of the operator $(I-\Delta)^s$.

Based on a straightforward computation of the Green's function $G(x,y)$, one can see that, when $s \le \frac{n}{2}$, $G(x,y)$ is singular along the diagonal, which means that a generic element of Θ^s is a generalized function (or in other words, a tempered distribution) which may not be pointwisely defined. In particular, when $s = \frac{n}{2}$, the Green's function of $(I-\Delta)^{\frac{n}{2}}$ on $\mathbb{R}^n$ has a *logarithmic singularity* and hence the corresponding GFF is logarithmically correlated; the dim-n order-$\frac{n}{2}$ GFF is also referred to as the *dim-n log-correlated GFF* or the *log-correlated GFF on* $\mathbb{R}^n$. When $s \in \frac{1}{2}\mathbb{N}$ and $s < \frac{n}{2}$, the Green's function of $(I-\Delta)^s$ on $\mathbb{R}^n$ has a *polynomial singularity* with degree $n-2s$ and thus the corresponding GFF is polynomially correlated. In this article, we will focus on treating the log-correlated GFFs, i.e., $s = \frac{n}{2}$. For further discussions of polynomially correlated GFFs, we refer interested readers to [**5**].

REMARK 2.1. There are two types of models that are generally considered in the study of GFFs: one is the *massless GFF*, which is the model associated with Δ^s (e.g., [**12**, **15**]), and the other is the *massive GFF* where the operator is $(I-\Delta)^s$, as we introduced above. Here we adopt the massive GFF model instead of the massless one mainly for technical reasons. The GFF associated with Δ^s has to be defined on a bounded domain, while the GFF associated with $(I-\Delta)^s$ can be defined on the entire $\mathbb{R}^n$. With the GFF defined on $\mathbb{R}^n$, we can use the Fourier transform tools to compute the covariance function in (2.2), and we do not have to worry about any boundary condition[2] in the computations. Furthermore, given s and n, since the Green's function of $(I-\Delta)^s$ on $\mathbb{R}^n$ and that of Δ^s on an $n-$dimensional domain have exactly the same singularity along the diagonal, the corresponding massive GFF and massless GFF should exhibit the same local behaviors. In particular, we expect that the results discussed in this article on the asymptotics of the regularization of the GFF will still apply if the massive GFF is replaced by the massless GFF.

3. Regularization of GFF on $\mathbb{R}^n$

As we have pointed out in the previous section, when $s = \frac{n}{2}$, a generic GFF element is only a generalized function. Therefore, in order to treat the GFF analytically, we need to carry out a regularization procedure. Various regularization approaches have been adopted in the study of singular GFFs. The common approaches, as we have mentioned in the Introduction, include the GMC theory, the eigenfunction expansion with random coefficients, and the approach of taking geometric averages. In this article, we will adopt a regularization based on spherical averages of the GFF.

[2]With the massless GFF model, it is often complicated to choose the proper boundary condition for Δ^s, especially when $s > 1$.

3.1. Spherical Averages of GFF on $\mathbb{R}^n$. Consider the Hilbert space $H^{\frac{n}{2}} := H^{\frac{n}{2}}(\mathbb{R}^n)$ with $n \geq 2$. As mentioned before, there exists a separable Banach space $\Theta^{\frac{n}{2}} := \Theta^{\frac{n}{2}}(\mathbb{R}^n)$ and a Gaussian measure $\mathcal{W}^{\frac{n}{2}} := \mathcal{W}^{\frac{n}{2}}(\mathbb{R}^n)$ such that the triple $\left(H^{\frac{n}{2}}, \Theta^{\frac{n}{2}}, \mathcal{W}^{\frac{n}{2}}\right)$ forms an AWS, and $\theta \in \Theta^{\frac{n}{2}}$ sampled under $\mathcal{W}^{\frac{n}{2}}$ is a generic element of the dim-n log-correlated GFF. Given $x \in \mathbb{R}^n$ and $\varepsilon > 0$, we denote by σ_ε^x the spherical average measure over $\partial B_\varepsilon(x)$[3], and compute its Fourier transform as

$$\hat{\sigma_\varepsilon^x}(\xi) = \frac{2^{\frac{n-2}{2}}\Gamma\left(\frac{n}{2}\right)}{(\varepsilon|\xi|)^{\frac{n-2}{2}}} J_{\frac{n-2}{2}}(\varepsilon|\xi|)\, e^{i(x,\xi)_{\mathbb{R}^n}} \text{ for every } \xi \in \mathbb{R}^n. \tag{3.1}$$

where $J_\nu(\cdot)$, $\nu \in \mathbb{R}$, refers to the Bessel function (of the first kind) of order ν. It is easy to check that, for every $x \in \mathbb{R}^n$ and every $\varepsilon > 0$, $\sigma_\varepsilon^x \in H^{-\frac{n}{2}}$ and hence $h_{\sigma_\varepsilon^x} := (I - \Delta)^{-\frac{n}{2}} \sigma_\varepsilon^x \in H^{\frac{n}{2}}$, which implies that the spherical average of the GFF $\theta \mapsto \mathcal{I}\left(h_{\sigma_\varepsilon^x}\right)(\theta)$ is well defined as a Gaussian random variable. Since $\sigma_\varepsilon^x \to \delta_x$ as $\varepsilon \downarrow 0$ in the sense of tempered distributions, we can view the random variable $\mathcal{I}\left(h_{\sigma_\varepsilon^x}\right)(\theta)$ as an approximation of $\theta(x)$ when ε is small. One can easily check that for $x \in \mathbb{R}^n$, the covariance function of $\left\{\mathcal{I}\left(h_{\sigma_\varepsilon^x}\right) : \varepsilon > 0\right\}$, based on (2.1) and (2.2), is given by

$$\mathbb{E}^{\mathcal{W}}\left[\mathcal{I}\left(h_{\sigma_{\varepsilon_1}^x}\right)\mathcal{I}\left(h_{\sigma_{\varepsilon_2}^x}\right)\right] = \frac{\Gamma\left(\frac{n}{2}\right)}{2\pi^{\frac{n}{2}}(\varepsilon_1\varepsilon_2)^{\frac{n-2}{2}}} \int_0^\infty \frac{r}{(1+r^2)^{\frac{n}{2}}} J_{\frac{n-2}{2}}(\varepsilon_1 r)\, J_{\frac{n-2}{2}}(\varepsilon_2 r)\, dr \tag{3.2}$$

for every $\varepsilon_1 \geq \varepsilon_2 > 0$. In fact, under certain circumstances, the right hand side of (3.2) can be evaluated explicitly. In later discussions, $I_\nu(\cdot)$ and $K_\nu(\cdot)$, $\nu \in \mathbb{R}$, refer to the modified Bessel functions of order ν (see, e.g. §3.7 of [**22**]).

First, let $n = 2$ and Θ be the dim-2 log-correlated GFF. Given $x \in \mathbb{R}^2$ and $\varepsilon_1 \geq \varepsilon_2 > 0$, we have that

$$\mathbb{E}^{\mathcal{W}}\left[\mathcal{I}\left(h_{\sigma_{\varepsilon_1}^x}\right)\mathcal{I}\left(h_{\sigma_{\varepsilon_2}^x}\right)\right] = \frac{1}{2\pi} K_0(\varepsilon_1)\, I_0(\varepsilon_2)$$

(see, e.g., Appendix of [**6**]). An immediate consequence of the formula above is that, for every $\varepsilon_1 \geq \varepsilon_2 \geq \varepsilon_3 > 0$,

$$\mathbb{E}\left[\mathcal{I}\left(h_{\sigma_{\varepsilon_1}^x}\right)\mathcal{I}\left(h_{\sigma_{\varepsilon_3}^x}\right)|\mathcal{I}\left(h_{\sigma_{\varepsilon_2}^x}\right)\right] = \mathbb{E}\left[\mathcal{I}\left(h_{\sigma_{\varepsilon_1}^x}\right)|\mathcal{I}\left(h_{\sigma_{\varepsilon_2}^x}\right)\right]\mathbb{E}\left[\mathcal{I}\left(h_{\sigma_{\varepsilon_3}^x}\right)|\mathcal{I}\left(h_{\sigma_{\varepsilon_2}^x}\right)\right],$$

which means that $\mathcal{I}\left(h_{\sigma_{\varepsilon_1}^x}\right)$ and $\mathcal{I}\left(h_{\sigma_{\varepsilon_3}^x}\right)$ are conditionally independent, conditioning on $\mathcal{I}\left(h_{\sigma_{\varepsilon_2}^x}\right)$. This is equivalent to saying that $\left\{\mathcal{I}\left(h_{\sigma_\varepsilon^x}\right) : \varepsilon > 0\right\}$ is a backward Markov Gaussian process (see, e.g., [**11**]). Furthermore, the formula of the covariance function leads to a natural normalization of the spherical average. Namely, if we define

$$\tilde{\theta}_\varepsilon(x) := \frac{\mathcal{I}\left(h_{\sigma_\varepsilon^x}\right)}{I_0(\varepsilon)} \text{ for } x \in \mathbb{R}^2 \text{ and } \varepsilon > 0,$$

then, for every $x \in \mathbb{R}^2$, the Gaussian process $\left\{\tilde{\theta}_\varepsilon(x) : \varepsilon > 0\right\}$ has the same distribution as the Brownian motion up to a deterministic time change. Meanwhile,

[3]Throughout the article, we denote by $B_\varepsilon(x)$ the open ball centered at x with radius ε and by $\partial B_\varepsilon(x)$ the sphere centered at x with radius ε.

$\lim_{\varepsilon\downarrow 0} I_0(\varepsilon) = 1$ and hence $\tilde{\theta}_\varepsilon(x)$ can still be viewed as an approximation of $\theta(x)$ when ε is small.

However, the (backward) Markov property fails for the spherical averages of the log-correlated GFF in higher dimensions. Now let $n = 4$ and Θ be the dim-4 log-correlated GFF. It is found in [**6**] that given $x \in \mathbb{R}^4$ and $\varepsilon_1 \geq \varepsilon_2 > 0$,

$$\mathbb{E}^{\mathcal{W}}\left[\mathcal{I}\left(h_{\sigma^x_{\varepsilon_1}}\right)\mathcal{I}\left(h_{\sigma^x_{\varepsilon_2}}\right)\right] = -\frac{1}{4\pi^2}\left[K_1'(\varepsilon_1)\frac{I_1(\varepsilon_2)}{\varepsilon_2} + I_1'(\varepsilon_2)\frac{K_1(\varepsilon_1)}{\varepsilon_1}\right]. \tag{3.3}$$

As one can see, the covariance function in (3.3) is no longer "separable" with respect to the radii ε_1 and ε_2, and hence $\left\{\mathcal{I}\left(h_{\sigma^x_\varepsilon}\right) : \varepsilon > 0\right\}$ no longer possesses the Markov property. Heuristically speaking, one can view the cause of this matter as being that the spherical average of the GFF alone does not contain enough information for the process to be Markovian. To overcome this matter, one may take one more piece of information given by the derivative $d\sigma^x_\varepsilon := \frac{d}{d\varepsilon}\sigma^x_\varepsilon$ of the average with respect to the radius, and consider $\left\{\left(\mathcal{I}\left(h_{\sigma^x_\varepsilon}\right), \mathcal{I}\left(h_{d\sigma^x_\varepsilon}\right)\right)^T : \varepsilon > 0\right\}$ as a vector-valued Gaussian process. We define

$$\mathbf{C}(\varepsilon) := \begin{pmatrix} \frac{I_1(\varepsilon)}{\varepsilon} & I_1'(\varepsilon) \\ \frac{I_2(\varepsilon)}{\varepsilon} & I_1''(\varepsilon) \end{pmatrix} \text{ and } \mathbf{V}^x_\varepsilon := \begin{pmatrix} \mathcal{I}\left(h_{\sigma^x_\varepsilon}\right) \\ \mathcal{I}\left(h_{d\sigma^x_\varepsilon}\right) \end{pmatrix} \text{ for } x \in \mathbb{R}^4 \text{ and } \varepsilon > 0.$$

It is shown in [**6**] that if we normalize the vector by setting

$$\mathbf{U}^x_\varepsilon := (\mathbf{C}(\varepsilon))^{-1}\mathbf{V}^x_\varepsilon,$$

then, for every $x \in \mathbb{R}^4$, the Gaussian family $\{\mathbf{U}^x_\varepsilon : \varepsilon > 0\}$ is a backward Markov Gaussian process. Furthermore, it leads to a linear combination of $\mathcal{I}\left(h_{\sigma^x_\varepsilon}\right)$ and $\mathcal{I}\left(h_{d\sigma^x_\varepsilon}\right)$ that becomes a suitable regularization of θ. To be specific, we can find two explicit functions f_1 and f_2 on $[0,\infty)$, such that if

$$\tilde{\theta}_\varepsilon(x) := f_1(\varepsilon)\mathcal{I}\left(h_{\sigma^x_\varepsilon}\right)(\theta) + f_2(\varepsilon)\mathcal{I}\left(h_{d\sigma^x_\varepsilon}\right)(\theta) \text{ for } x \in \mathbb{R}^4 \text{ and } \varepsilon > 0,$$

then, for every $x \in \mathbb{R}^4$, $\left\{\tilde{\theta}_\varepsilon(x) : \varepsilon > 0\right\}$ has the same distribution as the Brownian motion up to a deterministic time change.

Similarly, when Θ is the dim-$2n$ log-correlated GFF, $\{\mathcal{I}(h_{\sigma^x_\varepsilon}) : \varepsilon > 0\}$ fails to possess the Markov property. But one can extend the approach in [**6**] to this case by taking

$$\mathbf{V}^x_\varepsilon := \left(\mathcal{I}\left(h_{\sigma^x_\varepsilon}\right), \mathcal{I}\left(h_{d\sigma^x_\varepsilon}\right), \cdots, \mathcal{I}\left(h_{d^{\nu-1}\sigma^x_\varepsilon}\right)\right)^T \text{ for } x \in \mathbb{R}^{2n} \text{ and } \varepsilon > 0.$$

Then, a suitable regularization of θ can be obtained by a linear combination of $\mathcal{I}\left(h_{\sigma^x_\varepsilon}\right)$ and all its derivatives up to the order of $n-1$, following a similar procedure as that in $\mathbb{R}^4$.

It is clear that in even dimensions, by considering the spherical average and its derivatives, one can obtain a regularized GFF whose concentric family behaves just like a Brownian motion upon changing the time. However, this property no longer holds in odd dimensions. In $\mathbb{R}^{2n+1}$, a careful analysis shows that the covariance function of the spherical averages, as defined in (3.2), does not produce finitely many terms in the form of $a(\varepsilon_1)b(\varepsilon_2)$. For example, in $\mathbb{R}^3$ we have that, given $x \in \mathbb{R}^3$ and $\varepsilon_1 \geq \varepsilon_2 > 0$,

$$\mathbb{E}^{\mathcal{W}}\left[\mathcal{I}\left(h_{\sigma^x_{\varepsilon_1}}\right)\mathcal{I}\left(h_{\sigma^x_{\varepsilon_2}}\right)\right] = \frac{1}{4\pi^2}\left[\frac{\varepsilon_2 - \varepsilon_1}{\varepsilon_1\varepsilon_2}K_1(\varepsilon_1 - \varepsilon_2) - \frac{\varepsilon_2 + \varepsilon_1}{\varepsilon_1\varepsilon_2}K_1(\varepsilon_1 + \varepsilon_2)\right].$$

We expect that in higher odd dimensions, the formula of the covariance function contains similar structures. Specifically, not only the covariance function of spherical averages of the log-correlated GFF fails to be separable with respect to the radii, but it fails in a worse way than the situation in even dimensions, which makes it impossible to obtain a suitable regularization of the GFF based on the spherical averages and its derivatives of any finitely many orders. In our opinion, such a phenomenon is due to the fact that the operator $(I-\Delta)^{\frac{2n+1}{2}}$ associated with the log-correlated GFF on $\mathbb{R}^{2n+1}$ is non-local. Therefore, in order to obtain a suitable regularization to approximate the value of θ at a point x, we should gather information of θ from the entire space, instead of only exploring the local region near x. In particular, one of the possibilities is to examine the averages of θ over infinitely many spheres that propagate to infinity. We will connect the Fourier-Bessel expansion of certain function f with the averages of θ over infinitely many spheres centered at x whose radii increase to infinity. To this end, we consider actions that take the following form:

$$\mu_\varepsilon^x = \sum_{m=1}^{\infty} c_m \sigma_{r_m(\varepsilon)}^x \text{ for } x \in \mathbb{R}^n \text{ and } \varepsilon > 0,$$

where $r_m(\varepsilon) \uparrow \infty$ as $m \uparrow \infty$ for every $\varepsilon > 0$, and $\{c_m : m \geq 1\} \subseteq (0,\infty)$ are some proper coefficients.

3.2. Fourier-Bessel Series with Spherical Averages of the log-correlated GFF on $\mathbb{R}^3$. From now on, we will restrict our discussions to $\mathbb{R}^3$, but similar methods and arguments apply to any dimension. We write $\Theta := \Theta^{\frac{3}{2}}(\mathbb{R}^3)$ and $\mathcal{W} := \mathcal{W}^{\frac{3}{2}}(\mathbb{R}^3)$ for short, and $(\Theta, \mathcal{B}_\Theta, \mathcal{W})$ is the log-correlated GFF on $\mathbb{R}^3$. Let $x \in \mathbb{R}^3$ and $\varepsilon > 0$. Assume that

$$\mu_\varepsilon^x = \sum_{m=1}^{\infty} c_m \sigma_{j_m \varepsilon}^x,$$

where $j_m = m\pi$ for $m \geq 1$ are the positive zeros of the Bessel function $J_{\frac{1}{2}}$ in ascending order of magnitude. Our goal is to find out c_m such that

$$\hat{\mu_\varepsilon^x}(\xi) = 1_{[0,\frac{1}{\varepsilon})}(|\xi|)\, e^{i(x,\xi)_{\mathbb{R}^3}} = 1_{[0,1)}(\varepsilon|\xi|)\, e^{i(x,\xi)_{\mathbb{R}^3}} \text{ for every } \xi \in \mathbb{R}^3.$$

To this end, we observe that if $f(y) := y^{\frac{1}{2}} 1_{[0,1)}(y)$, then by [**22**, Chapter 18], the *Fourier-Bessel expansion*[4] of f is given by

$$f(y) = \sum_{m=1}^{\infty} \left[\frac{2}{J_{\frac{3}{2}}^2(j_m)} \int_0^1 t^{\frac{3}{2}} J_{\frac{1}{2}}(j_m t)\, dt \right] J_{\frac{1}{2}}(j_m y).$$

By (3.1), in order to have

$$\hat{\mu_\varepsilon^x}(\xi) = 1_{[0,1)}(\varepsilon|\xi|)\, e^{i(x,\xi)_{\mathbb{R}^3}} = \frac{f(\varepsilon|\xi|)}{\sqrt{\varepsilon|\xi|}} e^{i(x,\xi)_{\mathbb{R}^3}},$$

we require

$$c_m = \frac{1}{\Gamma\left(\frac{3}{2}\right)} \left(\frac{j_m}{2}\right)^{\frac{1}{2}} \frac{2}{J_{\frac{3}{2}}^2(j_m)} \int_0^1 t^{\frac{3}{2}} J_{\frac{1}{2}}(j_m t)\, dt = 2(-1)^{m+1} \text{ for every } m \in \mathbb{N}.$$

[4]The authors are grateful to Oliver Nadeau-Chamard for introducing this technical tool to the first author and for all the subsequent discussions.

Since $f(y)$ is bounded and continuous in $(0,1)$ and $\int_0^1 t^{\frac{1}{2}} f(t)\,dt < \infty$, the theory of the Fourier-Bessel expansion guarantees that

$$\sum_{m=1}^{\infty} c_m \widehat{\sigma^x_{j_m\varepsilon}}(\xi) = \frac{f(\varepsilon|\xi|)}{\sqrt{\varepsilon|\xi|}} e^{i(x,\xi)_{\mathbb{R}^3}} = \hat{\mu^x_\varepsilon}(\xi) \text{ for every } \xi \in \mathbb{R}^3.$$

Furthermore, one can easily check that $\mu^x_\varepsilon \in H^{-\frac{3}{2}}$ and $\mu^x_\varepsilon \to \delta_x$ as $\varepsilon \downarrow 0$ in the sense of tempered distributions. In other words, for each $x \in \mathbb{R}^3$ and $\varepsilon > 0$, by taking averages over a family of spheres centered at x with radii $j_m\varepsilon$ propagating to the entire $\mathbb{R}^3$, and by taking a proper linear combination of these spherical averages, we can design an action (that can be applied to θ) that has the desired Fourier transform. Besides, we claim that the following results hold.

LEMMA 3.1. *Given $x \in \mathbb{R}^3$ and $\varepsilon > 0$, let $S_N := \sum_{m=1}^N c_m \sigma^x_{j_m\varepsilon}$ for each positive integer N, where j_m and c_m are as defined above. Then S_N converges to μ^x_ε in $H^{-\frac{3}{2}}$ as $N \to \infty$.*

The proof of Lemma 3.1 consists of straightforward computations of the Bessel functions. We will not include the proof in this article. Upon establishing Lemma 3.1, one may take $\hat{\mu^x_\varepsilon}(\xi) = 1_{[0,\frac{1}{\varepsilon})}(|\xi|)\, e^{i(x,\xi)_{\mathbb{R}^3}}$ into the computation of the covariance function of $\{\mathcal{I}(h_{\mu^x_\varepsilon}) : x \in \mathbb{R}^3, \varepsilon > 0\}$. In this case, we discover the following favorable properties of this family.

THEOREM 3.2. *Let $\{\mathcal{I}(h_{\mu^x_\varepsilon}) : x \in \mathbb{R}^3, \varepsilon > 0\}$ be the family introduced above.*

(1) Given $x \in \mathbb{R}^3$ and $\varepsilon_1 \geq \varepsilon_2 > 0$,

$$\mathbb{E}^{\mathcal{W}}\left[\mathcal{I}\left(h_{\mu^x_{\varepsilon_1}}\right)\mathcal{I}\left(h_{\mu^x_{\varepsilon_2}}\right)\right] = \frac{1}{2\pi^2}\left[-\frac{1}{\sqrt{1+\varepsilon_1^2}} + \ln\left(1+\sqrt{1+\varepsilon_1^2}\right) - \ln\varepsilon_1\right] =: G(\varepsilon_1),$$

where the function $G(\varepsilon_1) \asymp$[5] *$-\frac{1}{2\pi^2}\ln\varepsilon_1$ when ε_1 is small. Moreover, $\{\mathcal{I}(h_{\mu^x_\varepsilon}) : \varepsilon > 0\}$ has the same distribution as a Brownian motion up to a deterministic time change.*

(2) Given $x \neq y$ and $\varepsilon_1 \geq \varepsilon_2 > 0$,

$$\mathbb{E}^{\mathcal{W}}\left[\mathcal{I}\left(h_{\mu^x_{\varepsilon_1}}\right)\mathcal{I}\left(h_{\mu^y_{\varepsilon_2}}\right)\right] = \frac{1}{(2\pi)^{\frac{3}{2}}\sqrt{|x-y|}}\int_0^{\frac{1}{\varepsilon_1}} \frac{r^{\frac{3}{2}}}{(1+r^2)^{\frac{3}{2}}} J_{\frac{1}{2}}(|x-y|\,r)\,dr.$$

If $0 < |x-y| \leq \varepsilon_1$, then as $\varepsilon_1 \downarrow 0$,

$$\mathbb{E}^{\mathcal{W}}\left[\mathcal{I}\left(h_{\mu^x_{\varepsilon_1}}\right)\mathcal{I}\left(h_{\mu^y_{\varepsilon_2}}\right)\right] = \frac{1}{2\pi^2}K_0(\varepsilon_1) + \mathcal{O}(\varepsilon_1) \asymp -\frac{1}{2\pi^2}\ln\varepsilon_1.$$

If $|x-y| > \varepsilon_1$, then as $|x-y| \downarrow 0$,

$$\mathbb{E}^{\mathcal{W}}\left[\mathcal{I}\left(h_{\mu^x_{\varepsilon_1}}\right)\mathcal{I}\left(h_{\mu^y_{\varepsilon_2}}\right)\right] = \frac{1}{2\pi^2}K_0(|x-y|) + \mathcal{O}(\varepsilon_1) \asymp -\frac{1}{2\pi^2}\ln|x-y|.$$

Again, we omit the proof of Theorem 3.2 since the results follow directly from Lemma 3.1. By now, one should believe that $\{\mathcal{I}(h_{\mu^x_\varepsilon}) : x \in \mathbb{R}^3, \varepsilon > 0\}$ is a suitable choice of regularization for the log-correlated GFF on $\mathbb{R}^3$, since it possesses the favorable properties as listed above. In fact, these properties are exactly the same as those established in [**12**] and [**6**] for the regularization of the log-correlated GFF on $\mathbb{R}^2$ and $\mathbb{R}^4$, respectively.

[5] "$\asymp$" means that one quantity is asymptotic to another.

4. Random Measure and KPZ in $\mathbb{R}^3$

In the rest of the article, we will outline the construction of the Liouville Quantum Gravity (LQG) measure and the proof of the Knizhnik-Polyakov-Zamolodchikov (KPZ) formula for the log-correlated GFF on $\mathbb{R}^3$ via the regularization we have introduced above. Based on the methods and the results in [**12**] and [**6**], we believe that the properties stated in Theorem 3.2 are the key ingredients that enable us to construct the LQG measure and to prove the KPZ formula. The specific technicalities needed to achieve our goal follow closely those developed in [**6**]. We will only explain the main ideas in this article, and leave the complete proofs to a future work.

4.1. Construction of Random Measure in $\mathbb{R}^3$. We are now ready to use the Gaussian family $\left\{\mathcal{I}\left(h_{\mu_\varepsilon^x}\right) : x \in \mathbb{R}^3, \varepsilon > 0\right\}$ to construct the LQG measure in $\mathbb{R}^3$. For the simplicity of notation, we again set $\tilde{\theta}_\varepsilon(x) := \mathcal{I}\left(h_{\mu_\varepsilon^x}\right)(\theta)$ for every $x \in \mathbb{R}^3$ and $\varepsilon > 0$. It is easy to see that, without loss of generality, we can assume that $(x, \varepsilon) \mapsto \tilde{\theta}_\varepsilon(x)$ is continuous for almost every $\theta \in \Theta$, and there is no doubt that $(\theta, x, \varepsilon) \mapsto \tilde{\theta}_\varepsilon(x)$ is measurable with respect to $\mathcal{B}_\Theta \times \mathcal{B}_{\mathbb{R}^3} \times \mathcal{B}_{(0,\infty)}$. Heuristically speaking, the LQG measure is the random measure on $\mathbb{R}^3$ which formally takes the representation of "$m^\theta(dx) = e^{\gamma\theta(x)}dx$" where γ is a positive constant. However, $m^\theta(dx)$ is not well-defined due to the singularity of θ. To make sense of $m^\theta(dx)$, we adopt the regularized family $\left\{\tilde{\theta}_\varepsilon(x) : x \in \mathbb{R}^3, \varepsilon > 0\right\}$ and for every $\varepsilon > 0$, define a random measure on $\mathbb{R}^3$ by

$$\theta \mapsto m_\varepsilon^\theta(dx) := E_\varepsilon^\theta(x)\,dx,$$

where

$$E_\varepsilon^\theta(x) := \exp\left(\gamma\tilde{\theta}_\varepsilon(x) - \frac{\gamma^2}{2}G(\varepsilon)\right).$$

The purpose of having the coefficient $e^{-\frac{\gamma^2}{2}G(\varepsilon)}$ is such that $\mathbb{E}^{\mathcal{W}}\left[E_\varepsilon^\theta(x)\right] = 1$ for every $x \in \mathbb{R}^3$ and $\varepsilon > 0$. For every $\varepsilon > 0$ and $B \subseteq \mathcal{B}_{\mathbb{R}^3}$, $\theta \mapsto m^\theta(B)$ is measurable and hence, by Fubini's theorem,

$$\mathbb{E}^{\mathcal{W}}\left[m_\varepsilon^\theta(B)\right] = \int_B \mathbb{E}^{\mathcal{W}}\left[E_\varepsilon^\theta(x)\right]dx = \mathrm{Vol}(B).$$

THEOREM 4.1. *If $0 < \gamma^2 < \pi^2$ and $\varepsilon_n := \varepsilon^{n^2}$ for $n \geq 1$ with $0 < \varepsilon < 1$, then for almost every $\theta \in \Theta$, there exists a non-negative Borel measure $m^\theta(dx)$ on $\mathbb{R}^3$ such that for every $f \in C_c\left(\mathbb{R}^3\right)$,*

$$\int_{\mathbb{R}^3} f(x)\,m_{\varepsilon_n}^\theta(dx) \longrightarrow \int_{\mathbb{R}^3} f(x)\,m^\theta(dx) \;\; as\ n \to \infty$$

almost surely as well as in $L^2(\mathcal{W})$.

The main idea of the proof is to control the L^2 difference between two consecutive terms along the concerned sequence. The result in Theorem 4.1 implies that the the LQG measure $m^\theta(dx)$ can be interpreted as the weak convergence limit of the sequence of random measures $m_{\varepsilon_n}^\theta(dx)$. It is not hard to see that the limiting random measure does not depend on the choice of the sequence $\{\varepsilon_n : n \geq 1\}$ in the sense that, if $\{\varepsilon'_n : n \geq 1\}$ is another sequence (decaying to 0) such that $m_{\varepsilon'_n}^\theta(dx)$

weakly converges to some Borel measure $m'^{\theta}(dx)$ almost surely, then $m'^{\theta}(dx)$ and $m^{\theta}(dx)$ are identical.

4.2. KPZ Formula on $\mathbb{R}^3$. A long celebrated work in the field of conformal field theory and quantum gravity is the Knizhnik-Polyakov-Zamolodchikov (KPZ) formula proposed by Knizhnik, Polyakov and Zamolodchikov in [**17**]. Under the specific setting of the LQG measure, the KPZ formula provides an exact correspondence between the scaling exponent of a set in the Euclidean geometry, and its counterpart under the random geometry associated with the LQG measure. In the original work [**17**], the KPZ formula was conjectured to hold for more general geometric parameters in more general settings. The first complete mathematical proof of the KPZ formula was presented in [**12**] for the LQG measure constructed with the 2D log-correlated GFF based on the circle averaging regularization of the GFF. Later the formula was also established with different choices of regularization under various conditions, e.g., for log-correlated GFFs in even dimensions based on the sphere averaging regularization ([**6**]), and for log-correlated GFFs whose covariance function is $\sigma-$positive definite in arbitrary dimensions based on the GMC theory. In this section, we investigate the possibility of proving the KPZ formula for the LQG measure constructed above. If this is achieved, then this work will provide an extension of the method and the results from [**12**] to arbitrary dimensions, and offer an alternative method from the GMC approach without assuming the $\sigma-$positive-definiteness of the covariance function. This will further extend the scope of the applicability of regularizations based on geometric averages of singular GFFs, and confirm that such a regularization is a good candidate for studying the geometry of singular GFFs.

For a bounded domain $D \subseteq \mathbb{R}^3$, we say that ρ is the *Euclidean scaling exponent* of D if

$$\lim_{\varepsilon\downarrow 0}\frac{\ln \operatorname{Vol}(D_\varepsilon)}{\ln \varepsilon^3}=\rho,$$

where $D_\varepsilon := \cup_{z\in D}B_\varepsilon(z)$ is the canonical ε-neighborhood of D for every $\varepsilon>0$. We note that D has Euclidean scaling exponent ρ if and only if D has Minkowski dimension $3-3\rho$ (see, e.g., §4.1 of [**18**]).

Now we will look at the notion of scaling exponent under the random geometry associated with $m^{\theta}(dx)$. Given $0<\gamma^2<\pi^2$ and $x\in\mathbb{R}^3$, it is easy to check that

$$\mathbb{E}^{\mathcal{W}}\left[\limsup_{n\to\infty} e^{6\gamma^2 G(\varepsilon_n)} m^{\theta}\left(\overline{B_{\varepsilon_n}(x)}\right)\right]=0.$$

Set

$$\Theta_x := \left\{\theta\in\Theta : \limsup_{n\to\infty} e^{6\gamma^2 G(\varepsilon_n)} m^{\theta}\left(\overline{B_{\varepsilon_n}(x)}\right)=0\right\}.$$

Then, Θ_x is a measurable set and $\mathcal{W}(\Theta_x)=1$. For $\omega>0$ and $\theta\in\Theta$, we set

$$R(x,\theta;\omega) := \begin{cases}\sup\left\{r>0 : m^{\theta}(B_r(x))\le\omega\right\}, & \text{if } \theta\in\Theta_x,\\ 0, & \text{otherwise.}\end{cases} \tag{4.1}$$

and define the *isothermal $\omega-$neighborhood* of D as

$$D^{\omega,\theta} := \{x\in D : R(x,\theta;\omega)=0\}\cup\left\{x\in\mathbb{R}^3 : R(x,\theta;\omega)>0,\ \operatorname{dist}(x,D)<R(x,\theta;\omega)\right\}.$$

It is not hard to justify that for every $\omega>0$, $(x,\theta)\mapsto R(x,\theta;\omega)$ is measurable with respect to $\mathcal{B}_{\mathbb{R}^3}\times\mathcal{B}_{\Theta}$, as well as that $\theta\mapsto D^{\omega,\theta}$ is measurable with respect to $\mathcal{B}_\Theta$ and for almost every $\theta \in \Theta$, $D^{\omega,\theta}$ is a Borel set on $\mathbb{R}^3$. We call Q the *quantum scaling exponent* of D if

$$\lim_{\omega \downarrow 0} \frac{\ln \mathbb{E}^{\mathcal{W}}\left[m^\theta\left(D^{\omega,\theta}\right)\right]}{\ln \omega} = Q. \tag{4.2}$$

Seeing from (4.1), it is clear that in order to study $R(x,\theta;\omega)$, which holds the key to understanding Q, we should consider $R(x,\theta;\omega)$ as a function of the pair $(x,\theta) \in \mathbb{R}^3 \times \Theta$. Our first step is to find, for every $(x,\theta) \in \mathbb{R}^3 \times \Theta$, a non-negative regular and σ-finite measure $\hat{m}^{\theta,x}(dy)$ on $\mathbb{R}^3$ such that for every $r > 0$, compact set $X \subseteq \mathbb{R}^3$ and $F \in C_0\left(\mathbb{R}^3 \times [0,\infty)\right)$,

$$\int_\Theta \int_X F\left(x, m^\theta\left(B_r(x)\right)\right) m^\theta(dx)\,\mathcal{W}(d\theta) = \int_X \int_\Theta F\left(x, \hat{m}^{\theta,x}\left(B_r(x)\right)\right) \mathcal{W}(d\theta)\,dx.$$

This is to say that the distribution of $\left(x, m^\theta\left(B_r(x)\right)\right)$ under $m^\theta(dx)\,\mathcal{W}(d\theta)$ is identical to that of $\left(x, \hat{m}^{\theta,x}\left(B_r(x)\right)\right)$ under $dx\mathcal{W}(d\theta)$. Thus, to analyze $R(x,\theta;\omega)$ under $m^\theta(dx)\,\mathcal{W}(d\theta)$, we may equivalently study $\hat{R}(x,\theta;\omega)$ under $dx\mathcal{W}(d\theta)$, where

$$\hat{R}(x,\theta;\omega) := \begin{cases} \sup\left\{r > 0 : \hat{m}^{\theta,x}\left(B_r(x)\right) \le \omega\right\}, & \text{if } \theta \in \Theta_x, \\ 0, & \text{otherwise.} \end{cases}$$

Further, for every $x \in \mathbb{R}^3$ and every $r > 0$, the distribution of $\hat{m}^{\theta,x}\left(B_r(x)\right)$ is independent of x. Thus, conditioning on x, the distribution of $\hat{R}(x,\theta;\omega)$ under $\mathcal{W}$ does not depend on x either. Without loss of generality, we will assume $x = 0$ and simplify the notations as $B_r := B_r(0)$, $\hat{m}^\theta(B_r) := \hat{m}^{\theta,0}(B_r(0))$ and $\hat{R}(\theta;\omega) := \hat{R}(0,\theta;\omega)$.

To proceed from here, we will need to find out the relation between $\hat{R}(\theta;\omega)$ and ω, which is entirely determined by the relation between $\hat{m}^\theta(B_r)$ and r. So, it suffices to study $\hat{m}^\theta(B_r)$ as a (random) function of r. To this end, we will approximate $\hat{m}^\theta(B_r)$ by conditioning on the average of θ over the sphere ∂B_r. Recall that if $r(t) := G^{-1}(t + G(1))$ for $t \ge 0$, then the process $\{X_t : t \ge 0\}$ with

$$X_t(\theta) := \tilde{\theta}_{r(t)}(0) - \tilde{\theta}_1(0)$$

has the same distribution as the standard Brownian motion. It is easy to see that, when t is sufficiently large, the conditional expectation $\mathbb{E}^{\mathcal{W}}\left[\hat{m}^\theta\left(B_{r(t)}\right)|X_t\right]$ is approximated by

$$Y_t(\theta) := \exp\left(\gamma X_t(\theta) - \left(6\pi^2 - \frac{\gamma^2}{2}\right)t\right).$$

In other words, when r is sufficiently small,

$$\hat{m}^\theta(B_r) \approx \mathbb{E}^{\mathcal{W}}\left[\hat{m}^\theta\left(B_{r(t)}\right)|X_t\right] \approx Y_t(\theta)$$

Each "$\approx$" relation here needs to be justified with rigorous arguments, which will be done in the future work.

Next, let $D \subseteq \mathbb{R}^3$ be a bounded Borel set with Euclidean scaling exponent $\rho \in [0,1]$. We will replace $\hat{m}^\theta(B_r)$ with $Y_t(\theta)$ in the covering of D. For each $\omega > 0$, we define the stopping time

$$T^*_\omega(\theta) := \inf\left\{t \ge 0 : Y_t(\theta) \le \omega\right\},$$

corresponding to which we have the random radius

$$r^*_\omega(\theta) := G^{-1}\left(T^*_\omega(\theta) + G(1)\right)$$

and the random neighborhood

$$D^{\omega^*,\theta} := \cup_{z\in D} B_{r^*_\omega(\theta)}(z).$$

For every $s \leq 0$, by Doob's stopping time theorem,

$$\left\{\exp\left[sY_{t\wedge T^*_\omega(\theta)}(\theta) - \frac{s^2}{2}\left(t\wedge T^*_\omega(\theta)\right)\right] : t \geq 0\right\}$$

is a uniformly bounded martingale. In addition, the continuity of Brownian motion implies that

$$Y_{T^*_\omega}(\theta) = \frac{\ln\omega}{\gamma} + \left(6\pi^2 - \frac{\gamma^2}{2}\right)\frac{T^*_\omega(\theta)}{\gamma}.$$

Therefore, we have that

$$\mathbb{E}^{\mathcal{W}}\left[\exp\left(-\frac{\gamma s^2 - 2s\left(6\pi^2 - \frac{\gamma^2}{2}\right)}{2\gamma}\right)T^*_\omega\right] = \omega^{-s/\gamma}, \tag{4.3}$$

and that is

$$\mathbb{E}^{\mathcal{W}}\left[\hat{m}^\theta\left(D^{\omega^*,\theta}\right)\right] \approx \mathbb{E}^{\mathcal{W}}\left[(r^*_\omega)^{3\rho}\right] \approx \mathbb{E}^{\mathcal{W}}\left[\exp\left(-6\pi^2\rho T^*_\omega\right)\right].$$

Again, each "$\approx$" relation requires rigorous proof.

Finally, let us choose $s \in [-\gamma, 0]$ such that $\frac{s^2}{2} - \frac{s}{\gamma}\left(6\pi^2 - \frac{\gamma^2}{2}\right) = 6\pi^2\rho$. Then, according to (4.3), we have that

$$\begin{aligned}\lim_{\omega\downarrow 0}\frac{\ln \mathbb{E}^{\mathcal{W}}\left[\hat{m}^\theta\left(D^{\omega^*,\theta}\right)\right]}{\ln\omega} &= \lim_{\omega\downarrow 0}\frac{\ln \mathbb{E}^{\mathcal{W}}\left[\exp\left(-6\pi^2\rho T^*_\omega\right)\right]}{\ln\omega}\\ &= \lim_{\omega\downarrow 0}\frac{\ln\omega^{-s/\gamma}}{\ln\omega} = -\frac{s}{\gamma}.\end{aligned}$$

By identifying Q with $-\frac{s}{\gamma}$, we see that

$$\lim_{\omega\downarrow 0}\frac{\ln \mathbb{E}^{\mathcal{W}}\left[\hat{m}^\theta\left(D^{\omega^*,\theta}\right)\right]}{\ln\omega} = Q,$$

and $Q \in [0,1]$ is determined by the following quadratic relation with ρ:

$$\rho = \frac{\gamma^2}{12\pi^2}Q^2 + \left(1 - \frac{\gamma^2}{12\pi^2}\right)Q. \tag{4.4}$$

Following the steps outlined above, we claim the following result is true.

THEOREM 4.2. *For every $D \subseteq \mathbb{R}^3$ bounded Borel set with Euclidean scaling exponent $\rho \in [0,1]$, D has quantum scaling exponent $Q \in [0,1]$ as defined in (4.2), where Q is related to ρ by (4.4).*

(4.4) is the KPZ formula that the LQG measure $m^\theta(dx)$ constructed in §3 satisfies, and it gives the exact description of how the scaling property of a set transforms from the Euclidean geometry to the random geometry associated with $m^\theta(dx)$.

References

[1] Julien Barral, Xiong Jin, Rémi Rhodes, and Vincent Vargas, *Gaussian multiplicative chaos and KPZ duality*, Comm. Math. Phys. **323** (2013), no. 2, 451–485, DOI 10.1007/s00220-013-1769-z. MR3096527

[2] David D. Bleecker, *Nonperturbative conformal quantum gravity*, Classical Quantum Gravity **4** (1987), no. 4, 827–849. MR895904

[3] Yaiza Canzani, Dmitry Jakobson, and Igor Wigman, *Scalar curvature and Q-curvature of random metrics*, J. Geom. Anal. **24** (2014), no. 4, 1982–2019, DOI 10.1007/s12220-013-9406-9. MR3261729

[4] S. Chatterjee, A. Dembo, and J. Ding. On level sets of gaussian fields. *Preprint*, pages 1–6, 2013. arXiv:1310.5175v1.

[5] Linan Chen, *Thick points of high-dimensional Gaussian free fields* (English, with English and French summaries), Ann. Inst. Henri Poincaré Probab. Stat. **54** (2018), no. 3, 1492–1526, DOI 10.1214/17-AIHP846. MR3825889

[6] Linan Chen and Dmitry Jakobson, *Gaussian free fields and KPZ relation in* $\mathbb{R}^4$, Ann. Henri Poincaré **15** (2014), no. 7, 1245–1283, DOI 10.1007/s00023-013-0277-1. MR3225731

[7] Alessandra Cipriani and Rajat Subhra Hazra, *Thick points for a Gaussian free field in 4 dimensions*, Stochastic Process. Appl. **125** (2015), no. 6, 2383–2404, DOI 10.1016/j.spa.2015.01.004. MR3322868

[8] B. Clarke, D. Jakobson, N. Kamran, L. Silberman, and J. Taylor. Manifold of metrics with fixed volume form. pages 1–14. With an appendix by Y. Canzani, D. Jakobson and L. Silberman. arxiv:1309.1348.

[9] Jian Ding and Ofer Zeitouni, *Extreme values for two-dimensional discrete Gaussian free field*, Ann. Probab. **42** (2014), no. 4, 1480–1515, DOI 10.1214/13-AOP859. MR3262484

[10] Jian Ding, Rishideep Roy, and Ofer Zeitouni, *Convergence of the centered maximum of log-correlated Gaussian fields*, Ann. Probab. **45** (2017), no. 6A, 3886–3928, DOI 10.1214/16-AOP1152. MR3729618

[11] J. L. Doob, *Stochastic processes*, Wiley Classics Library, John Wiley & Sons, Inc., New York, 1990. Reprint of the 1953 original; A Wiley-Interscience Publication. MR1038526

[12] Bertrand Duplantier and Scott Sheffield, *Liouville quantum gravity and KPZ*, Invent. Math. **185** (2011), no. 2, 333–393, DOI 10.1007/s00222-010-0308-1. MR2819163

[13] Bertrand Duplantier, Rémi Rhodes, Scott Sheffield, and Vincent Vargas, *Renormalization of critical Gaussian multiplicative chaos and KPZ relation*, Comm. Math. Phys. **330** (2014), no. 1, 283–330, DOI 10.1007/s00220-014-2000-6. MR3215583

[14] Leonard Gross, *Abstract Wiener spaces*, Proc. Fifth Berkeley Sympos. Math. Statist. and Probability (Berkeley, Calif., 1965/66), Univ. California Press, Berkeley, Calif., 1967, pp. 31–42. MR0212152

[15] Xiaoyu Hu, Jason Miller, and Yuval Peres, *Thick points of the Gaussian free field*, Ann. Probab. **38** (2010), no. 2, 896–926, DOI 10.1214/09-AOP498. MR2642894

[16] Jean-Pierre Kahane, *Some random series of functions*, 2nd ed., Cambridge Studies in Advanced Mathematics, vol. 5, Cambridge University Press, Cambridge, 1985. MR833073

[17] V. G. Knizhnik, A. M. Polyakov, and A. B. Zamolodchikov, *Fractal structure of 2D-quantum gravity*, Modern Phys. Lett. A **3** (1988), no. 8, 819–826, DOI 10.1142/S0217732388000982. MR947880

[18] Peter Mörters and Yuval Peres, *Brownian motion*, Cambridge Series in Statistical and Probabilistic Mathematics, vol. 30, Cambridge University Press, Cambridge, 2010. With an appendix by Oded Schramm and Wendelin Werner. MR2604525

[19] Rémi Rhodes and Vincent Vargas, *Multidimensional multifractal random measures*, Electron. J. Probab. **15** (2010), no. 9, 241–258, DOI 10.1214/EJP.v15-746. MR2609587

[20] Rémi Rhodes and Vincent Vargas, *KPZ formula for log-infinitely divisible multifractal random measures*, ESAIM Probab. Stat. **15** (2011), 358–371, DOI 10.1051/ps/2010007. MR2870520

[21] Rémi Rhodes and Vincent Vargas, *Gaussian multiplicative chaos and applications: a review*, Probab. Surv. **11** (2014), 315–392, DOI 10.1214/13-PS218. MR3274356

[22] G. N. Watson, *A treatise on the theory of Bessel functions*, Cambridge Mathematical Library, Cambridge University Press, Cambridge, 1995. Reprint of the second (1944) edition. MR1349110

805 Sherbrooke St. West, Department of Mathematics and Statistics, McGill University, Montreal, QC, H3A 0B9, Canada.

805 Sherbrooke St. West, Department of Mathematics and Statistics, McGill University, Montreal, QC, H3A 0B9, Canada.

Contemporary Mathematics
Volume **739**, 2019
https://doi.org/10.1090/conm/739/14892

Tangent nodal sets for random spherical harmonics

Suresh Eswarathasan

ABSTRACT. In this note, we consider a fixed vector field V on S^2 and study the distribution of points which lie on the nodal set (of a random spherical harmonic) where V is also tangent. We show that the expected value of the corresponding counting function is asymptotic to the eigenvalue with a leading coefficient that is independent of the vector field V. This demonstrates, in some form, a universality for vector fields up to lower order terms.

1. Introduction

1.1. Nodal Sets. In the 18th century Ernest Chladni first described nodal sets during his study of modes of vibration on a rigid surface: the observed nodal pattern corresponds to the sets that remain stationary during vibrations. To some extent, the manner in which these patterns develop as the frequency of the mode becomes larger has remained an enigma to mathematicians.

Consider a compact boundaryless Riemannian manifold (M, g) and the spectrum of the corresponding Laplace-Beltrami operator $-\Delta_g$ which we order as $\lambda_0^2 = 0 < \lambda_1^2 \leq \lambda_2^2 \leq \ldots$ and tends to infinity. Let us denote a corresponding L^2 orthonormal basis by $\{\varphi_j\}_j$ and let $Z(\varphi_j) = \{x \in M : \varphi_j(x) = 0\}$ be its nodal set. Courant demonstrated in the 1920s that the number of connected components of $M \backslash Z(\varphi_j)$, commonly referred to as *nodal domains*, is bounded above by a uniform constant times j. The study of the nodal set and nodal domains of eigenfunctions under various assumptions is a well-developed area of research and has demonstrated a number of connections to other areas of mathematics and mathematical physics; for more information, see the ICM article of Nazarov-Sodin [**NS10**].

In particular, much more can be said in settings which exhibit some rigid structure such as the torus or the sphere. In this note, we will consider an aspect of nodal sets (motivated by a question pertaining to nodal domains) for certain kinds of eigenfunctions on the 2-sphere S^2.

1.2. Spherical Harmonics. Consider the 2-dimensional sphere S^2, its positive Laplace-Beltrami operator $-\Delta_g$ where g is the round metric, and normalized volume measure dV_g. Consider the eigenfunction equation

$$-\Delta_g \varphi_l = \lambda_l^2 \varphi_l,$$

where $l \in \mathbb{N}$. Let E_l denote the eigenspace for the eigenvalue λ_l^2. We note that the eigenvalues λ_l on the sphere are explicit and have large multiplicities, with the formulas for them being

$$\lambda_l = l(l+1) \text{ and } \mathcal{N}_l = \dim E_{\lambda_l} = 2l+1.$$

Given l, we fix an L^2-orthonormal basis for E_l which we denote by $\{\varphi_{l,k}\}_{k=1}^{\mathcal{N}_l}$ and results in the identification $\mathbb{R}^{\mathcal{N}_l} \simeq E_l$. In particular, using standard spherical coordinates on $S^2 \subset \mathbb{R}^3$, we set $\varphi_{l,k} := e^{ik\theta} P_l^k(\cos\phi)$ where P_l^m is the associated Legendre polynomial of degree (l,m); this basis is commonly known as *ultraspherical*. For further reading on spherical harmonics we refer the reader to [**AAR99**].

1.3. Random Model. Next, we consider random eigenfunctions, that is, functions of the form

$$f_l(x) = \sqrt{\frac{1}{\mathcal{N}_l}} \sum_{k=1}^{\mathcal{N}_l} a_k \varphi_{l,k}(x)$$

where a_k are Gaussian $N(0,1)$ i.i.d. random variables. Thanks to our identification of E_l with $\mathbb{R}^{\mathcal{N}_l}$, we can put a Gaussian measure ν on E_n with the expression

$$d\nu_l(f) = \exp(-\|\overrightarrow{a}\|^2/2) \frac{da_1 da_2 \cdots da_{\mathcal{N}_l}}{(2\pi)^{\mathcal{N}_l/2}},$$

where $\overrightarrow{a} = (a_k)_{k=1}^{\mathcal{N}_l} \in \mathbb{R}^{\mathcal{N}_l}$. Note that the measure ν does not depend on our chosen basis $\{\varphi_{l,k}\}_k$. Moreover, we see that $\mathbb{E}_l\left[f_l^2\right] = 1$ as an immediate consequence of the addition theorem for spherical harmonics.

Finally, we can consider the product measure $\nu := \otimes_{l=1}^{\infty} \nu_l$ on the space

$$\Omega := \oplus_{l=1}^{\infty} E_l \tag{1.3.1}$$

which can be seen as a probability space of random sequences of spherical harmonics.

1.4. Tangent Nodal Sets. The inspiration for this note is the interaction between tangent/normal spaces to nodal sets of random eigenfunctions and various geometric quantities of these submanifolds, a notion that was also considered by Gayet-Welschinger who give upper and lower bounds on the expected Betti numbers (a natural step to understand the singularities of a vector field) for elliptic pseudodifferential operators [**GW14**], Dang-Rivère (who themselves were motivated by the work [**GW14**]) who give asymptotics pertaining to the equidistribution of normal cocycles for Laplace eigenfunctions on general compact manifolds [**DR17**], and Rudnick-Wigman who consider fixed normal directions to nodal sets on the flat torus [**RW18**]. Note that on the sphere, there is no non-vanishing continuous vector thanks to the hairy ball theorem.

We seek in some sense a more singular cocycle to that of Dang-Rivière in that we would like to understand the current which places weight on the intersection of the normal bundle and that of the natural section generated by a general V along the nodal set *without* taking into account orientation (or rather, we would like to put an absolute value on the current considered in [**DR17**] in turn adding another layer of complication). This work is a step towards this albeit in a very specific setting.

Let $\mathcal{V}(S^2)$ be the set of smooth vector fields on S^2 and consider a fixed $V \in \mathcal{V}(S^2)$. Given an eigenfunction $f_l \in C^\infty(S^2)$, we can consider the action of the vector field V on f_l through the following formula,

$$Vf_l(x) = \langle \nabla f, V \rangle_{g(x)},$$

where $\langle , \rangle_{g(x)}$ is the inner product on T_xS^2 and ∇f_l is the gradient of f with respect to the round metric g. Hence $\langle \nabla f_l, V\rangle_{g(x)} = 0$ is equivalent to $V(x) \in T_x\{f_l = 0\}$. That is, $V(f_l) = 0$ implies that $V \in T_x\{f_l = 0\}$.

For the set of regular points in the nodal set $Z(f_l)$ (which is ν_l-almost surely true thanks to Bulinskaya's Lemma; see the standard text [**AW09**]), we would like to understand the statistics of when $V \in T_x(Z(f_l))$ or more succinctly the so-called *V-tangent nodal set* $Z_V(f_l) = \{x \in S^2 : f_l(x) = 0, Vf_l(x) = 0\}$ for a random eigenfunction f_l. We will later show that $Z_V(f_l)$ is almost surely finite as well in Section 3.1. In particular, we would like to understand the large-l behavior of the expected value of $Z_V(f_l)$.

1.5. Main Result. The main result of this note is the following:

THEOREM 1.5.1. *Let $V \in \mathcal{V}(S^2)$ be fixed and have finitely many zeroes. Let m be the maximal order of vanishing amongst all the zeroes.*

We have the following asymptotic for the expected value:

$$\mathbb{E}_l[\#Z_V(f_l)] = \frac{\sqrt{2}}{4\pi^2}\, l^2 + \mathcal{O}_V(l^{1+\frac{3m}{3m+2}}),$$

which holds for all $l \geq l_0(V)$. Furthermore, the remainder term has a bounded dependence on V and its derivatives.

In the case that 0 is a regular value of V and of order m, $\{V = 0\}$ is a smooth curve of finite length and we get a similar asymptotic but with a remainder term of $\mathcal{O}_V(l^{1+\frac{3m}{3m+1}})$

REMARK 1.5.2. The slightly larger than normal error term is due to weak singularities of the first intensity which arise from the zeroes of V, which are unavoidable due to topological reasons. See Section 3.6 and proceeding comments for a detailed explanation.

REMARK 1.5.3. This form of independence contrasts that of Theorems 1.2 and 1.3 in the work of Rudnick-Wigman [**RW18**] where they consider the number of points in $Z(f_l)$ with a fixed normal direction ζ on the flat torus of dimension d; Rudnick-Wigman obtain an exact expected value where the angle ζ appears explictly as well as an upper bound for sequences of certain deterministic eigenfunctions.

1.6. Current work in progress. This calculation stems from current work in progress on nodal domains where we have the following:

GOAL 1.6.1. *Find an explicit relationship, or some kind of substantial lack of relation, between $\mathbb{E}_l[\#Z_V(f_l)]$ and $\mathbb{E}_l$[number of connected components of $Z(f_l)$]. More specifically, we would like to relate the constant $\frac{\sqrt{2}}{4\pi^2}$ to C_{NS} where the latter is the Nazarov-Sodin constant.*

The works of Nazarov-Sodin [**NS1, NS2**] determine an asymptotic law on the counting function for the number of connected components of the nodal set of a random spherical harmonic f_l. Let us state this specific result for reference purposes:

THEOREM 1.6.2. [**NS1**] *There exists a positive constant C_{NS} (depending only on the dimension of S^2 and not its geometry) such that with probability tending to*

1 as $l \to \infty$,

$$\text{the number of connected components of } Z(f_l) = C_{NS}\frac{\omega_2}{(2\pi)^2}l^2 + o(l^2),$$

where ω_2 *is the volume of the unit ball in 2 dimensions.*

In the process of establishing this result (with the latter work actually addressing the more general hypotheses of Gaussian processes), the authors provide many fundamental ideas and powerful techniques which have allowed for the study of the topology of the connected nodal components of functions coming from Gaussian ensembles; see the papers of Canzani-Sarnak [**CS18**] and Sarnak-Wigman [**SW18**] as well as the series of works by Gayet-Welschinger (see [**GW14**] and the references therein).

The intuition relating $\#Z_V(f_l)$ to the number of connected components of $Z(f_l)$ is the following: Nazarov and Sodin show that "most" nodal components have small diameter and on the length-scale of l^{-1} where this observation can be made, the vector V is "locally" straight and we therefore expect "most" small components to have two points where V is tangent. Rigorizing this intuition and using it to gain a better understanding of C_{NS} is work in-progress.

2. Geometric Preliminaries

2.1. Coordinates & bases. Consider $S^2 \subset \mathbb{R}^3$ and take $x \in S^2$. Throughout this note, we consider spherical coordinates at x given by

$$(\sin\phi_x\cos\theta_x, \sin\phi_x\sin\theta_x, \cos\phi_x)$$

where $\theta_x \in [0, 2\pi)$ and $\phi_x \in (0,\pi)$. Using this system of coordinates, our metric on the set $[0, 2\pi) \times (0, \pi)$ becomes

$$g(\theta_x, \phi_x) = \begin{pmatrix} \sin^2\phi_x & 0 \\ 0 & 1 \end{pmatrix}.$$

Throughout our computation, we will frequently use the orthogonal (instead of an orthonormal) basis $\{\frac{\partial}{\partial\theta_x}, \frac{\partial}{\partial\phi_x}\}$ due to the coordinate singularity at $\phi_x = 0$.

REMARK 2.1.1. Although the subscripts of x, which are meant to signify our coordinate representation is attached to the x-variable, may seem tedious notation-wise, it will become useful when calculating entries of the covariance matrix as we must consider various derivatives in x and y of the spectral projector $P_l(d(x, y))$ before setting $x = y$.

2.2. Vector field V. We now let V be a smooth vector field on S^2. We note that in our chosen spherical coordinates, $V = v_1(\theta_x, \phi_x)\frac{\partial}{\partial\theta_x} + v_2(\theta_x, \phi_x)\frac{\partial}{\partial\phi_x}$ and define $V^\perp := v_2(\theta_x, \phi_x)\frac{\partial}{\partial\theta_x} + (-\sin^2\phi_x)v_1(\theta, \phi_x)\frac{\partial}{\partial\phi_x}$. We choose $V^\perp$ in this particular way so that the ordered set $\{V, V^\perp\}$ has positive orientation and is orthogonal. In two dimensions, we have the following consequence: given V, any other choice of $\widetilde{V^\perp}$ for which $\{V, \widetilde{V^\perp}\}$ is orthogonal and positively orientated is just a (variable) rescaling of $V^\perp$ at each point of S^2.

In regards to the action of V on a smooth function f in local coordinates denoted by x, after slightly abused notation, we have that

$$V f_l(x) = \langle\nabla f_l, V\rangle_{g(x)} = \nabla f_l(x)^T \cdot g\,V =: d(f_l)_x(V) = \nabla_{\theta_x,\phi_x} f_l \cdot V$$

where $\nabla_{\theta_x,\phi_x} f_l$ is the Euclidean gradient in the coordinates θ_x, ϕ_x and have expressed the metric inner product through the multiplication of matrices; notice that we have used the geometric definition of the gradient in order to relate the metric gradient to the Euclidean gradient. Hence we have the local expression $\nabla f_l(x) = g^{-1}\nabla_{\theta_x,\phi_x} f_l(x)$.

We conclude with the quick observation that

$$\|V^{\perp}\|_g(x) = \|V\|_g(x)\sin\phi_x = \|V\|_g(x)\sqrt{\det g(x)},$$

a fact which we will use in our calculations in Section 3.4.

3. Calculating the Expectation

3.1. Preparing the probability space. We must first verify the conditions of Theorem 6.2 in Azaš-Wchebor [**AW09**], particularly points (iii) and (iv). The following lemma succinctly addresses these two conditions:

LEMMA 3.1.1. *For the centered Gaussian field $F_l = (f_l, Vf_l)$, we have the distribution of F_l is non-degenerate away from (fixed neighborhoods of) the zeroes of V and that*

$$\mathbb{P}_l[NonSing_{V,l}^{\complement} := \{\omega \in E_l : \exists x \in S^2 - \{V = 0\}$$

such that

$$f_l(x) = 0, Vf_l(x) = 0, \text{ and } (DF_l)_x(x) \text{ is not invertible}\}] = 0.$$

PROOF. As calculated below, we find that the determinant of the covariance matrix for F_l equals $\sqrt{2\pi}\|V\|_g\sqrt{l^2+1}$ and the subsequent probability mass function is uniformly bounded in fixed neighbourhoods away from $\{V = 0\}$; this important fact will play a crucial role in a later part of this note when we localize away from these zeroes on scales depending on eigenvalue parameter l.

The verification of the measure 0 property follows directly from a generalized version of Bulinskaya's Lemma in higher dimension as stated in Proposition 6.5 of [**AW09**]. Given that F_l is C^2 in fixed neighbourhoods (to be specified later) away from $\{V = 0\}$, we immediately obtain our desired result.

□

Hence, we can now identify Ω with $\oplus_l NonSing_{V,l}$. It is important to notice that we have not yet made any assumptions on the structure of the vanishing set $\{V(x) = 0\}$. This will only play a role in the penultimate step of the proof of Theorem 1.5.1 in Section 3.6.

3.2. Preparing the orthogonal determinant. We will employ the Kac-Rice formula (see [**AW09**] Chapter 6) and compute the quantity

$$\int_{S^2} \Phi_l(0,0)\,\mathbb{E}\left[|\overset{\perp}{\det} DF_l(x)|\big|f_l(x) = 0, Vf_l(x) = 0\right] dV_g(x)$$

where $F_l = (f_l, Vf_l)$ and Φ_l is the Gaussian probability density function of F_l. The quantity $\det^{\perp} DF_l$ is the orthogonal determinant, which is defined as the determinant of the map $DF_l^* DF_l$. Note that in some parts of the literature on random waves, the orthogonal determinant is labelled as the determinant of J_{F_l}.

It follows that evaluating this determinant at the orthogonal basis $\{V, V^{\perp}\}$ (i.e. expressing our coordinate basis vector fields with respect to the proposed orthogonal basis), at least away from the zeroes of V, with respect to spherical coordinates gives us

$$\stackrel{\perp}{\det} DF_l = \left(\frac{a_V^{\theta_x}}{\|V\|_g^2} V f_l + \frac{a_{V^\perp}^{\theta_x}}{\|V^\perp\|_g^2} V^\perp f_l\right)\left(\frac{a_V^{\phi_x}}{\|V\|_g^2} VV f_l + \frac{a_{V^\perp}^{\phi_x}}{\|V^\perp\|_g^2} V^\perp V f_l\right) \tag{3.2.1}$$

$$- \left(\frac{a_V^{\phi_x}}{\|V\|_g^2} V f_l + \frac{a_{V^\perp}^{\phi_x}}{\|V^\perp\|_g^2} V^\perp f_l\right)\left(\frac{a_V^{\theta_x}}{\|V\|_g^2} VV f_l + \frac{a_{V^\perp}^{\theta_x}}{\|V^\perp\|_g^2} V^\perp V f_l\right). \tag{3.2.2}$$

Here, $a_V^{\theta_x} := \langle \frac{\partial}{\partial\theta_x}, V\rangle_g = (\sin^2\phi_x)v_1$, $a_V^{\phi_x} := \langle \frac{\partial}{\partial\phi_x}, V\rangle_g = v_2$, $a_{V^\perp}^{\theta_x} := \langle \frac{\partial}{\partial e_x}, V^\perp\rangle_g = (\sin^2\phi_x)v_2$, and $a_{V^\perp}^{\phi_x} := \langle \frac{\partial}{\partial\phi_x}, V^\perp\rangle_g = (-\sin^2\phi_x)v_1$; these coefficients follow immediately from our expression of V and $V^\perp$ with respect to our coordinates.

Using the conditioning, the absolute value of the orthogonal determinant reduces to

$$\begin{aligned}
&\frac{1}{\|V\|_g^2\|V^\perp\|_g^2}\left|\left(a_{V^\perp}^{\theta_x}a_V^{\phi_x} - a_{V^\perp}^{\phi_x}a_V^{\theta_x}\right)(V^\perp f_l)(VVf_l)\right| \\
&= \frac{1}{\|V\|_g^2\|V^\perp\|_g^2}(\sin^2\phi_x)(v_2^2 + \sin^2\phi_x v_1^2)\left|(V^\perp f_l)(VVf_l)\right| \\
&= \frac{1}{\|V\|_g^2\|V^\perp\|_g^2}\cdot\|V\|_g^2\det(g)\cdot\left|(V^\perp f_l)(VVf_l)\right| \\
&= \frac{1}{\|V^\perp\|_g^2}\det(g)\cdot\left|(V^\perp f_l)(VVf_l)\right| \\
&= \frac{1}{\|V\|_g^2}\cdot\left|(V^\perp f_l)(VVf_l)\right| \qquad (3.2.3)
\end{aligned}$$

where g is the round metric in coordinates. Hence, we take the Gaussian field $(f_l, Vf_l, V^\perp f_l, VVf_l)$ and compute the corresponding conditional covariance matrix. That is, we compute the covariance for the field $X_2 := (V^\perp f_l, VVf_l)$ conditioned on the event that $X_1 := (f_l, Vf_l) = 0$.

3.3. Entries of the Covariance Matrix. We list the coefficients of the full covariance matrix for the field $(f_l, Vf_l, V^\perp f_l, VVf_l)$. Note that due to symmetry after restricting to the diagonal, i.e. setting $\theta_x = \theta_y$ and $\phi_x = \phi_y$, we only need to compute the entries on the diagonal and above. These calculations are done in full detail in Appendix - Section A.

For the convenience of the reader, we now list the final form of the entries:

- $a_{11} = P_l(h(x,y))_{|x=y} = P_l(1)$
- $a_{12} = V_y P_l(h(x,y))_{|x=y} = 0$
- $a_{13} = V_y^\perp P_l(h(x,y))_{|x=y} = 0$
- $a_{14} = V_y V_y P_l(h(x,y))_{|x=y} = -\|V\|_{g(x)}^2 P_l'(1)$
- $a_{22} = V_x V_y P_l(h(x,y))_{|x=y} = \|V\|_{g(x)}^2 P_l'(1)$
- $a_{23} = V_x V_y^\perp P_l(h(x,y))_{|x=y} = 0$
- $a_{24} = V_x V_y V_y P_l(h(x,y)_{|x=y} =$

$$\begin{aligned}
&\Big((v_1^x)^2\frac{\partial v_1^x}{\partial\theta_x}\sin^2\phi_x + v_1^x\frac{\partial v_1^x}{\partial\phi_x}v_2^x\sin^2\phi_x + (v_1^x)^2 v_2^x\sin\phi_x\cos\phi_x \\
&+ v_1^x v_2^x\frac{\partial v_2^x}{\partial\theta_x} + (v_2^x)^2\frac{\partial v_2^x}{\partial\phi_x}\Big)P_l'(1) =: \tilde{a}_{24}(\theta_x,\phi_x)P_l'(1)
\end{aligned}$$

• $a_{33} = V_x^\perp V_y^- P_l(h(x,y))_{|x=y} = \|V^\perp\|^2_{g(x)} P_l'(1)$

• $a_{34} = V_x^\perp V_y V_y P_l(h(x,y))_{|x=y} =$

$$\Big(v_2^x v_1^x \frac{\partial v_1^x}{\partial \theta_x} \sin^2\phi_x + (v_2^x)^2 v_1^x \sin\phi_x\cos\phi_x + (v_2^x)^2 \frac{\partial v_1^x}{\partial \phi_x}\sin^2\phi_x \\ + (v_2^x)^2 v_1^x \sin\phi_x \cos\phi_x + (v_1^x)^3 \cos\phi_x \sin^3\phi_x \\ - (v_1^x)^2 \frac{\partial v_2^x}{\partial\theta_x}\sin^2\phi_x - v_1^x v_2^x \frac{\partial v_2^x}{\partial \phi_x}\sin^2\phi_x \Big) P_l'(1) =: \tilde{a}_{34}(\theta_x,\phi_x) P_l'(1)$$

• $a_{44} = V_x V_x V_y V_y P_l(h(x,y))_{|x=y} = \frac{3}{8}\|V\|^4_{g(x)} l^4 + \frac{6}{8}\|V\|^4_{g(x)} l^3 + \frac{1}{2}\tilde{a}^1_{44}(\theta_x,\phi_x) l^2 + (\frac{1}{2}\tilde{a}^1_{44}(\theta_x,\phi_x) - \frac{3}{8}\|V\|^4) l$ where

$$\tilde{a}^1_{44}(\theta_x,\phi_x) = v_1^x \frac{\partial v_1^x}{\partial\theta_x} v_1^x \frac{\partial v_1^x}{\partial \theta_x}\sin^2\phi_x \\ + (v_1^x)^4 \sin^2\phi_x - (v_1^x)^3 \frac{\partial v_2^x}{\partial \theta_x}\sin\phi_x\cos\phi_x + (v_1^x)^2 \frac{\partial v_1^x}{\partial\theta_x} v_2^x \sin\phi_x\cos\phi_x \\ + v_1^x \frac{\partial v_1^x}{\partial \theta_x}\frac{\partial v_1^x}{\partial\phi_x} v_2^x \sin^2\phi_x + (v_1^x)^2 \frac{\partial v_1^x}{\partial\theta_x} v_2^x \sin\phi_x\cos\phi_x - (v_1^x)^2 v_2^x \frac{\partial v_2^x}{\partial\phi_x}\sin\phi_x\cos\phi_x \\ + (v_1^x)^2 (v_2^x)^2 \sin^2\phi_x + (v_1^x)^2 \frac{\partial v_1^x}{\partial\theta_x} v_2^x \cos\phi_x \sin\phi_x \\ - (v_1^x)^3 \frac{\partial v_2^x}{\partial\theta_x}\cos\phi_x\sin\phi_x + (v_1^x)^2 \left(\frac{\partial v_2^x}{\partial\theta_x}\right)^2 \\ + (v_1^x)^2 (v_2^x)^2 \cos^2\phi_x + v_1^x \frac{\partial v_1^x}{\partial\phi_x}(v_2^x)^2 \cos\phi_x\sin\phi_x \\ + (v_1^x)^2 (v_2^x)^2 \cos^2\phi_x + v_1^x \frac{\partial v_2^x}{\partial\theta_x} v_2^x \frac{\partial v_2^x}{\partial \phi_x} \\ + v_2^x \frac{\partial v_1^x}{\partial\phi_x} v_1^x \frac{\partial v_1^x}{\partial\theta_x}\sin^2\phi_x + v_2^x (v_1^x)^2 \frac{\partial v_1^x}{\partial\theta_x}\cos\phi_x\sin\phi_x \\ + (v_1^x)^2 (v_2^x)^2 \cos^2\phi_x + v_1^x \frac{\partial v_1^x}{\partial\phi_x}(v_2^x)^2 \sin\phi_x\cos\phi_x + v_1^x \frac{\partial v_1^x}{\partial\phi_x}(v_2^x)^2\cos\phi_x\sin\phi_x$$

$$+ \left(\frac{\partial v_1^x}{\partial\phi_x}\right)^2 (v_2^x)^2 \sin^2\phi_x + v_1^x \frac{\partial v_1^x}{\partial\phi_x}(v_2^x)^2 \sin\phi_x\cos\phi_x \\ + (v_1^x)^2 (v_2^x)^2 \cos^2\phi_x - (v_1^x)^2 v_2^x \frac{\partial v_2^x}{\partial\phi_x}\cos\phi_x\sin\phi_x \\ + (v_1^x)^2(v_2^x)^2\sin^2\phi_x + v_1^x v_2^x \frac{\partial v_2^x}{\partial\phi_x}\frac{\partial v_2^x}{\partial\theta_x} + \left(\frac{\partial v_2^x}{\partial\phi_x}\right)^2 (v_2^x)^2 \\ + (v_2^x)^4.$$

3.4. Conditional covariance matrix. In this section, we would like to compute the covariance matrix for $(V^\perp f_l, VVf_l)$ conditioned on the random vector (f_l, Vf_l), particularly at the value $(0,0)$. Given that we started with the Gaussian vector field $(f_l, Vf_l, V^\perp f_l, VVf_l)$, we can conveniently apply formulas found in [**AT00**] Section 1.2 for our desired matrix.

The 4×4-matrix total covariance matrix whose entries we computed in the previous section allows us to explicit calculate

$$\underbrace{\begin{pmatrix} a_{33} & a_{34} \\ a_{43} & a_{44} \end{pmatrix}}_{=:M_1} - \underbrace{\begin{pmatrix} a_{31} & a_{32} \\ a_{41} & a_{42} \end{pmatrix}}_{=:M_2} \cdot \underbrace{\begin{pmatrix} a_{11} & a_{12} \\ a_{21} & a_{22} \end{pmatrix}^{-1}}_{=:M_3} \cdot \underbrace{\begin{pmatrix} a_{13} & a_{14} \\ a_{23} & a_{24} \end{pmatrix}}_{=:M_4}.$$

We note that the probability mass function $\Phi_F(0,0)$ of the random field $F = (f_l, Vf_l)$ is $\frac{1}{(2\pi)\sqrt{\det C_{11}}}$ where $\det C_{11} = \frac{\|V\|^2}{2}(l^2+l)$. For the sake of clarity, let us write out these individual matrices:

$$M_1 = \begin{pmatrix} \|V^\perp\|^2_{g(x)} \frac{l^2+l}{2} & \tilde{a}_{34}(\theta_x,\phi_x)\frac{l^2+l}{2} \\ \tilde{a}_{34}(\theta_x,\phi_x)\frac{l^2+l}{2} & \begin{matrix}\left[\frac{3}{8}\|V\|^4_{g(x)}l^4 + \frac{6}{8}\|V\|^4_{g(x)}l^3 + \frac{1}{2}\tilde{a}^1_{44}(\theta_x,\phi_x)l^2 + \right. \\ \left. (\frac{1}{2}\tilde{a}^1_{44}(\theta_x,\phi_x) - \frac{3}{8}\|V\|^4)l\right]\end{matrix} \end{pmatrix}$$

$$M_2 = \begin{pmatrix} 0 & 0 \\ -\|V\|^2_{g(x)}\frac{l^2+l}{2} & \tilde{a}_{24}(\theta_x,\phi_x)\frac{l^2+l}{2} \end{pmatrix}$$

$$M_3 = \begin{pmatrix} 1 & 0 \\ 0 & \frac{2}{\|V\|^2_g(l^2+l)} \end{pmatrix}$$

$$M_4 = \begin{pmatrix} 0 & -\|V\|^2_{g(x)}\frac{l^2+l}{2} \\ 0 & \tilde{a}_{24}(\theta_x,\phi_x)\frac{l^2+l}{2} \end{pmatrix}$$

where $\lambda_l^2 = l^2 + l$ is our Laplace eigenvalue. This leads to

$$M_2 \cdot M_3 \cdot M_4 = \begin{pmatrix} 0 & 0 \\ 0 & \|V\|^4_{g(x)}(\frac{l^2+l}{2})^2 + \frac{(\tilde{a}_{24}(\theta_x,\phi_x))^2}{\|V\|^2_{g(x)}}(\frac{l^2+l}{2}) \end{pmatrix}.$$

Finally, we obtain the symmetric 4×4 matrix $M_1 - M_2M_3M_4$, whose entries $m_{i,j}$ are the following

$$\begin{aligned} m_{1,1} &= \|V^\perp\|^2_{g(x)}\frac{l^2+l}{2} = \frac{\|V^\perp\|^2_{g(x)}}{2}l^2 + \mathcal{O}(l) \\ m_{1,2} &= \tilde{a}_{34}(\theta_x,\phi_x)\frac{l^2+l}{2} = \frac{\tilde{a}_{34}(\theta_x,\phi_x)}{2}l^2 + \mathcal{O}(l) \\ m_{2,2} &= \left(\frac{3}{8}\|V\|^4_{g(x)} - \frac{1}{4}\|V\|^4_{g(x)}\right)l^4 + \left(\frac{6}{8}\|V\|^4_{g(x)} - \frac{1}{2}\|V\|^4_{g(x)}\right)l^3 \\ &+ \left(\frac{1}{2}\tilde{a}^1_{44}(\theta_x,\phi_x) - \frac{1}{4}\|V\|^4_{g(x)} - \frac{(\tilde{a}_{24}(\theta_x,\phi_x))^2}{2\|V\|^2_{g(x)}}\right)l^2 \\ &+ \left(\frac{1}{2}\tilde{a}^1_{44}(\theta_x,\phi_x) - \frac{3}{8}\|V\|^4 - \frac{(\tilde{a}_{24}(\theta_x,\phi_x))^2}{2\|V\|^2_{g(x)}}\right)l \\ &= \frac{\|V\|^4}{8}l^4 + \frac{\|V\|^4}{4}l^3 + \frac{1}{\|V\|^2_{g(x)}}\mathcal{O}(l^2). \end{aligned}$$

Note that the implicit constants appearing in our big-O notation are uniformly bounded in ϕ_x, θ_x. Let us refer to our resulting conditional covariance matrix, with these particular entries, as

$$\mathbf{\Delta}_l(\theta_x, \phi_x) := \begin{pmatrix} m_{1,1} & m_{1,2} \\ m_{1,2} & m_{2,2} \end{pmatrix}.$$

3.5. Evaluating the first intensity. Notice that $\det \mathbf{\Delta}_l(\theta_x, \phi_x)$ is possibly singular (i.e. blows up) when $(v_2^x)^2 + \sin^2\phi_x (v_1^x)^2 = \langle V, V\rangle_g = 0$ which is equivalent to having $V = 0$. Given that every continuous vector field on S^2 must have at least one zero, it is natural that we place some restrictions on how V vanishes. However, we will show that having a singular determinant in this sense is not the case via an explicit calculation.

As $\operatorname{tr}(\mathbf{\Delta}_l(\theta_x, \phi_x))$ and $\det(\mathbf{\Delta}_l(\theta_x, \phi_x))$ play an important role in the some upcoming calculations, we write out these quantities explicitly for sake of reference:

$$\operatorname{tr}(\mathbf{\Delta}_l(\theta_x, \phi_x)) = \left(\frac{3}{8}\|V\|_{g(x)}^4 - \frac{1}{4}\|V\|_{g(x)}^4\right) l^4 + \left(\frac{6}{8}\|V\|_{g(x)}^4 - \frac{1}{2}\|V\|_{g(x)}^4\right) l^3$$
$$+ \left(\frac{1}{2}\tilde{a}_{44}^1(\theta_x, \phi_x) - \frac{1}{4}\|V\|_{g(x)}^4 - \frac{(\tilde{a}_{24}(\theta_x, \phi_x))^2}{2\|V\|_{g(x)}^2} + \frac{1}{2}\|V^\perp\|^2\right) l^2$$
$$+ \left(\frac{1}{2}\tilde{a}_{44}^1(\theta_x, \phi_x) - \frac{3}{8}\|V\|^4 - \frac{(\tilde{a}_{24}(\theta_x, \phi_x))^2}{2\|V\|_{g(x)}^2} + \frac{1}{2}\|V^\perp\|^2\right) l.$$

Using the following formula,

$$\det(\mathbf{\Delta}_l(\theta_x, \phi_x)) =$$
$$\left(\frac{\|V^\perp\|_g^2(x) \cdot \|V\|_g^4(x)}{16}\right) l^6$$
$$+ \frac{\|V^\perp\|_g^2(x)}{2}\left(\frac{3\|V\|_g^4(x)}{16}\right) l^5$$
$$+ \left(\frac{\|V^\perp\|_g^2(x)}{2}\left(\frac{\tilde{a}_{44}^1(\theta_x, \phi_x)}{2} - \frac{(\tilde{a}_{24}(\theta_x, \phi_x))^2}{2\|V\|_g^2}\right) - \frac{(\tilde{a}_{34}(\theta_x, \phi_x))^2}{4}\right) l^4$$
$$+ \left(\frac{\|V^\perp\|_g^2(x)}{2}\left(\frac{\tilde{a}_{44}^1(\theta_x, \phi_x)}{2} - \frac{3\|V\|_g^4(x)}{8} - \frac{(\tilde{a}_{24}(\theta_x, \phi_x))^2}{2\|V\|_g^2}\right) - \frac{(\tilde{a}_{34}(\theta_x, \phi_x))^2}{2}\right) l^3$$
$$+ \left(\frac{\|V^\perp\|_g^2(x)}{2}\left(\frac{\tilde{a}_{44}^1(\theta_x, \phi_x)}{2} - \frac{3\|V\|_g^4(x)}{8} + \frac{(\tilde{a}_{24}(\theta_x, \phi_x))^2}{2\|V\|_g^2}\right) - \frac{(\tilde{a}_{34}(\theta_x, \phi_x))^2}{4}\right) l^2,$$

we get that $\det(\mathbf{\Delta}_l(\theta_x, \phi_x)) = \frac{\|V\|^4\|V^\perp\|^2}{16} l^6 + \mathcal{O}(l^5)$ and the remainder term is uniformly bounded in θ_x, ϕ_x thanks to $\|V^\perp\|_{g(x)} = \|V\|_{g(x)}\sqrt{\det g(x)}$ cancelling out the length factor $\|V\|_{g(x)}$ in the denominators. We record this observation in the following:

LEMMA 3.5.1. *The conditional covariance has the determinant*

$$\det(\mathbf{\Delta}_l(\theta_x, \phi_x)) = \frac{\|V\|^6 \det(g(x))}{16} l^6 + \mathcal{O}(l^5)$$

with a uniformly bounded remainder. Moreover, $V(\theta_x, \phi_x) = 0$ *for some* (θ_x, ϕ_x) *if and only if* $\det(\mathbf{\Delta}_l(\theta_x, \phi_x)) = 0$ *for the same* (θ_x, ϕ_x)*, uniformly for all* $l \geq l_0(V)$.

Recall $\det^{\perp} DF_l = \frac{1}{\|V\|_g^2} \cdot \left|(V^{\perp} f_l)(VVf_l)\right|$ from equation (3.2.3). And since

$$\Phi_l(0,0)\,\mathbb{E}\left[|\overset{\perp}{\det} DF_l(x)|\big|f_l(x)=0, Vf_l(x)=0\right] =$$

$$\frac{1}{2\pi\sqrt{\det C_{11}}}\frac{1}{\|V\|_g^2}\cdot\mathbb{E}\left[\left|(V^{\perp}f_l)(VVf_l)\right|\,\big|f_l(x)=0, Vf_l(x)=0\right] \tag{3.5.2}$$

where $\det C_{11} = \frac{\|V\|^2}{2}(l^2+l)$, we are now lead to our section's main proposition:

PROPOSITION 3.5.3. *For $\alpha \in (\frac{5}{27}, \frac{1}{3})$, define the set $U := \{x \in S^2 : \|V\| \geq l^{-\alpha}\}$. Then the 1st intensity satisfies the following asymptotic on U:*

$$K_V(x) = \Phi_l(0,0)\,\mathbb{E}\left[|\overset{\perp}{\det} DF_l(x)|\big|f_l(x)=0, Vf_l(x)=0\right] = \frac{\sqrt{2}}{4\pi^2}l^2 + \mathcal{O}_V(l^{1+3\alpha})$$

as $l \to \infty$, where the remainder terms are uniformly bounded in $x \in U$ but have a dependence on V and possibly its derivatives. Over $U^{\complement}$, we have $K_V(x) = o_V(l^2)$ where once again the subscript notation of V denotes a bounded dependence on the derivatives of V.

PROOF. (of Proposition 3.5.3)

The crux of our proof is in precisely estimating the Gaussian integral

$$\frac{1}{2\pi\sqrt{\det \mathbf{\Delta}_l}}\int_{\mathbb{R}^2}|t_1t_2|\exp\left(-\frac{1}{2}\vec{t}\,\det\mathbf{\Delta_l}^{-1}\vec{t}\right)dt_1dt_2, \tag{3.5.4}$$

where

$$\mathbf{\Delta_l}^{-1} = \frac{1}{\det\mathbf{\Delta}_l}\begin{pmatrix} \frac{\|V\|^4}{8}l^4+\frac{\|V\|^4}{4}l^3+\frac{1}{\|V\|^2}\mathcal{O}(l^2) & -\frac{\tilde{a_{34}}(\theta_x,\phi_x)}{2}l^2-\frac{\tilde{a_{34}}(\theta_x,\phi_x)}{2}l \\ -\frac{\tilde{a_{34}}(\theta_x,\phi_x)}{2}l^2-\frac{\tilde{a_{34}}(\theta_x,\phi_x)}{2}l & \frac{\|V^{\perp}\|^2}{2}l^2+\frac{\|V^{\perp}\|^2}{2}l\end{pmatrix}.$$

Equation (3.5.4) follows directly from the Kac-Rice formula, the orthogonal determinant computed in equation (3.2.3), and expected value formulas for Gaussian random variables.

We first prove the bound for the first intensity over $U^{\complement}$. Performing the sequence of transformations:

$$\vec{t} = \sqrt{\det\mathbf{\Delta}_l}\,\vec{s} \Rightarrow \vec{s} = \frac{l^2}{\|V\|}(r_1,r_2)$$

we are then left with the quadratic form being the

$$\begin{pmatrix}\frac{\|V\|^6}{8}+\mathcal{O}(l^{-1}) & \|V\|^2\mathcal{O}(l^{-2}) \\ \|V\|^2\mathcal{O}(l^{-2}) & \frac{\sin\phi_x^2}{2}\frac{1}{l^2}+\mathcal{O}(l^{-3}).\end{pmatrix}$$

We can make use of our parameter α which dictates our localization around the vanishing set of V to make this quadratic form "almost" diagonal with positive eigenvalues. Hence, we set $\alpha < \frac{1}{3}$ in order to make the top-right entry be $> Cl^{-2}$. Let $\varepsilon(\alpha) = \frac{1}{3} - \alpha$. A direct calculation along with Lemma 3.5.1 then shows us that equation (3.5.4) is

$$\mathcal{O}(\|V\|^{13}l^{5-4\varepsilon(\alpha)}) = \mathcal{O}(l^{-13\alpha}l^{5-4\varepsilon(\alpha)})$$

which itself is little-o of l^2 if $\alpha > \frac{5}{27} = \frac{1}{3} - \frac{4}{27}$.

Let us localize away from the zero set $\{V = 0\}$ of our vector field V, specifically onto the set $U = \{x \in S^2 : \|V\| \geq l^{-\alpha}\}$ where α is as above. Here, $l \geq l_0(V)$ as established in Lemma 3.5.1. Working over this subset of S^2 allows us to perform our usual algebraic manipulations in the calculation to follow.

Thanks to the explicit form of the inverse of our conditional covariance matrix, we can perform the following sequence of transformations

$$\vec{t} = \sqrt{\det \mathbf{\Delta}_l}\,\vec{s} \Rightarrow \vec{s} = \left(\frac{r_1}{l^2}, \frac{r_2}{l}\right) \Rightarrow \vec{r} = \left(\frac{2\sqrt{2}}{\|V\|^2}u_1, \frac{\sqrt{2}}{\|V^\perp\|}u_2\right)$$

to reduce the quadratic form appearing in the exponential to

$$\begin{pmatrix} 1+\frac{2}{l}+\frac{1}{\|V\|^6}\mathcal{O}(\frac{1}{l^2}) & -\frac{2\tilde{a_{34}}(\theta_x,\phi_x)}{\|V\|^2\|V^\perp\|}\frac{1}{l}-\frac{2\tilde{a_{34}}(\theta_x,\phi_x)}{\|V\|^2\|V^\perp\|}\frac{1}{l^2} \\ -\frac{2\tilde{a_{34}}(\theta_x,\phi_x)}{\|V\|^2\|V^\perp\|}\frac{1}{l}-\frac{2\tilde{a_{34}}(\theta_x,\phi_x)}{\|V\|^2\|V^\perp\|}\frac{1}{l^2} & 1+\frac{1}{l} \end{pmatrix}$$

$$=\begin{pmatrix} 1+\frac{2}{l} & 0 \\ 0 & 1+\frac{1}{l}\end{pmatrix}+\begin{pmatrix} \frac{1}{\|V\|^6}\mathcal{O}(\frac{1}{l^2}) & -\frac{2\tilde{a_{34}}(\theta_x,\phi_x)}{\|V\|^2\|V^\perp\|}\frac{1}{l}-\frac{2\tilde{a_{34}}(\theta_x,\phi_x)}{\|V\|^2\|V^\perp\|}\frac{1}{l^2} \\ -\frac{2\tilde{a_{34}}(\theta_x,\phi_x)}{\|V\|^2\|V^\perp\|}\frac{1}{l}-\frac{2\tilde{a_{34}}(\theta_x,\phi_x)}{\|V\|^2\|V^\perp\|}\frac{1}{l^2} & 0 \end{pmatrix}$$

The quadratic polynomial $q(u_1,u_2)$ in u_1,u_2 which is generated by this sum of matrices is

$$\begin{aligned} q(u_1,u_2) &= \left(1+\frac{2}{l}+\frac{1}{\|V\|^6}\mathcal{O}\left(\frac{1}{l^2}\right)\right)u_1^2+\left(1+\frac{1}{l}\right)u_2^2 \\ &+\left(-\frac{4\tilde{a_{34}}(\theta_x,\phi_x)}{\|V\|^2\|V^\perp\|}\frac{1}{l}-\frac{4\tilde{a_{34}}(\theta_x,\phi_x)}{\|V\|^2\|V^\perp\|}\frac{1}{l^2}\right)u_1u_2. \end{aligned}$$

On the set $U\subset S^2$, in the region $|u_1|<2|u_2|$, $q(u_1,u_2)=u_1^2+(1+\mathcal{O}(l^{-(1-3\alpha)}))u_2^2$ which follows from the assumption that V must not be allowed to become too small in l which in turn requires that $\frac{\tilde{a_{34}}(\theta_x,\phi_x)}{\|V\|^2\|V^\perp\|l} = \mathcal{O}(\frac{1}{\|V\|^3 l}) = \mathcal{O}\left(\frac{1}{l^{1-3\alpha}}\right)$; the uniform boundedness in (θ_x,ϕ_x) follows as $\frac{\tilde{a_{34}}(\theta_x,\phi_x)}{\sin\phi_x}\in C^\infty(S^2)$ after using the explicit form of the entries of the conditional covariance matrix as given in Remark 3.3.

In the complementary region $|u_1|\geq 2|u_2|$, we have that

$$q(u_1,u_2) = \left(1+\mathcal{O}(l^{-(1-3\alpha)})\right)u_1^2+(1+l^{-1})u_2.$$

Therefore, on all of $\mathbb{R}^2$, we have that $q(u_1,u_2) = \left(1+\mathcal{O}(l^{-(1-3\alpha)})\right)u_1^2+(1+\mathcal{O}(l^{-(1-3\alpha)}))u_2^2$. Thus, after a series of reductions, we are left to estimating the quantity

$$\begin{aligned} &\frac{(\det\mathbf{\Delta}_l)^{3/2}}{\pi}\frac{8}{\|V\|^4\|V^\perp\|^2}l^{-6}\int_{\mathbb{R}^2}|u_1u_2|\exp\left(-\frac{1}{2}q(u_1,u_2)\right)du_1\,du_2 \\ &=\frac{(\det\mathbf{\Delta}_l)^{3/2}}{\pi}\frac{8}{\|V\|^4\|V^\perp\|^2}l^{-6}\times \\ &\int_{\mathbb{R}^2}|u_1u_2|\exp\left(-\frac{1}{2}\left(\left(1+\mathcal{O}(l^{-(1-3\alpha)})\right)u_1^2+(1+\mathcal{O}(l^{-(1-3\alpha)}))u_2^2\right)\right)du_1\,du_2. \end{aligned}$$

Hence, our leading term in our asymptotic will come from the expression

$$\frac{(\det\mathbf{\Delta}_l)^{3/2}}{\pi}\frac{8}{\|V\|^4\|V^\perp\|^2}l^{-6}\int_{R^2}|u_1u_2|\exp\left(-\frac{1}{2}\left(u_1^2+u_2^2\right)\right)du_1\,du_2.$$

Using that $\det(\mathbf{\Delta}_l(\theta_x,\phi_x)) = \frac{\|V\|^4\|V^\perp\|^2}{16}l^6+\mathcal{O}(l^5)$, the simple calculation

$$\int_{R^2}|u_1u_2|\exp\left(-\frac{1}{2}(u_1^2+u_2^2)\right)du_1\,du_2 = \left(2\int_0^\infty -\frac{d}{du_1}\exp\left(-\frac{1}{2}u_1^2\right)du_1\right)^2 = 4,$$

and the remaining factors in equation (3.5.2) for the 1st intensity, we find that the asymptotic of the 1st intensity (on our neighborhood U) equals

$$= \Phi_l(0,0)\,\mathbb{E}\left[|\overset{\perp}{\det}\, DF_l(x)||f_l(x)=0, Vf_l(x)=0\right]$$

$$= \frac{\sqrt{2}}{4\pi^2}\,l^2 + \mathcal{O}_V\left(l^{1+3\alpha}\right)$$

where the big-O terms have implicit constants which are uniformly bounded in θ_x, ϕ_x.

REMARK 3.5.5. We observe throughout our calculation that $\frac{5}{27} < \alpha < 1/3$ is necessary for suitable remainder terms.

□

3.6. Proof of Theorem 1.5.1. We conclude our note with the proof of the proposed expected value asymptotic:

PROOF. We remind ourselves that $U := \{x \in S^2 : \|V\| \geq l^{-\alpha}\}$. Given Proposition 3.5.3, we can now proceed to integrating over $S^2 - \{V = 0\}$; recall that dV_g is normalized. Note also that we now have an approximate expression for the truncated first intensity of the form:

$$\left(\frac{\sqrt{2}}{4\pi^2}\,l^2 + \mathcal{O}_V\left(l^{1+3\alpha}\right)\right)\mathbf{1}_U(x) + \left(o_V(l^2)\right)\mathbf{1}_{U^\complement}(x)$$

In the case of $\{V = 0\}$ being finite, we know that the volume of each zero's neighborhood is asymptotic to $l^{-2\frac{\alpha}{m}}$ where m is the smallest order amongst all the zeroes of V. Indeed, let $x \in \{V = 0\}$ with V having vanishing order m, then one can use finitely many balls of radius $r = Cl^{-\alpha/m}$ to cover $\{V = 0\}$ and therefore estimate the area of U accordingly. Comparing the integrated bound from near the zeroes of $l^{2-2\frac{\alpha}{m}}\log(l)$ (and then summing over the zeroes) and the corresponding bound from away of $\frac{\sqrt{2}}{4\pi^2}l^2 + \mathcal{O}\left(l^{1+3\alpha}\right) + \mathcal{O}\left(l^{2-2\frac{\alpha}{m}}\right)$, we find that setting $\frac{5}{54} < \alpha = \frac{1}{3+\frac{2}{m}} < 1/3$ gives us our best remainder estimate.

In the second case of $\{V = 0\}$ being a smooth curve of finite length (thanks to 0 being a regular value of V and us being in a compact setting) and V vanishes to order m in the normal directions, we obtain an integrated bound from near the zeroes of $l^{2-\frac{\alpha}{m}}$; proceeding similarly as above we obtain that $\frac{5}{54} < \alpha = \frac{1}{3+\frac{1}{m}} < 1/3$ gives the best remainder estimate. □

Appendix A. Computation of Covariance Matrix

Before we begin, let us remind ourselves that $P_l(t)$ is the standard l-th degree Legendre polynomial and

$$h(x,y) := \tilde{h}(\theta_x, \theta_y, \phi_x, \phi_y) = \cos\phi_x \cos\phi_y + \sin\phi_x \sin\phi_y \cos(\theta_x - \theta_y).$$

We remind ourselves of the coordinate representations

$$V_y = v_1(\theta_y, \phi_y)\frac{\partial}{\partial\theta_y} + v_2(\theta_y, \phi_y)\frac{\partial}{\partial\phi_y}$$

and $V_y^\perp = v_2(\theta_y, \phi_y)\frac{\partial}{\partial\theta_y} - \sin^2\phi_y v_1(\theta_y, \phi_y)\frac{\partial}{\partial\phi_y}$. Similar definitions hold for V_x and $V_x^\perp$.

For the ease of exposition, let us list the specific formulas for the entries of the full covariance matrix for the random field $(f_l, Vf_l, V^{\perp}f_l, VVf_l) \in \mathbb{R}^4$:

- $a_{11} = P_l(h(x,y))_{|x=y}$
- $a_{12} = V_y P_l(h(x,y))_{|x=y}$
- $a_{13} = V_y^{\perp} P_l(h(x,y))_{|x=y}$
- $a_{14} = V_y V_y P_l(h(x,y))_{|x=y}$
- $a_{22} = V_x V_y P_l(h(x,y))_{|x=y}$
- $a_{23} = V_x V_y^{\perp} P_l(h(x,y))_{|x=y}$
- $a_{24} = V_x V_y V_y P_l(h(x,y))_{|x=y}$
- $a_{33} = V_x^{\perp} V_y^{\perp} P_l(h(x,y))_{|x=y}$
- $a_{34} = V_x^{\perp} V_y V_y P_l(h(x,y))_{|x=y}$
- $a_{44} = V_x V_x V_y V_y P_l(h(x,y))_{|x=y}$

We now proceed to calculating these entries along the diagonal:

<u>a_{11}</u>:

$$P_l(h)_{|\theta_x=\theta_y,\phi_x=\phi_y} = P_l(1)$$

<u>a_{12}</u>:

$$\begin{aligned}&v_1^y P_l'(h)(\sin\phi_x \sin\phi_y \sin(\theta_x-\theta_y))\\ &+v_2^y P_l'(h)(-\cos\phi_x \sin\phi_y + \sin\phi_x \cos\phi_y cos(\theta_x-\theta_y)_{|\theta_x=\theta_y,\phi_x=\phi_y} = 0\end{aligned}$$

<u>a_{13}</u>:

$$\begin{aligned}&v_2^y P_l'(h)(\sin\phi_x \sin\phi_y \sin(\theta_x-\theta_y)) + (-\sin^2\phi_y)v_1^y P_l'(h)(-\cos\phi_x \sin\phi_y\\ &+\sin\phi_x \cos\phi_y cos(\theta_x-\theta_y))_{|\theta_x=\theta_y,\phi_x=\phi_y} = 0\end{aligned}$$

<u>a_{14}</u>: We set $T_1 := v_1^y(\theta_y,\phi_y)\frac{\partial}{\partial\theta_y}$ and $T_2 := v_2^y(\theta_y,\phi_y)\frac{\partial}{\partial\phi_y}$. This will allow us to organize our derivative calculations more easily. We continue working with the general formula, pre-evaluation at $x = y$, appearing for a_{12}. We break this calculation into blocks arising from different applications of the vector fields T_1 and T_2.

$$\begin{aligned}
\text{block } 1 = T_1(S_1):\quad & v_1^y \frac{\partial v_1^y}{\partial\theta_y} P_l'(h)(\sin\phi_x\sin\phi_y\sin(\theta_x-\theta_y)) + (v_1^y)^2 P_l''(h)\\
&\cdot(\sin\phi_x\sin\phi_y\sin(\theta_x-\theta_y))^2\\
&+(v_1^y)^2 P_l'(h)(-\sin\phi_x\sin\phi_y\cos(\theta_x-\theta_y))\\
+\text{block } 2 = T_1(S_2):\quad & v_1^y \frac{\partial v_2^y}{\partial\theta_y} P_l'(h)(-\cos\phi_x\sin\phi_y + \sin\phi_x\cos\phi_y\cos(\theta_x-\theta_y)) +\\
& v_1^y v_2^y P_l''(h)(\sin\phi_x\sin\phi_y\sin(\theta_x-\theta_y))\\
&\cdot(-\cos\phi_x\sin\phi_y + \sin\phi_x\cos\phi_y\cos(\theta_x-\theta_y)) +\\
& v_1^y v_2^y P_l'(h)(\sin\phi_x\cos\phi_y\sin(\theta_x-\theta_y))\\
+\text{block } 3 = T_2(S_1):\quad & \frac{\partial v_1^y}{\partial\phi_y} v_2^y P_l'(h)(\sin\phi_x\sin\phi_y\sin(\theta_x-\theta_y)) +\\
& v_1^y v_2^y P_l''(h)(-\cos\phi_x\sin\phi_y + \sin\phi_x\cos\phi_y\cos(\theta_x-\theta_y))\\
&\cdot(\sin\phi_x\sin\phi_y\sin(\theta_x-\theta_y)) +\\
& v_1^y v_2^y P_l'(h)(\sin\phi_x\cos\phi_y\sin(\theta_x-\theta_y))
\end{aligned}$$

$$+\text{block } 4 = T_2(S_2): \quad v_2^y \frac{\partial v_2^y}{\partial \phi_y} P_l'(h)(-\cos\phi_x \sin\phi_y + \sin\phi_x \cos\phi_y \cos(\theta_x - \theta_y)) + (v_2^y)^2 P_l''(h)(-\cos\phi_x \sin\phi_y + \sin\phi_x \cos\phi_y \cos(\theta_x - \theta_y))^2 + (v_2^y)^2 P_l'(h) \cdot(-\cos\phi_x \cos\phi_y - \sin\phi_x \sin\phi_y \cos(\theta_x - \theta_y))_{|\theta_x=\theta_y, \phi_x=\phi_y}$$

$$= \text{final:} \quad -(v_1^x)^2 P_l'(1)(\sin^2\phi_x) - (v_2^x)^2 P_l'(1) = -\|V\|^2_{g(x)} P_l'(1)$$

$\underline{a_{22}}$:

$$\text{block 1:} \quad -v_1^x v_1^y P_l''(h)(\sin\phi_x \sin\phi_y \sin(\theta_x - \theta_y))^2 + v_1^x v_1^y P_l'(h)\,(\sin\phi_x \sin\phi_y \cos(\theta_x - \theta_y))$$

$$+\text{block 2:} \quad v_1^x v_2^y P_l''(h)(-\sin\phi_x \sin\phi_y \sin(\theta_x - \theta_y)) \cdot(-\cos\phi_x \sin\phi_y + \sin\phi_x \cos\phi_y \cos(\theta_x - \theta_y)) + v_1^x v_2^y P_l'(h)(-\sin\phi_x \cos\phi_y \sin(\theta_x - \theta_y))$$

$$+\text{block 3:} \quad v_1^y v_2^x P_l''(h)(-\sin\phi_x \cos\phi_y + \cos\phi_x \sin\phi_y \cos(\theta_x - \theta_y)) \cdot(\sin\phi_x \sin\phi_y \sin(\theta_x - \theta_y)) + v_1^y v_2^x P_l'(h)(\cos\phi_x \sin\phi_y \sin(\theta_x - \theta_y))$$

$$+\text{block 4:} \quad v_2^x v_2^y P_l''(-\sin\phi_x \cos\phi_y + \cos\phi_y \sin\phi_y \cos(\theta_x - \theta_y)) \cdot(-\cos\phi_x \sin\phi_y + \sin\phi_x \cos\phi_y \cos(\theta_x - \theta_y)) + v_2^x v_2^y P_l'(\sin\phi_x \sin\phi_y + \cos\phi_x \cos\phi_y \cos(\theta_x - \theta_y))_{|\theta_x=\theta_y, \phi_x=\phi_y}$$

$$= \text{final:} \quad (v_1^x)^2 P_l'(1)(\sin^2\phi_x) + (v_2^x)^2 P_l'(1) = \|V\|^2_{g(x)} P_l'(1)$$

$\underline{a_{23}}$:

$$\text{block 1:} \quad -v_1^x v_2^y P_l''(h)(\sin\phi_x \sin\phi_y \sin(\theta_x - \theta_y))^2 + v_1^x v_2^y P_l'(h)(\sin\phi_x \sin\phi_y \cos(\theta_x - \theta_y))$$

$$+\text{block 2:} \quad -v_1^x(-\sin^2\phi_y) v_1^y P_l''(h)(\sin\phi_x \sin\phi_y \sin(\theta_x - \theta_y)) \cdot(-\cos\phi_x \sin\phi_y + \sin\phi_x \cos\phi_y \cos(\theta_x - \theta_y)) - v_1^x(-\sin^2\phi_y) v_1^y P_l'(h)(\sin\phi_x \cos\phi_y \sin(\theta_x - \theta_y))$$

$$+\text{block 3:} \quad v_2^x v_2^y P_l''(h)(-\sin\phi_x \cos\phi_y + \cos\phi_x \sin\phi_y \cos(\theta_x - \theta_y)) \cdot(\sin\phi_x \sin\phi_y \sin(\theta_x - \theta_y)) + v_2^x v_2^y P_l'(h)(\cos\phi_x \sin\phi_y \sin(\theta_x - \theta_y))$$

$$+\text{block 4:} \quad v_2^x(-\sin^2\phi_y) v_1^y P_l''(h)(-\sin\phi_x \cos\phi_y + \cos\phi_x \sin\phi_y \cos(\theta_x - \theta_y)) \cdot(-\cos\phi_x \sin\phi_y + \sin\phi_x \cos\phi_y \cos(\theta_x - \theta_y)) + v_2^x(-\sin^2\phi_y) v_1^y P_l'(h)(\sin\phi_x \sin\phi_y + \cos\phi_x \cos\phi_y \cos(\theta_x - \theta_y))_{|\theta_x=\theta_y, \phi_x=\phi_y}$$

$$= \text{final:} \quad v_1^x v_2^x(\sin^2\phi_x) P_l'(1) + (-\sin^2\phi_x) v_1^x v_2^x P_l'(1) = 0$$

$\underline{a_{33}}$: We continue working with the general formula, pre-evaluation at $x = y$, appearing for a_{13}. We break this calculation into blocks arising from different applications of the components of the vector fields $V^\perp$.

$$\begin{aligned}
\text{block 1:}\quad & v_2^x v_2^y P_l''(h)(-\sin\phi_x \sin\phi_y \sin(\theta_x-\theta_y))(\sin\phi_x\sin\phi_y\sin(\theta_x-\theta_y)) \\
& +v_2^x v_2^y P_l'(h)(\sin\phi_x\sin\phi_y\cos(\theta_x-\theta_y))
\end{aligned}$$

$$\begin{aligned}
+\text{block 2:}\quad & -\sin^2\phi_y v_2^x v_1^y P_l''(h) \\
& (-\sin\phi_x\sin\phi_y\sin(\theta_x-\theta_y))(-\cos\phi_x\sin\phi_y+\sin\phi_x\cos\phi_y\cos(\theta_x-\theta_y)) \\
& -\sin^2\phi_y v_2^x v_1^y P_l'(h)(-\sin\phi_x\cos\phi_y\sin(\theta_x-\theta_y))
\end{aligned}$$

$$\begin{aligned}
+\text{block 3:}\quad & -\sin^2\phi_x v_1^x v_2^y P_l''(h)(-\sin\phi_x\cos\phi_y+\cos\phi_x\sin\phi_y\cos(\theta_x-\theta_y)) \\
& (\sin\phi_x\sin\phi_y\sin(\theta_x-\theta_y)) \\
& -\sin^2\phi_x v_1^x v_2^y P_l'(h)(\cos\phi_x\sin\phi_y\sin(\theta_x-\theta_y))
\end{aligned}$$

$$\begin{aligned}
+\text{block 4:}\quad & (-\sin^2\phi_x)(-\sin^2\phi_y)v_1^x v_1^y P_l''(h) \\
& \cdot(-\sin\phi_x\cos\phi_y+\cos\phi_x\sin\phi_y\cos(\theta_x-\theta_y)) \\
& \cdot(-\cos\phi_x\sin\phi_y+\sin\phi_x\cos\phi_y\cos(\theta_x-\theta_y)) \\
& +(-\sin^2\phi_x)(-\sin^2\phi_y)v_1^x v_1^y P_l'(h) \\
& \cdot(\sin\phi_x\sin\phi_y+\cos\phi_x\cos\phi_y\cos(\theta_x-\theta_y))_{|\theta_x=\theta_y,\phi_x=\phi_y}
\end{aligned}$$

$$=\text{final:}\quad \sin^2\phi_x(v_2^x)^2P_l'(1)+\sin^4\phi_x(v_1^x)^2P_l'(1) = \|V^\perp\|^2_{g(x)}P_l'(1)$$

$\underline{a_{24}}$: Let $T_1 = v_1^x \frac{\partial}{\partial\theta_x}$ and $T_2 = v_2^x\frac{\partial}{\partial\phi_x}$. We continue working with the general formula, pre-evaluation at $x = y$, appearing for a_{14}. We break this calculation into blocks arising from different applications of the vector fields T_1 and T_2.

$$\begin{aligned}
T_1\text{ (block 1), S1:}\quad & v_1^x v_1^y \frac{\partial v_1^y}{\partial\theta_y} P_l''(h)(-\sin\phi_x\sin\phi_y\sin(\theta_x-\theta_y)) \\
& \cdot(\sin\phi_x\sin\phi_y\sin(\theta_x-\theta_y)) \\
& +v_1^x v_1^y \frac{\partial v_1^y}{\partial\theta_y} P_l'(h)(\sin\phi_x\sin\phi_y\cos(\theta_x-\theta_y))
\end{aligned}$$

$$\begin{aligned}
+T_1\text{ (block 1), S2:}\quad & v_1^x (v_1^y)^2 P_l'''(h)(-\sin\phi_x\sin\phi_y\sin(\theta_x-\theta_y)) \\
& \cdot(\sin\phi_x\sin\phi_y\sin(\theta_x-\theta_y))^2 \\
& +2v_1^x(v_1^y)^2P_l''(h)(\sin\phi_x\sin\phi_y\sin(\theta_x-\theta_y)) \\
& \cdot(\sin\phi_x\sin\phi_y\cos(\theta_x-\theta_y))
\end{aligned}$$

$$\begin{aligned}
+T_1\text{ (block 1), S3:}\quad & v_1^x(v_1^y)^2P_l''(h)(-\sin\phi_x\sin\phi_y\sin(\theta_x-\theta_y)) \\
& \cdot(-\sin\phi_x\sin\phi_y\cos(\theta_x-\theta_y)) \\
& +v_1^x(v_1^y)^2P_l'(h)(\sin\phi_x\sin\phi_y\sin(\theta_x-\theta_y))
\end{aligned}$$

$$\begin{aligned}
+T_1\text{ (block 2), S1:}\quad & v_1^x v_1^y\frac{\partial v_2^y}{\partial\theta_y}P_l''(h)(-\sin\phi_x\sin\phi_y\sin(\theta_x-\theta_y)) \\
& \cdot(-\cos\phi_x\sin\phi_y+\sin\phi_x\cos\phi_y\cos(\theta_x-\theta_y)) \\
& +v_1^x v_1^y\frac{\partial v_2^y}{\partial\theta_y}P_l'(h)(-\sin\phi_x\cos\phi_y\sin(\theta_x-\theta_y))
\end{aligned}$$

$$+T_1 \text{ (block 2), S2:}\quad \begin{aligned} & v_1^x v_1^y v_2^y P_l'''(h)(-\sin\phi_x \sin\phi_y \sin(\theta_x - \theta_y)) \\ & \cdot(\sin\phi_x \sin\phi_y \sin(\theta_x - \theta_y)) \\ & \cdot(-\cos\phi_x \sin\phi_y + \sin\phi_x \cos\phi_y \cos(\theta_x - \theta_y)) \\ & + v_1^x v_1^y v_2^y P_l''(h)(\sin\phi_x \sin\phi_y \cos(\theta_x - \theta_y)) \\ & \cdot(-\cos\phi_x \sin\phi_y + \sin\phi_x \cos\phi_y \cos(\theta_x - \theta_y)) \\ & + v_1^x v_1^y v_2^y P_l''(h)(\sin\phi_x \sin\phi_y \sin(\theta_x - \theta_y)) \\ & \cdot(-\sin\phi_x \cos\phi_y \sin(\theta_x - \theta_y)) \end{aligned}$$

$$+T_1 \text{ (block 2), S3:}\quad \begin{aligned} & v_1^x v_1^y v_2^y P_l''(h)(-\sin\phi_x \sin\phi_y \sin(\theta_x - \theta_y)) \\ & \cdot(\sin\phi_x \cos\phi_y \sin(\theta_x - \theta_y)) \\ & + v_1^x v_1^y v_2^y P_l'(h)(\sin\phi_x \cos\phi_y \cos(\theta_x - \theta_y)) \end{aligned}$$

$$+T_1 \text{ (block 3), S1:}\quad \begin{aligned} & v_1^x \frac{\partial v_1^y}{\partial \phi_y} v_2^y P_l''(h)(-\sin\phi_x \sin\phi_y \sin(\theta_x - \theta_y)) \\ & \cdot(\sin\phi_x \sin\phi_y \sin(\theta_x - \theta_y)) + \\ & v_1^x \frac{\partial v_1^y}{\partial \phi_y} v_2^y P_l'(h)(\sin\phi_x \sin\phi_y \cos(\theta_x - \theta_y)) \end{aligned}$$

$$+T_1 \text{ (block 3), S2:}\quad \begin{aligned} & v_1^x v_1^y v_2^y P_l'''(h)(-\sin\phi_x \sin\phi_y \sin(\theta_x - \theta_y)) \\ & \cdot(-\cos\phi_x \sin\phi_y + \sin\phi_x \cos\phi_y \cos(\theta_x - \theta_y)) \\ & \cdot(\sin\phi_x \sin\phi_y \sin(\theta_x - \theta_y)) \\ & + v_1^x v_1^y v_2^y P_l''(h)(-\sin\phi_x \cos\phi_y \sin(\theta_x - \theta_y)) \\ & \cdot(\sin\phi_x \sin\phi_y \sin(\theta_x - \theta_y)) \\ & + v_1^x v_1^y v_2^y P_l''(h)(-\cos\phi_x \sin\phi_y + \sin\phi_x \cos\phi_y \cos(\theta_x - \theta_y)) \\ & \cdot(\sin\phi_x \sin\phi_y \cos(\theta_x - \theta_y)) \end{aligned}$$

$$+T_1 \text{ (block 3), S3:}\quad \begin{aligned} & + v_1^x v_1^y v_2^y P_l''(h)(-\sin\phi_x \sin\phi_y \sin(\theta_x - \theta_y)) \\ & \cdot(\sin\phi_x \cos\phi_y \sin(\theta_x - \theta_y)) \\ & + v_1^x v_1^y v_2^y P_l'(h)(\sin\phi_x \cos\phi_y \cos(\theta_x - \theta_y)) \end{aligned}$$

$$+T_1 \text{ (block 4), S1:}\quad \begin{aligned} & v_1^x v_2^y \frac{\partial v_2^y}{\partial \phi_y} P_l''(h)(-\sin\phi_x \sin\phi_y \sin(\theta_x - \theta_y))(-\cos\phi_x \sin\phi_y \\ & + \sin\phi_x \cos\phi_y \cos(\theta_x - \theta_y)) \\ & + v_1^x v_2^y \frac{\partial v_2^y}{\partial \phi_y} P_l'(h)(-\sin\phi_x \cos\phi_y \sin(\theta_x - \theta_y)) \end{aligned}$$

$$+T_1 \text{ (block 4), S2:}\quad \begin{aligned} & v_1^x (v_2^y)^2 P_l'''(h)(-\sin\phi_x \sin\phi_y \sin(\theta_x - \theta_y)) \\ & \cdot(-\cos\phi_x \sin\phi_y + \sin\phi_x \cos\phi_y \cos(\theta_x - \theta_y))^2 \\ & + 2v_1^x (v_2^y)^2 P_l''(h)(-\cos\phi_x \sin\phi_y + \sin\phi_x \cos\phi_y \cos(\theta_x - \theta_y)) \\ & \cdot(-\sin\phi_x \cos\phi_y \sin(\theta_x - \theta_y)) \end{aligned}$$

T_1 (block 4), S3: $$v_1^x (v_2^y)^2 P_l''(h)(-\sin\phi_x \sin\phi_y \sin(\theta_x-\theta_y))$$
$$\cdot(-\cos\phi_x\cos\phi_y - \sin\phi_x\sin\phi_y\cos(\theta_x-\theta_y))$$
$$+v_1^x (v_2^y)^2 P_l'(h)(\sin\phi_x\sin\phi_y\sin(\theta_x-\theta_y))$$

$+T_2$ (block 1), S1: $$v_2^x v_1^y \frac{\partial v_1^y}{\partial \theta_y} P_l''(h)(-\sin\phi_x\cos\phi_y + \cos\phi_x\sin\phi_y\cos(\theta_x-\theta_y))$$
$$\cdot(\sin\phi_x\sin\phi_y\sin(\theta_x-\theta_y)) + v_2^x v_1^y \frac{\partial v_1^y}{\partial \theta_y} P_l'(h)$$
$$\cdot(\cos\phi_x\sin\phi_y\sin(\theta_x-\theta_y))$$

$+T_2$ (block 1), S2: $$v_2^x (v_1^y)^2 P_l'''(h)(-\sin\phi_x\cos\phi_y + \cos\phi_x\sin\phi_y\cos(\theta_x-\theta_y))$$
$$\cdot(\sin\phi_x\sin\phi_y\sin(\theta_x-\theta_y))^2$$
$$+2v_2^x (v_1^y)^2 P_l''(h)(\sin\phi_x\sin\phi_y\sin(\theta_x-\theta_y))$$
$$\cdot(\cos\phi_x\sin\phi_y\sin(\theta_x-\theta_y))$$

$+T_2$ (block 1), S3: $$v_2^x (v_1^y)^2 P_l''(h)(-\sin\phi_x\cos\phi_y + \cos\phi_x\sin\phi_y\cos(\theta_x-\theta_y))$$
$$\cdot(-\sin\phi_x\sin\phi_y\cos(\theta_x-\theta_y))$$
$$+v_2^x (v_1^y)^2 P_l'(h)(-\cos\phi_x\sin\phi_y\cos(\theta_x-\theta_y))$$

$+T_2$ (block 2), S1: $$v_2^x v_1^y \frac{\partial v_2^y}{\partial \theta_y} P_l''(h)(-\sin\phi_x\cos\phi_y + \cos\phi_x\sin\phi_y\cos(\theta_x-\theta_y))$$
$$\cdot(-\cos\phi_x\sin\phi_y + \sin\phi_x\cos\phi_y\cos(\theta_x-\theta_y))$$
$$+v_2^x v_1^y \frac{\partial v_2^y}{\partial \theta_y} P_l'(h)\cdot(\sin\phi_x\sin\phi_y + \cos\phi_x\cos\phi_y\cos(\theta_x-\theta_y))$$

$+T_2$ (block 2), S2: $$v_2^x v_1^y v_2^y P_l'''(h)(-\sin\phi_x\cos\phi_y + \cos\phi_x\sin\phi_y\cos(\theta_x-\theta_y))$$
$$\cdot(\sin\phi_x\sin\phi_y\sin(\theta_x-\theta_y))$$
$$\cdot(-\cos\phi_x\sin\phi_y + \sin\phi_x\cos\phi_y\cos(\theta_x-\theta_y))$$
$$+v_2^x v_1^y v_2^y P_l''(h)(\cos\phi_x\sin\phi_y\sin(\theta_x-\theta_y))$$
$$\cdot(-\cos\phi_x\sin\phi_y + \sin\phi_x\cos\phi_y\cos(\theta_x-\theta_y))$$
$$+v_2^x v_1^y v_2^y P_l''(h)(\sin\phi_x\sin\phi_y\sin(\theta_x-\theta_y))$$
$$\cdot(\sin\phi_x\sin\phi_y + \cos\phi_x\cos\phi_y\cos(\theta_x-\theta_y))$$

$+T_2$ (block 2), S3: $$v_2^x v_1^y v_2^y P_l''(h)(-\sin\phi_x\cos\phi_y + \cos\phi_x\sin\phi_y\cos(\theta_x-\theta_y))$$
$$\cdot(\sin\phi_x\cos\phi_y\sin(\theta_x-\theta_y))$$
$$+v_2^x v_1^y v_2^y P_l'(h)(\cos\phi_x\cos\phi_y\sin(\theta_x-\theta_y))$$

$+T_2$ (block 3), S1: $$v_2^x \frac{\partial v_1^y}{\partial \phi_y} v_2^y P_l''(h)(-\sin\phi_x\cos\phi_y + \cos\phi_x\sin\phi_y\cos(\theta_x-\theta_y))$$
$$\cdot(\sin\phi_x\sin\phi_y\sin(\theta_x-\theta_y))$$
$$+v_2^x \frac{\partial v_1^y}{\partial \phi_y} v_2^y P_l'(h)(\cos\phi_x\sin\phi_y\sin(\theta_x-\theta_y))$$

$+T_2$ (block 3), S2:
$$v_1^y v_2^x v_2^y P_l'''(h)(-\sin\phi_x\cos\phi_y+\cos\phi_x\sin\phi_y\cos(\theta_x-\theta_y))$$
$$\cdot(-\cos\phi_x\sin\phi_y+\sin\phi_x\cos\phi_y\cos(\theta_x-\theta_y))$$
$$\cdot(\sin\phi_x\sin\phi_y\sin(\theta_x-\theta_y))$$
$$+v_1^y v_2^x v_2^y P_l''(h)(\sin\phi_x\sin\phi_y+\cos\phi_x\cos\phi_y\cos(\theta_x-\theta_y))$$
$$\cdot(\sin\phi_x\sin\phi_y\sin(\theta_x-\theta_y))$$
$$+v_1^y v_2^x v_2^y P_l''(h)(-\cos\phi_x\sin\phi_y+\sin\phi_x\cos\phi_y\cos(\theta_x-\theta_y))$$
$$\cdot(\cos\phi_x\sin\phi_y\sin(\theta_x-\theta_y))$$

T_2 (block 3), S3:
$$v_2^x v_1^y v_2^y P_l''(h)(-\sin\phi_x\cos\phi_y+\cos\phi_x\sin\phi_y\cos(\theta_x-\theta_y))$$
$$\cdot(\sin\phi_x\cos\phi_y\sin(\theta_x-\theta_y))$$
$$+v_2^x v_1^y v_2^y P_l'(h)(\cos\phi_x\cos\phi_y\sin(\theta_x-\theta_y))$$

$+T_2$ (block 4), S1:
$$v_2^x v_2^y \frac{\partial v_2^y}{\partial\phi_y} P_l''(h)(-\sin\phi_x\cos\phi_y+\cos\phi_x\sin\phi_y\cos(\theta_x-\theta_y))$$
$$\cdot(-\cos\phi_x\sin\phi_y+\sin\phi_x\cos\phi_y\cos(\theta_x-\theta_y))$$
$$+v_2^x v_2^y \frac{\partial v_2^y}{\partial\phi_y} P_l'(h)(\sin\phi_x\sin\phi_y+\cos\phi_x\cos\phi_y\cos(\theta_x-\theta_y))$$

$+T_2$ (block 4), S2:
$$v_2^x (v_2^y)^2 P_l'''(h)(-\sin\phi_x\cos\phi_y+\cos\phi_x\sin\phi_y\cos(\theta_x-\theta_y))$$
$$\cdot(-\cos\phi_x\sin\phi_y+\sin\phi_x\cos\phi_y\cos(\theta_x-\theta_y))^2$$
$$+2v_2^x (v_2^y)^2 P_l''(h)(-\cos\phi_x\sin\phi_y+\sin\phi_x\cos\phi_y\cos(\theta_x-\theta_y))$$
$$\cdot(\sin\phi_x\sin\phi_y+\cos\phi_x\cos\phi_y\cos(\theta_x-\theta_y))$$

$+T_2$ (block 4), S3:
$$v_2^x (v_2^y)^2 P_l''(h)(-\sin\phi_x\cos\phi_y+\cos\phi_x\sin\phi_y\cos(\theta_x-\theta_y))$$
$$\cdot(-\cos\phi_x\cos\phi_y-\sin\phi_x\sin\phi_y\cos(\theta_x-\theta_y))$$
$$+v_2^x (v_2^y)^2 P_l'(h)$$
$$\cdot(\sin\phi_x\cos\phi_y-\cos\phi_x\sin\phi_y\cos(\theta_x-\theta_y))_{|\theta_x=\theta_y,\phi_x=\phi_y}$$

$$= \text{final} : \Big((v_1^x)^2\frac{\partial v_1^x}{\partial\theta_x}\sin^2\phi_x+(v_1^x)^2 v_2^x\sin\phi_x\cos\phi_x+v_1^x\frac{\partial v_1^x}{\partial\phi_x}v_2^x\sin^2\phi_x$$
$$+(v_1^x)^2 v_2^x\sin\phi_x\cos\phi_x$$
$$-v_2^x(v_1^x)^2\sin\phi_x\cos\phi_x+v_1^x v_2^x\frac{\partial v_2^x}{\partial\theta_x}+(v_2^x)^2\frac{\partial v_2^x}{\partial\phi_x}\Big)P_l'(1)$$

$$=\Big((v_1^x)^2\frac{\partial v_1^x}{\partial\theta_x}\sin^2\phi_x+v_1^x\frac{\partial v_1^x}{\partial\phi_x}v_2^x\sin^2\phi_x+(v_1^x)^2v_2^x\sin\phi_x\cos\phi_x$$
$$+v_1^x v_2^x\frac{\partial v_2^x}{\partial\theta_x}+(v_2^x)^2\frac{\partial v_2^x}{\partial\phi_x}\Big)P_l'(1)$$

$\underline{a_{34}}$: We notice this entry is exactly the same as a_{24} except that for every occurence of v_1^x and v_2^x, we substitute with $-\sin^2\phi_x v_2^x$ and v_1^x, respectively. Let $T_1 = v_2^x\frac{\partial}{\partial\theta_x}$ and $T_2 = -\sin^2\phi_x v_1^x\frac{\partial}{\partial\phi_x}$

T_1 (block 1), S1:

$$v_2^x v_1^y \frac{\partial v_1^y}{\partial \theta_y} P_l''(h)(-\sin\phi_x \sin\phi_y \sin(\theta_x-\theta_y)) \cdot(\sin\phi_x \sin\phi_y \sin(\theta_x-\theta_y)) + v_2^x v_1^y \frac{\partial v_1^y}{\partial \theta_y} P_l'(h)(\sin\phi_x \sin\phi_y \cos(\theta_x-\theta_y))$$

$+T_1$ (block 1), S2:

$$v_2^x (v_1^y)^2 P_l'''(h)(-\sin\phi_x \sin\phi_y \sin(\theta_x-\theta_y)) \cdot(\sin\phi_x \sin\phi_y \sin(\theta_x-\theta_y))^2 + 2v_2^x (v_1^y)^2 P_l''(h)(\sin\phi_x \sin\phi_y \sin(\theta_x-\theta_y)) \cdot(\sin\phi_x \sin\phi_y \cos(\theta_x-\theta_y))$$

$+T_1$ (block 1), S3:

$$v_2^x (v_1^y)^2 P_l''(h)(-\sin\phi_x \sin\phi_y \sin(\theta_x-\theta_y)) \cdot(-\sin\phi_x \sin\phi_y \cos(\theta_x-\theta_y)) + v_2^x (v_1^y)^2 P_l'(h)(\sin\phi_x \sin\phi_y \sin(\theta_x-\theta_y))$$

$+T_1$ (block 2), S1:

$$v_2^x v_1^y \frac{\partial v_2^y}{\partial \theta_y} P_l''(h)(-\sin\phi_x \sin\phi_y \sin(\theta_x-\theta_y)) \cdot(-\cos\phi_x \sin\phi_y + \sin\phi_x \cos\phi_y \cos(\theta_x-\theta_y)) + v_2^x v_1^y \frac{\partial v_2^y}{\partial \theta_y} P_l'(h)(-\sin\phi_x \cos\phi_y \sin(\theta_x-\theta_y))$$

$+T_1$ (block 2), S2:

$$v_2^x v_1^y v_2^y P_l'''(h)(-\sin\phi_x \sin\phi_y \sin(\theta_x-\theta_y)) \cdot(\sin\phi_x \sin\phi_y \sin(\theta_x-\theta_y)) \cdot(-\cos\phi_x \sin\phi_y + \sin\phi_x \cos\phi_y \cos(\theta_x-\theta_y)) + v_2^x v_1^y v_2^y P_l''(h)(\sin\phi_x \sin\phi_y \cos(\theta_x-\theta_y)) \cdot(-\cos\phi_x \sin\phi_y + \sin\phi_x \cos\phi_y \cos(\theta_x-\theta_y)) + v_2^x v_1^y v_2^y P_l''(h)(\sin\phi_x \sin\phi_y \sin(\theta_x-\theta_y)) \cdot(-\sin\phi_x \cos\phi_y \sin(\theta_x-\theta_y))$$

$+T_1$ (block 2), S3:

$$v_2^x v_1^y v_2^y P_l''(h)(-\sin\phi_x \sin\phi_y \sin(\theta_x-\theta_y)) \cdot(\sin\phi_x \cos\phi_y \sin(\theta_x-\theta_y)) + v_2^x v_1^y v_2^y P_l'(h)(\sin\phi_x \cos\phi_y \cos(\theta_x-\theta_y))$$

$+T_1$ (block 3), S1:

$$v_2^x \frac{\partial v_1^y}{\partial \phi_y} v_2^y P_l''(h)(-\sin\phi_x \sin\phi_y \sin(\theta_x-\theta_y)) \cdot(\sin\phi_x \sin\phi_y \sin(\theta_x-\theta_y)) + v_2^x \frac{\partial v_1^y}{\partial \phi_y} v_2^y P_l'(h)(\sin\phi_x \sin\phi_y \cos(\theta_x-\theta_y))$$

$+T_1$ (block 3), S2:
$$v_2^x v_1^y v_2^y P_l'''(h)(-\sin\phi_x \sin\phi_y \sin(\theta_x - \theta_y))$$
$$\cdot(-\cos\phi_x \sin\phi_y + \sin\phi_x \cos\phi_y \cos(\theta_x - \theta_y))$$
$$\cdot(\sin\phi_x \sin\phi_y \sin(\theta_x - \theta_y))$$
$$+v_2^x v_1^y v_2^y P_l''(h)(-\sin\phi_x \cos\phi_y \sin(\theta_x - \theta_y))$$
$$\cdot(\sin\phi_x \sin\phi_y \sin(\theta_x - \theta_y))$$
$$+v_2^x v_1^y v_2^y P_l''(h)(-\cos\phi_x \sin\phi_y + \sin\phi_x \cos\phi_y \cos(\theta_x - \theta_y))$$
$$\cdot(\sin\phi_x \sin\phi_y \cos(\theta_x - \theta_y))$$

$+T_1$ (block 3), S3:
$$+v_2^x v_1^y v_2^y P_l''(h)(-\sin\phi_x \sin\phi_y \sin(\theta_x - \theta_y))$$
$$\cdot(\sin\phi_x \cos\phi_y \sin(\theta_x - \theta_y))$$
$$+v_2^x v_1^y v_2^y P_l'(h)(\sin\phi_x \cos\phi_y \cos(\theta_x - \theta_y))$$

$+T_1$ (block 4), S1:
$$v_2^x v_2^y \frac{\partial v_2^y}{\partial \phi_y} P_l''(h)(-\sin\phi_x \sin\phi_y \sin(\theta_x - \theta_y)(-\cos\phi_x \sin\phi_y$$
$$+\sin\phi_x \cos\phi_y \cos(\theta_x - \theta_y)) +$$
$$v_2^x v_2^y \frac{\partial v_2^y}{\partial \phi_y} P_l'(h)(-\sin\phi_x \cos\phi_y \sin(\theta_x - \theta_y))$$

T_1 (block 4), S2:
$$v_2^x (v_2^y)^2 P_l'''(h)(-\sin\phi_x \sin\phi_y \sin(\theta_x - \theta_y))$$
$$\cdot(-\cos\phi_x \sin\phi_y + \sin\phi_x \cos\phi_y \cos(\theta_x - \theta_y))^2$$
$$+2v_2^x (v_2^y)^2 P_l''(h)(-\cos\phi_x \sin\phi_y + \sin\phi_x \cos\phi_y \cos(\theta_x - \theta_y))$$
$$\cdot(-\sin\phi_x \cos\phi_y \sin(\theta_x - \theta_y))$$

T_1 (block 4), S3:
$$v_2^x (v_2^y)^2 P_l''(h)(-\sin\phi_x \sin\phi_y \sin(\theta_x - \theta_y))$$
$$\cdot(-\cos\phi_x \cos\phi_y - \sin\phi_x \sin\phi_y \cos(\theta_x - \theta_y))$$
$$+v_2^x (v_2^y)^2 P_l'(h)(\sin\phi_x \sin\phi_y \sin(\theta_x - \theta_y))$$

$+T_2$ (block 1), S1:
$$(-\sin^2\phi_x v_1^x) v_1^y \frac{\partial v_1^y}{\partial \theta_y} P_l''(h)$$
$$\cdot(-\sin\phi_x \cos\phi_y + \cos\phi_x \sin\phi_y \cos(\theta_x - \theta_y))$$
$$\cdot(\sin\phi_x \sin\phi_y \sin(\theta_x - \theta_y)) + (-\sin^2\phi_x v_1^x) v_1^y \frac{\partial v_1^y}{\delta \theta_y} P_l'(h)$$
$$\cdot(\cos\phi_x \sin\phi_y \sin(\theta_x - \theta_y))$$

T_2 (block 1), S2:
$$(-\sin^2\phi_x v_1^x)(v_1^y)^2 P_l'''(h)$$
$$\cdot(-\sin\phi_x \cos\phi_y + \cos\phi_x \sin\phi_y \cos(\theta_x - \theta_y))$$
$$\cdot(\sin\phi_x \sin\phi_y \sin(\theta_x - \theta_y))^2$$
$$+2(-\sin^2\phi_x v_1^x)(v_1^y)^2 P_l''(h)(\sin\phi_x \sin\phi_y \sin(\theta_x - \theta_y))$$
$$\cdot(\cos\phi_x \sin\phi_y \sin(\theta_x - \theta_y))$$

T_2 (block 1), S3:

$$\begin{aligned}
&(-\sin^2\phi_x v_1^x)(v_1^y)^2 P_l''(h)\\
&\cdot(-\sin\phi_x\cos\phi_y+\cos\phi_x\sin\phi_y\cos(\theta_x-\theta_y))\\
&\cdot(-\sin\phi_x\sin\phi_y\cos(\theta_x-\theta_y))\\
&+(-\sin^2\phi_x v_1^x)(v_1^y)^2 P_l'(h)(-\cos\phi_x\sin\phi_y\cos(\theta_x-\theta_y))
\end{aligned}$$

T_2 (block 2), S1:

$$\begin{aligned}
&(-\sin^2\phi_x v_1^x)v_1^y\frac{\partial v_2^y}{\partial\theta_y}P_l''(h)\\
&\cdot(-\sin\phi_x\cos\phi_y+\cos\phi_x\sin\phi_y\cos(\theta_x-\theta_y))\\
&\cdot(-\cos\phi_x\sin\phi_y+\sin\phi_x\cos\phi_y\cos(\theta_x-\theta_y))\\
&+(-\sin^2\phi_x v_1^x)v_1^y\frac{\partial v_2^y}{\partial\theta_y}P_l'(h)\\
&\cdot(\sin\phi_x\sin\phi_y+\cos\phi_x\cos\phi_y\cos(\theta_x-\theta_y))
\end{aligned}$$

T_2 (block 2), S2:

$$\begin{aligned}
&(-\sin^2\phi_x v_1^x)v_1^y v_2^y P_l'''(h)\\
&\cdot(-\sin\phi_x\cos\phi_y+\cos\phi_x\sin\phi_y\cos(\theta_x-\theta_y))\\
&\cdot(\sin\phi_x\sin\phi_y\sin(\theta_x-\theta_y))\\
&\cdot(-\cos\phi_x\sin\phi_y+\sin\phi_x\cos\phi_y\cos(\theta_x-\theta_y))\\
&+(-\sin^2\phi_x v_1^x)v_1^y v_2^y P_l''(h)(\cos\phi_x\sin\phi_y\sin(\theta_x-\theta_y))\\
&\cdot(-\cos\phi_x\sin\phi_y+\sin\phi_x\cos\phi_y\cos(\theta_x-\theta_y))\\
&+(-\sin^2\phi_x v_1^x)v_1^y v_2^y P_l''(h)(\sin\phi_x\sin\phi_y\sin(\theta_x-\theta_y))\\
&\cdot(\sin\phi_x\sin\phi_y+\cos\phi_x\cos\phi_y\cos(\theta_x-\theta_y))
\end{aligned}$$

T_2 (block 2), S3:

$$\begin{aligned}
&(-\sin^2\phi_x v_1^x)v_1^y v_2^y P_l''(h)\\
&\cdot(-\sin\phi_x\cos\phi_y+\cos\phi_x\sin\phi_y\cos(\theta_x-\theta_y))\\
&\cdot(\sin\phi_x\cos\phi_y\sin(\theta_x-\theta_y))\\
&+(-\sin^2\phi_x v_1^x)v_1^y v_2^y P_l'(h)(\cos\phi_x\cos\phi_y\sin(\theta_x-\theta_y))
\end{aligned}$$

T_2 (block 3), S1:

$$\begin{aligned}
&(-\sin^2\phi_x v_1^x)\frac{\partial v_1^y}{\partial\phi_y}v_2^y P_l''(h)\\
&\cdot(-\sin\phi_x\cos\phi_y+\cos\phi_x\sin\phi_y\cos(\theta_x-\theta_y))\\
&\cdot(\sin\phi_x\sin\phi_y\sin(\theta_x-\theta_y))\\
&+(-\sin^2\phi_x v_1^x)\frac{\partial v_1^y}{\partial\phi_y}v_2^y P_l'(h)(\cos\phi_x\sin\phi_y\sin(\theta_x-\theta_y))
\end{aligned}$$

T_2 (block 3), S2:
$$\begin{aligned}&v_1^y(-\sin^2\phi_x v_1^x)v_2^y P_l'''(h)\\&\cdot(-\sin\phi_x\cos\phi_y+\cos\phi_x\sin\phi_y\cos(\theta_x-\theta_y))\\&\cdot(-\cos\phi_x\sin\phi_y+\sin\phi_x\cos\phi_y\cos(\theta_x-\theta_y))\\&\cdot(\sin\phi_x\sin\phi_y\sin(\theta_x-\theta_y))\\&+v_1^y(-\sin^2\phi_x v_1^x)v_2^y P_l''(h)\\&\cdot(\sin\phi_x\sin\phi_y+\cos\phi_x\cos\phi_y\cos(\theta_x-\theta_y))\\&\cdot(\sin\phi_x\sin\phi_y\sin(\theta_x-\theta_y))\\&+v_1^y(-\sin^2\phi_x v_1^x)v_2^y P_l''(h)\\&\cdot(-\cos\phi_x\sin\phi_y+\sin\phi_x\cos\phi_y\cos(\theta_x-\theta_y))\\&\cdot(\cos\phi_x\sin\phi_y\sin(\theta_x-\theta_y))\end{aligned}$$

T_2 (block 3), S3:
$$\begin{aligned}&(-\sin^2\phi_x v_1^x)v_1^y v_2^y P_l''(h)\\&\cdot(-\sin\phi_x\cos\phi_y+\cos\phi_x\sin\phi_y\cos(\theta_x-\theta_y))\\&\cdot(\sin\phi_x\cos\phi_y\sin(\theta_x-\theta_y))\\&+(-\sin^2\phi_x v_1^x)v_1^y v_2^y P_l'(h)(\cos\phi_x\cos\phi_y\sin(\theta_x-\theta_y))\end{aligned}$$

T_2 (block 4), S1:
$$\begin{aligned}&(-\sin^2\phi_x v_1^x)v_2^y\frac{\partial v_2^y}{\partial\phi_y}P_l''(h)\\&\cdot(-\sin\phi_x\cos\phi_y+\cos\phi_x\sin\phi_y\cos(\theta_x-\theta_y))\\&\cdot(-\cos\phi_x\sin\phi_y+\sin\phi_x\cos\phi_y\cos(\theta_x-\theta_y))\\&+(-\sin^2\phi_x v_1^x)v_2^y\frac{\partial v_2^y}{\partial\phi_y}P_l'(h)\\&\cdot(\sin\phi_x\sin\phi_y+\cos\phi_x\cos\phi_y\cos(\theta_x-\theta_y))\end{aligned}$$

T_2 (block 4), S2:
$$\begin{aligned}&(-\sin^2\phi_x v_1^x)(v_2^y)^2 P_l'''(h)\\&\cdot(-\sin\phi_x\cos\phi_y+\cos\phi_x\sin\phi_y\cos(\theta_x-\theta_y))\\&\cdot(-\cos\phi_x\sin\phi_y+\sin\phi_x\cos\phi_y\cos(\theta_x-\theta_y))^2\\&+2(-\sin^2\phi_x v_1^x)(v_2^y)^2 P_l''(h)\\&\cdot(-\cos\phi_x\sin\phi_y+\sin\phi_x\cos\phi_y\cos(\theta_x-\theta_y))\\&\cdot(\sin\phi_x\sin\phi_y+\cos\phi_x\cos\phi_y\cos(\theta_x-\theta_y))\end{aligned}$$

T_2 (block 4), S3:
$$\begin{aligned}&(-\sin^2\phi_x v_1^x)(v_2^y)^2 P_l''(h)\\&\cdot(-\sin\phi_x\cos\phi_y+\cos\phi_x\sin\phi_y\cos(\theta_x-\theta_y))\\&\cdot(-\cos\phi_x\cos\phi_y-\sin\phi_x\sin\phi_y\cos(\theta_x-\theta_y))\\&+(-\sin^2\phi_x v_1^x)(v_2^y)^2 P_l'(h)\\&\cdot(\sin\phi_x\cos\phi_y-\cos\phi_x\sin\phi_y\cos(\theta_x-\theta_y))\end{aligned}$$

$$= \text{final} : \Big(v_2^x v_1^x \frac{\partial v_1^x}{\partial \theta_x} \sin^2 \phi_x$$
$$+ (v_2^x)^2 v_1^x \sin\phi_x \cos\phi_x + (v_2^x)^2 \frac{\partial v_1^x}{\partial \phi_x} \sin^2 \phi_x$$
$$+ (v_2^x)^2 v_1^x \sin\phi_x \cos\phi_x + (v_1^x)^3 \cos\phi_x \sin^3 \phi_x$$
$$- (v_1^x)^2 \frac{\partial v_2^x}{\partial \theta_x} \sin^2 \phi_x - v_1^x v_2^x \frac{\partial v_2^x}{\partial \phi_x} \sin^2 \phi_x \Big) P_l'(1)$$

<u>a_{44}</u>: Again, we set $T_1 = v_1^x \frac{\partial}{\partial \theta_x}$ and $T_2 = v_2^x \frac{\partial}{\partial \phi_x}$. As this calculaton is dependent on that for a_{24}, we continue to use the previous labeling. Due to the complexity of this entry, *for each resulting block of terms, we only keep those which will not vanish after restricting to the diagonal.*

$$T_1\ (T_1\ (\text{block 1})\ \text{S1}) : \qquad v_1^x \frac{\partial v_1^x}{\partial \theta_x} v_1^y \frac{\partial v_1^y}{\partial \theta_y} P_l'(h)(\sin\phi_x \sin\phi_y \cos(\theta_x - \theta_y))$$

$$T_1\ (T_1\ (\text{block 1})\ \text{S2}) : \qquad 2(v_1^x)^2 (v_1^y)^2 P_l''(h)(\sin\phi_x \sin\phi_y \cos(\theta_x - \theta_y))^2$$

$$\begin{aligned} T_1\ (T_1\ (\text{block 1})\ \text{S3}) : \qquad & (v_1^x)^2 (v_1^y)^2 P_l''(h)(-\sin\phi_x \sin\phi_y \cos(\theta_x - \theta_y)) \\ & \cdot(-\sin\phi_x \sin\phi_y \cos(\theta_x - \theta_y)) \\ & + (v_1^x)^2 (v_1^y)^2 P_l'(h)(\sin\phi_x \sin\phi_y \cos(\theta_x - \theta_y)) \end{aligned}$$

$$T_1\ (T_1\ (\text{block 2})\ \text{S1}) : \qquad (v_1^x)^2 v_1^y \frac{\partial v_2^y}{\partial \theta_y} P_l'(h)(-\sin\phi_x \cos\phi_y \cos(\theta_x - \theta_y))$$

$$T_1\ (T_1\ (\text{block 2})\ \text{S2}) : \qquad \text{completely vanishes}$$

$$T_1\ (T_1\ (\text{block 2})\ \text{S3}) : \qquad v_1^x \frac{\partial v_1^x}{\partial \theta_x} v_1^y v_2^y P_l'(h)(\sin\phi_x \cos\phi_y \cos(\theta_x - \theta_y))$$

$$T_1\ (T_1\ (\text{block 3})\ \text{S1}) : \qquad v_1^x \frac{\partial v_1^x}{\partial \theta_x} \frac{\partial v_1^y}{\partial \phi_y} v_2^y P_l'(h)(\sin\phi_x \sin\phi_y \cos(\theta_x - \theta_y))$$

$$T_1\ (T_1\ (\text{block 3})\ \text{S2}) : \qquad \text{completely vanishes}$$

$$T_1\ (T_1\ (\text{block 3})\ \text{S3}) : \qquad v_1^x \frac{\partial v_1^x}{\partial \theta_x} v_1^y v_2^y P_l'(h)(\sin\phi_x \cos\phi_y \cos(\theta_x - \theta_y))$$

$$T_1\ (T_1\ (\text{block 4})\ \text{S1}) : \qquad (v_1^x)^2 v_2^y \frac{\partial v_2^y}{\partial \phi_y} P_l'(h)(-\sin\phi_x \cos\phi_y \cos(\theta_x - \theta_y))$$

$$T_1\ (T_1\ (\text{block 4})\ \text{S2}) : \qquad \text{completely vanishes}$$

$$\begin{aligned} T_1\ (T_1\ (\text{block 4})\ \text{S3}) : \qquad & (v_1^x)^2 (v_2^y)^2 P_l''(h)(-\sin\phi_x \sin\phi_y \cos(\theta_x - \theta_y)) \\ & \cdot(-\cos\phi_x \cos\phi_y - \sin\phi_x \sin\phi_y \cos(\theta_x - \theta_y)) \\ & + (v_1^x)^2 (v_2^y)^2 P_l'(h)(\sin\phi_x \sin\phi_y \cos(\theta_x - \theta_y)) \end{aligned}$$

$$T_1\ (T_2\ (\text{block 1})\ \text{S1}) : \qquad v_1^x v_2^x v_1^y \frac{\partial v_1^y}{\partial \theta_y} P_l'(h)(\cos\phi_x \sin\phi_y \cos(\theta_x - \theta_y))$$

$$T_1\ (T_2\ (\text{block 1})\ \text{S2}) : \qquad \text{completely vanishes}$$

$$T_1\ (T_2\ (\text{block 1})\ \text{S3}):\qquad v_1^x \frac{\partial v_2^x}{\partial \theta_x}(v_1^y)^2 P_l'(h)(-\cos\phi_x \sin\phi_y \cos(\theta_x-\theta_y))$$

$$T_1\ (T_2\ (\text{block 2})\ \text{S1}):\qquad v_1^x \frac{\partial v_2^x}{\partial \theta_x} v_1^y \frac{\partial v_2^y}{\partial \theta_y} P_l'(h) \cdot(\sin\phi_x\sin\phi_y + \cos\phi_x\cos\phi_y\cos(\theta_x-\theta_y))$$

$$T_1\ (T_2\ (\text{block 2})\ \text{S2}):\qquad v_1^x v_2^x v_1^y v_2^y P_l''(h)(\sin\phi_x\sin\phi_y\cos(\theta_x-\theta_y)) \cdot(\sin\phi_x\sin\phi_y + \cos\phi_x\cos\phi_y\cos(\theta_x-\theta_y))$$

$$T_1\ (T_2\ (\text{block 2})\ \text{S3}):\ v_1^x v_2^x v_1^y v_2^y P_l'(h)(\cos\phi_x\cos\phi_y\cos(\theta_x-\theta_y))$$

$$T_1\ (T_2\ (\text{block 3})\ \text{S1}):\qquad v_1^x v_2^x \frac{\partial v_1^y}{\partial \phi_y} v_2^y P_l'(h)(\cos\phi_x\sin\phi_y\cos(\theta_x-\theta_y))$$

$$T_1\ (T_2\ (\text{block 3})\ \text{S2}):\qquad v_1^x v_1^y v_2^x v_2^y P_l''(h)(\sin\phi_x\sin\phi_y + \cos\phi_x\cos\phi_y\cos(\theta_x-\theta_y)) \cdot(\sin\phi_x\sin\phi_y\cos(\theta_x-\theta_y))$$

$$T_1\ (T_2\ (\text{block 3})\ \text{S3}):\qquad v_1^x v_2^x v_1^y v_2^y P_l'(h)(\cos\phi_x\cos\phi_y\cos(\theta_x-\theta_y))$$

$$T_1\ (T_2\ (\text{block 4})\ \text{S1}):\quad v_1^x \frac{\partial v_2^x}{\partial \theta_x} v_2^y \frac{\partial v_2^x}{\partial \phi_y} P_l'(h)(\sin\phi_x\sin\phi_y + \cos\phi_x\cos\phi_y\cos(\theta_x-\theta_y))$$

T_1 (T_2 (block 4) S2) : completely vanishes

T_1 (T_2 (block 4) S3) : completely vanishes

$$T_2\ (T_1\ (\text{block 1})\ \text{S1}):\qquad v_2^x \frac{\partial v_1^x}{\partial \phi_x} v_1^y \frac{\partial v_1^y}{\partial \theta_y} P_l'(h)(\sin\phi_x\sin\phi_y\cos(\theta_x-\theta_y)) + v_2^x v_1^x v_1^y \frac{\partial v_1^y}{\partial \theta_y} P_l'(h)(\cos\phi_x\sin\phi_y\cos(\theta_x-\theta_y))$$

T_2 (T_1 (block 1) S2) : completely vanishes

T_2 (T_1 (block 1) S3) : completely vanishes

T_2 (T_1 (block 2) S1) : completely vanishes

$$T_2\ (T_1\ (\text{block 2})\ \text{S2}):\qquad v_2^x v_1^x v_1^y v_2^y P_l''(h)(\sin\phi_x\sin\phi_y\cos(\theta_x-\theta_y)) \cdot(\sin\phi_x\sin\phi_y + \cos\phi_x\cos\phi_y\cos(\theta_x-\theta_y))$$

$$T_2\ (T_1\ (\text{block 2})\ \text{S3}):\qquad v_2^x v_1^x v_1^y v_2^y P_l'(h)(\cos\phi_x\cos\phi_y\cos(\theta_x-\theta_y)) + v_2^x \frac{\partial v_1^x}{\partial \phi_x} v_1^y v_2^y P_l'(h)(\sin\phi_x\cos\phi_y\cos(\theta_x-\theta_y))$$

$$T_2\ (T_1\ (\text{block 3})\ \text{S1}):\qquad v_2^x v_1^x \frac{\partial v_1^y}{\partial \phi_y} v_2^y P_l'(h)(\cos\phi_x\sin\phi_y\cos(\theta_x-\theta_y)) + v_2^x \frac{\partial v_1^x}{\partial \phi_x}\frac{\partial v_1^y}{\partial \phi_y} v_2^y P_l'(h)(\sin\phi_x\sin\phi_y\cos(\theta_x-\theta_y))$$

T_2 (T_1 (block 3) S2) : $\quad v_2^x v_1^x v_1^y v_2^y P_l''(h)(\sin\phi_x \sin\phi_y + \cos\phi_x \cos\phi_y \cos(\theta_x - \theta_y))$
$\cdot(\sin\phi_x \sin\phi_y \cos(\theta_x - \theta_y))$

T_2 (T_1 (block 3) S3) : $\quad v_2^x \dfrac{\partial v_1^x}{\partial \phi_x} v_1^y v_2^y P_l'(h)(\sin\phi_x \cos\phi_y \cos(\theta_x - \theta_y))$
$+v_2^x v_1^x v_1^y v_2^y P_l'(h)(\cos\phi_x \cos\phi_y \cos(\theta_x - \theta_y))$

T_2 (T_1 (block 4) S1) : completely vanishes

T_2 (T_1 (block 4) S2) : completely vanishes

T_2 (T_1 (block 4) S3) : completely vanishes

T_2 (T_2 (block 1) S1) : completely vanishes

T_2 (T_2 (block 1) S2) : completely vanishes

T_2 (T_2 (block 1) S3) : $\quad (v_2^x)^2 (v_1^y)^2 P_l''(h)(-\cos\phi_x \cos\phi_y - \sin\phi_x \sin\phi_y \cos(\theta_x - \theta_y))$
$\cdot(-\sin\phi_x \sin\phi_y \cos(\theta_x - \theta_y))$
$+v_2^x \dfrac{\partial v_2^x}{\partial \phi_x} (v_1^y)^2 P_l'(h)(-\cos\phi_x \sin\phi_y \cos(\theta_x - \theta_y))$
$+(v_2^x)^2 (v_1^y)^2 P_l'(h)(\sin\phi_x \sin\phi_y \cos(\theta_x - \theta_y))$

T_2 (T_2 (block 2) S1) : $\quad v_2^x \dfrac{\partial v_2^x}{\partial \phi_x} v_1^y \dfrac{\partial v_2^y}{\partial \theta_y} P_l'(h)(\sin\phi_x \sin\phi_y + \cos\phi_x \cos\phi_y \cos(\theta_x - \theta_y))$

T_2 (T_2 (block 2) S2) : completely vanishes

T_2 (T_2 (block 2) S3) : completely vanishes

T_2 (T_2 (block 3) S1) : completely vanishes

T_2 (T_2 (block 3) S2) : completely vanishes

T_2 (T_2 (block 3) S3) : completely vanishes

T_2 (T_2 (block 4) S1) : $\quad v_2^x \dfrac{\partial v_2^x}{\partial \phi_x} \dfrac{\partial v_2^y}{\partial \phi_y} v_2^y P_l'(h)(\sin\phi_x \sin\phi_y + \cos\phi_x \cos\phi_y \cos(\theta_x - \theta_y))$

T_2 (T_2 (block 4) S2) : $\quad 2(v_2^x)^2 (v_2^y)^2 P_l''(h)(\sin\phi_x \sin\phi_y + \cos\phi_x \cos\phi_y \cos(\theta_x - \theta_y))^2$

T_2 (T_2 (block 4) S3) : $\quad (v_2^x)^2 (v_2^y)^2 P_l''(h)(-\cos\phi_x \cos\phi_y - \sin\phi_x \sin\phi_y \cos(\theta_x - \theta_y))$
$\cdot(-\cos\phi_x \cos\phi_y - \sin\phi_x \sin\phi_y \cos(\theta_x - \theta_y))$
$+(v_2^x)^2 (v_2^y)^2 P_l'(h)(\sin\phi_x \sin\phi_y + \cos\phi_x \cos\phi_y \cos(\theta_x - \theta_y))$

$$= \text{final} : v_1^x \frac{\partial v_1^x}{\partial \theta_x} v_1^x \frac{\partial v_1^x}{\partial \theta_x} \sin^2\phi_x P_l'(1) + 2(v_1^x)^4 \sin^4\phi_x P_l''(1) + (v_1^x)^4 \sin^4\phi_x P_l''(1)$$
$$+(v_1^x)^4 \sin^2\phi_x P_l'(1) - (v_1^x)^3 \frac{\partial v_2^x}{\partial \theta_x} \sin\phi_x \cos\phi_x P_l'(1) + (v_1^x)^2 \frac{\partial v_1^x}{\partial \theta_x} v_2^x \sin\phi_x \cos\phi_x P_l'(1)$$
$$+v_1^x \frac{\partial v_1^x}{\partial \theta_x}\frac{\partial v_1^x}{\partial \phi_x} v_2^x \sin^2\phi_x P_l'(1) + (v_1^x)^2 \frac{\partial v_1^x}{\partial \theta_x} v_2^x \sin\phi_x \cos\phi_x P_l'(1)$$
$$-(v_1^x)^2 v_2^x \frac{\partial v_2^x}{\partial \phi_x} \sin\phi_x \cos\phi_x P_l'(1)$$
$$+(v_1^x)^2 (v_2^x)^2 \sin^2\phi_x P_l''(1) + (v_1^x)^2 (v_2^x)^2 \sin^2\phi_x P_l'(1)$$
$$+(v_1^x)^2 \frac{\partial v_1^x}{\partial \theta_x} v_2^x \cos\phi_x \sin\phi_x P_l'(1)$$
$$-(v_1^x)^3 \frac{\partial v_2^x}{\partial \theta_x} \cos\phi_x \sin\phi_x P_l'(1) + (v_1^x)^2 \left(\frac{\partial v_2^x}{\partial \theta_x}\right)^2 P_l'(1) + (v_1^x)^2 (v_2^x)^2 \sin^2\phi_x P_l''(1)$$
$$+(v_1^x)^2 (v_2^x)^2 \cos^2\phi_x P_l'(1) + v_1^x \frac{\partial v_1^x}{\partial \phi_x} (v_2^x)^2 \cos\phi_x \sin\phi_x P_l'(1)$$
$$+(v_1^x)^2 (v_2^x)^2 \sin^2\phi_x P_l''(1)$$
$$+(v_1^x)^2 (v_2^x)^2 \cos^2\phi_x P_l'(1) + v_1^x \frac{\partial v_2^x}{\partial \theta_x} v_2^x \frac{\partial v_2^x}{\partial \phi_x} P_l'(1) \text{ (from the T1 portion)}$$
$$+v_2^x \frac{\partial v_1^x}{\partial \phi_x} v_1^x \frac{\partial v_1^x}{\partial \theta_x} \sin^2\phi_x P_l'(1) + v_2^x (v_1^x)^2 \frac{\partial v_1^x}{\partial \theta_x} \cos\phi_x \sin\phi_x P_l'(1)$$
$$+(v_1^x)^2 (v_2^x)^2 \sin^2\phi_x P_l''(1)$$
$$+(v_1^x)^2 (v_2^x)^2 \cos^2\phi_x P_l'(1) + v_1^x \frac{\partial v_1^x}{\partial \phi_x} (v_2^x)^2 \sin\phi_x \cos\phi_x P_l'(1)$$
$$+v_1^x \frac{\partial v_1^x}{\partial \phi_x} (v_2^x)^2 \cos\phi_x \sin\phi_x P_l'(1)$$
$$+\left(\frac{\partial v_1^x}{\partial \phi_x}\right)^2 (v_2^x)^2 \sin^2\phi_x P_l'(1) + (v_1^x)^2 (v_2^x)^2 \sin^2\phi_x P_l''(1)$$
$$+v_1^x \frac{\partial v_1^x}{\partial \phi_x} (v_2^x)^2 \sin\phi_x \cos\phi_x P_l'(1)$$
$$+(v_1^x)^2 (v_2^x)^2 \cos^2\phi_x P_l'(1) + (v_2^x)^2 (v_1^x)^2 \sin^2\phi_x P_l''(1)$$
$$-(v_1^x)^2 v_2^x \frac{\partial v_2^x}{\partial \phi_x} \cos\phi_x \sin\phi_x P_l'(1)$$
$$+(v_1^x)^2 (v_2^x)^2 \sin^2\phi_x P_l'(1) + v_1^x v_2^x \frac{\partial v_2^x}{\partial \phi_x}\frac{\partial v_2^x}{\partial \theta_x} P_l'(1)$$
$$+\left(\frac{\partial v_2^x}{\partial \phi_x}\right)^2 (v_2^x)^2 P_l'(1) + 2(v_2^x)^4 P_l''(1)$$
$$+(v_2^x)^4 P_l''(1) + (v_2^x)^4 P_l'(1) \text{ (from the T2 portion).}$$

Acknowledgements: S.E. would like to thank I. Wigman for numerous helpful discussions and Z. Rudnick for suggesting the collaboration with Wigman which eventually led to this note and upcoming work.

References

[AT00] Robert J. Adler and Jonathan E. Taylor, *Random fields and geometry*, Springer Monographs in Mathematics, Springer, New York, 2007. MR2319516

[AAR99] George E. Andrews, Richard Askey, and Ranjan Roy, *Special functions*, Encyclopedia of Mathematics and its Applications, vol. 71, Cambridge University Press, Cambridge, 1999. MR1688958

[AW09] Jean-Marc Azaïs and Mario Wschebor, *Level sets and extrema of random processes and fields*, John Wiley & Sons, Inc., Hoboken, NJ, 2009. MR2478201

[CS18] Yaiza Canzani and Peter Sarnak, *Topology and nesting of the zero set components of monochromatic random waves*, Comm. Pure Appl. Math. **72** (2019), no. 2, 343–374, DOI 10.1002/cpa.21795. MR3896023

[DR17] Nguyen Viet Dang and Gabriel Rivière, *Equidistribution of the conormal cycle of random nodal sets*, J. Eur. Math. Soc. (JEMS) **20** (2018), no. 12, 3017–3071, DOI 10.4171/JEMS/828. MR3871498

[GW14] Damien Gayet and Jean-Yves Welschinger, *Betti numbers of random nodal sets of elliptic pseudo-differential operators*, Asian J. Math. **21** (2017), no. 5, 811–839, DOI 10.4310/AJM.2017.v21.n5.a2. MR3767266

[NS1] Fedor Nazarov and Mikhail Sodin, *On the number of nodal domains of random spherical harmonics*, Amer. J. Math. **131** (2009), no. 5, 1337–1357, DOI 10.1353/ajm.0.0070. MR2555843

[NS10] Fedor Nazarov and Mikhail Sodin, *Random complex zeroes and random nodal lines*, Proceedings of the International Congress of Mathematicians. Volume III, Hindustan Book Agency, New Delhi, 2010, pp. 1450–1484. MR2827851

[NS2] F. Nazarov and M. Sodin, *Asymptotic laws for the spatial distribution and the number of connected components of zero sets of Gaussian random functions*, Zh. Mat. Fiz. Anal. Geom. **12** (2016), no. 3, 205–278, DOI 10.15407/mag12.03.205. MR3522141

[RW18] Z. Rudnick, I. Wigman, *Points on nodal lines with given direction*, preprint, arXiv: 1802.09603 (2018).

[SW18] Peter Sarnak and Igor Wigman, *Topologies of nodal sets of random band-limited functions*, Comm. Pure Appl. Math. **72** (2019), no. 2, 275–342, DOI 10.1002/cpa.21794. MR3896022

(Permanent address) School of Mathematics, Cardiff University, Senghennyd Road, Cardiff, Wales CF24 4AG United Kingdom

Email address: eswarathasans@cardiff.ac.uk

Current address: Department of Mathematics and Statistics, McGill University, Burnside Hall Rm 1208, 805 Sherbrooke Street West, Montréal, Québec H3A 0B9 Canada

Email address: suresh.eswarathasan@mcgill.ca

Contemporary Mathematics
Volume **739**, 2019
https://doi.org/10.1090/conm/739/14893

Formal Zeta function expansions and the frequency of Ramanujan graphs

Joel Friedman

ABSTRACT We show that logarithmic derivative of the Zeta function of any regular graph is given by a power series about infinity whose coefficients are given in terms of the traces of powers of the graph's Hashimoto matrix.

We then consider the expected value of this power series over random, d-regular graph on n vertices, with d fixed and n tending to infinity. Under rather speculative assumptions, we make a formal calculation that suggests that for fixed d and n large, this expected value should have simple poles of residue $-1/2$ at $\pm(d-1)^{-1/2}$. We shall explain that calculation suggests that for fixed d there is an $f(d) > 1/2$ such that a d-regular graph on n vertices is Ramanujan with probability at least $f(d)$ for n sufficiently large.

Our formal computation has a natural analogue when we consider random covering graphs of degree n over a fixed, regular "base graph." This again suggests that for n large, a strict majority of random covering graphs are relatively Ramanujan.

We do not regard our formal calculations as providing overwhelming evidence regarding the frequency of Ramanujan graphs. However, these calculations are quite simple, and yield intiguing suggestions which we feel merit further study.

CONTENTS

Research supported in part by an NSERC grant.

1. Introduction

In this paper we shall give a calculation that suggests that there should be many Ramanujan graphs of any fixed degree and any sufficiently large number of vertices. Our calculation is quite speculative in that it formally sums an infinite number of asymptotic expansions term by term, only finitely many of which hold for graphs of a fixed number of vertices. However, the infinite sum is a simple calculation and suggests intriguing information regarding the fraction of graphs that are Ramanujan. It therefore seems reasonable to study convergent variants of our inifinte formal sums to see if they can yield provable results.

In more detail, we show that the logarithmic derivative of the Zeta function of a regular graph has a simple power series expansion at infinity. The coefficients of this power series invovle traces of successively larger powers of the Hashimoto matrix of a graph. For a standard model of a d-regular graph on n vertices, we consider the expected value of this logarithmic derivative for fixed d and large n. We make a number of assumptions to make a formal computation which suggests that for fixed d. the expected number of real poles near each of $\pm(d-1)^{-1/2}$ tends to $1/2$ as n tends to infinity. Assuming this $1/2$ is caused entirely by real poles of the Zeta function, then the expected number of positive or negative adjacency eigenvalues of absolute value in the interval $[2(d-1)^{1/2}, d)$ would be $1/4$ for both cases, positive and negative. Under the likely assumption that for fixed d, a random, d-regular graph on n vertices will have two or more such eigenvalues with probability bounded away from zero, then for sufficiently large n, a strict majority of random graphs are Ramanujan. We emphasize that all these conclusions are quite speculative.

Our computation involves a contour integral of the expected logarithmic derivative near $u = (d-1)^{-1/2}$, where our formal computation suggests that this expected logarithmic derivative has a pole. We write the contour in a particular way that assumes that the residue at this pole arises entirely from real eigenvalues of the Hashimoto matrix. However it is conceivable that some of the complex Hashimoto eigenvalues near $\pm 2(d-1)^{1/2}$ also contribute to this pole, in which case (if our other assumptions are correct) the limiting expected number of positive and negative real Zeta function poles may each be less than $1/4$.

We caution the reader that part of our contour passes through the open ball $|u| < (d-1)^{-1/2}$, which is one serious issue in the above computation. Indeed, for fixed d and n large, the Zeta function of a random, d-regular graph on n vertices has poles throughout the circle $|u| = (d-1)^{-1/2}$; this follows easily from the fact that a random such graph has a bounded expected number of cycles of any fixed length, and so its adjacency eigenvalue distribution tends to the Kesten-McKay distribution (see [**McK81**]). Hence, for large n the true expected Zeta function should have a pole distribution throughout $|u| = (d-1)^{-1/2}$, which makes it highly speculative to work in any part of the region $|u| < (d-1)^{-1/2}$. At the same time, this makes the residue of $-1/2$ at $\pm(d-1)^{-1/2}$ in the formal calculation all the more interesting, and is why we may conjecture that the imaginary poles near $\pm(d-1)^{-1/2}$ may possibly contribute this $-1/2$, if there is any true sense to this residue.

It is interesting that the $-1/2$ comes from a computation on the trace method for regular graphs that was essentially done by Broder and Shamir [**BS87**], although more justification for our computation comes from the asymptotic expansions of expected trace powers given later improvements of the Broder-Shamir method, namely [**Fri91**, **Fri08**, **FK14**].

As far as we know, this is the first direct application of Zeta functions per se to graph theory. Zeta functions of graphs arose first in the study of $\mathfrak{p}$-adic groups [**Iha66**, **Ser03**], and developed for general graphs by Sunada, Hashimoto, and Bass (see [**Ter11**], beginning of Part 2). Ihara's determinantal formula gave rise to what is now often called the *Hashimoto matrix* of a graph, which can be used to count *strictly non-backtracking* closed walks in a graph. Although the underlying graph of the Hashimoto matrix appeared in graph theory in the 1940's and 1960's (see [**dB46**, **HN60**, **Knu67**]), the study of its spectral properties seems largely inspired by the above work on graph Zeta functions. Friedman and Kohler [**FK14**] note that in the trace method for random, regular graphs, one gets better adjacency eigenvalue bounds if one first gets analogous trace estimates for the Hashimoto matrix. Furthermore, the solution to the Alon Second Eigenvalue Conjecture [**Fri08**] involved trace methods for the Hashimoto matrix rather than for the adjacency matrix. Hence the Hashimoto matrix—and therfore, by implication, also Zeta functions—have played a vital role in graph theory. However, we know of no previous direct applications of Zeta functions to obtain new theorems or conjectures in graph theory. Our theorems and conjectures—although they involve the Hashimoto matrix in their expansion at infinity—seem to fundamentally involve Zeta functions.

We remark that the expected traces of Hashimoto matrix powers are difficult to study directly. Indeed, these expected traces are complicated by *tangles* [**Fri08**, **FK14**], which are—roughly speaking—low probability events that force a graph to have large, positive real Hashimoto eigenvalues. It is known that such tangles must be removed to prove the Alon conjecture [**Fri08**] or its relativization [**FK14**]; furthermore by modifying trace powers to eliminate the pathological effect of tangles, the asymptotic expansions of expected trace powers become much simpler. For this reason we introduce a second formal power series, whose terms are a variant of the above terms, such that (1) it is probably simpler to understand the terms of this second formal power series, and (2) we believe that the second set of terms contain very similar information to the first.

We can generalize the above discussion to random covering maps of degree n over a fixed, regular "base graph." Doing so gives the analogous formal computation that suggests that for large n we expect that a majority of random cover maps to be *relatively Ramanujan*. Again, the $1/4$ we get (or $1/2$ for random, bipartite, regular graphs) comes from the analogue of the Broder-Shamir computation [**Fri03**] for random covering maps, but is further justified by higher order expansions [**LP10**, **Pud15**, **FK14**].

The rest of this paper is organized as follows. In Section 2 we desribe our main theorems and conjectures. In Section 3 we describe our terminology regarding graphs and Zeta functions. In Section 4 we prove our expansion near infinity of the logarithmic derivative of the Zeta function of a graph. In Section 5 we make a formal calculation of the expected above logarithmic derivative and make numerous conjetures. In Section 6 we describe variants of this formal computation which we believe will be easier to study, and yet will contain essentially the same information. In Section 7 we desribe other models of random graphs, especially covering maps of degree n over a fixed, regular base graph. In Section 8 we briefly describe our numerical experiments and what previous experiments in the literature have suggested.

2. Main Results

In this section we state our main results, although we will use some terminology to be made precise in later sections. Let us begin with the notion of a random graph that we use. For positive integers n and $d \geq 3$ we consider a random d-regular graph on n vertices. It is simplest to think of d as an even integer with a random graph generated by $d/2$ permutations on $\{1, \ldots, n\}$, which we denote $\mathcal{G}_{n,d}$, as was used in [**BS87**, **Fri91**, **Fri08**]; our models therefore allow for multiple edges and self-loops. We remark that there are similar "algebraic" models for d and n of any parity [**Fri08**] which we shall describe in Section 7.

We will give a formal calculation that indicates that for fixed d the expected number of adjacency eigenvalues of a graph in $\mathcal{G}_{n,d}$ of absolute value in $[2(d-1)^{1/2}, d)$ is one-half, $1/4$ positive and $1/4$ negative; a more conservative conjecture is that $1/4$ is an upper bound on each side. We also conjecture that the probability that for fixed d and $n \to \infty$, a graph of $\mathcal{G}_{n,d}$ has at least two such eigenvalues is bounded from below by a positive constant. These two conjectures—if true—imply that for any fixed d, for all n sufficiently large a strict majority of graphs in $\mathcal{G}_{n,d}$ are Ramanujan.

After explaining our conjecture, we will comment on generalizations to random covering maps of degree n to a fixed, regular graph. Then we will describe some numerical experiments we made to test our conjecture; although these calculations suggest that our formal computation may be close to the correct answer, our calculations are done on graphs with under one million vertices (and assume that certain software is computing correctly); it may be that one needs more vertices to see the correct trend, and it is commonly believed that there are fewer Ramanujan graphs than our formal calculation suggests; see [**MNS08**].

Let us put our conjectures in a historical context. For a graph, G, on n vertices, we let

$$\lambda_1(G) \geq \lambda_2(G) \geq \cdots \geq \lambda_n(G)$$

be the n eignevalues of A_G, the adjacency matrix of G. In [**Alo86**], Noga Alon conjectured that for fixed integer $d \geq 3$ and $\epsilon > 0$, as n tends to infinity, the probability that a random d-regular graph, G, on n has

$$\lambda_2(G) \leq 2(d-1)^{-1/2} + \epsilon$$

tends to one. Alon's interest in the above conjecture was that the above condition on $\lambda_2(G)$ implies ([**GG81**, **AM85**, **Tan84**]) that G has a number of interesting isoperimetric or "expansion" properties. Broder and Shamir [**BS87**] introduced a trace method to study the above question; [**BS87**, **FKS89**, **Fri91**] gave high probability bounds on $\lambda_2(G)$ with $2(d-1)^{-1/2}$ replaced with a larger constant, and [**Fri08**] finally settled the original conjecture. We remark that all the aforementioned papers actually give stronger bounds, namely with $\lambda_2(G)$ replaced with

$$\rho(G) \stackrel{\text{def}}{=} \max_{i \geq 2} |\lambda_2(G)|.$$

For many applications, it suffices to specify one particular graph, $G \in \mathcal{G}_{n,d}$, which satisfies the bound in Alon's conjecture. Such graphs were given in [**LPS88**,

Mar88, Mor94]; [**LPS88**] coined the term *Ramanujan graph* to describe a d-regular graph, G, satisfying

$$\lambda_i(G) \in \{d, -d\} \cup [-2(d-1)^{1/2}, 2(d-1)^{1/2}]$$

for all i. This is stronger than Alon's conjecture in that the ϵ of Alon becomes zero; it is weaker than what trace methods prove, in that $-d$ is permitted as an eigenvalue, i.e., the graph may be bipartite. Recently [**MSS13**] have proven the existence of a sequence of bipartite Ramanujan graphs of any degree d for a sequence of n's tending to infinity; [**LPS88, Mar88, Mor94**] constructed sequences for d such that $d-1$ is a prime power, but the [**LPS88, Mar88, Mor94**] are more explicit—contructible in polynomial of $\log n$ and d—than those of [**MSS13**]—constructible in polynomial of n and d. Our results do not suggest any obvious method of constructing Ramanujan graphs.

Our formal calculation for $\mathcal{G}_{n,d}$ is based on two results: Ihara Zeta functions of graphs (see [**Ter11**]), and a trace estimate essentially known since [**BS87**]. We give a stronger conjecture based on the trace methods and estimates of [**Fri91, Fri08, FK14**], which give somewhat more justification for the formal calculation we describe.

Let us roughly describe our methods. We will consider the function

$$\mathbb{E}_{G\in\mathcal{G}_{n,d}}\left[\zeta_G'(u)/\zeta_G(u)\right],$$

i.e., the expected logarithmic derivative of $\zeta_G(u)$; using the results of [**Fri08, FK14**] regarding the Alon conjecture, we can write the expected number of eigenvalues equal to or greater than $2(d-1)^{1/2}$ as half of a contour integral of the above logarithmic derivative near $2(d-1)^{-1/2}$ (really minus the logarithmic derivative, since we are counting poles). This contour integral is essentially unchanged if we replace minus this logarithmic derivative by the simpler expression

$$\mathcal{L}_G(u) = \sum_{k=0}^{\infty} u^{-1-k}\,\mathrm{Tr}(H_G^k)(d-1)^{-k},$$

where H_G denotes the Hashimoto matrix of a graph, G. We recall [**Fri08, FK14**] (although essentially from [**Fri91**]) that we can estimate the expected traces of Hashimoto matrices as

$$(1)\qquad \mathbb{E}_{G\in\mathcal{G}_{n,d}}\left[\mathrm{Tr}(H_G^k)\right] = P_0(k) + P_1(k)n^{-1} + \cdots + P_{r-1}(k)n^{1-r} + \mathrm{err}_{n,k,r}$$

where the $P_i(k)$ are functions of k and d, and for any fixed r we have

$$|\mathrm{err}_{n,k,r}| \le Ck^C n^{-r}(d-1)^k$$

for some $C = C(r)$. The leading term, $P_0(k)$, has been essentially known since [**BS87**] (see [**Fri91, Fri08, FK14**]) to be

$$P_0(k) = (d-1)^k + (d-1)^{k/2}\mathbb{I}_{\mathrm{even}}(k) + O(k)\,(d-1)^{k/3},$$

where $\mathbb{I}_{\mathrm{even}}(k)$ is the indicator function that k is even. If we assume that we may evaluate the expected value of $\mathcal{L}_G(u)$ by writing is as a formal sum, term by term, using (1), then the

$$(d-1)^{k/2}\mathbb{I}_{\mathrm{even}}(k)$$

term of $P_0(k)$ gives a term of the expected value of $\mathcal{L}_G(u)$ equal to

$$\frac{u}{u^2-(d-1)},$$

whose residue at $u = \pm(d-1)^{-1/2}$ is $1/2$. If this exchange of summation gives the correct asymptotics, i.e., the sum of the terms corresponding to $P_i(k)n^{-i}$ for $i \geq 1$ tends to zero as $n \to \infty$, then as $n \to \infty$ we would have that the expected number of positive real poles and negative real poles is $1/2$ each (with the $\pm(d-1)^{-1/2}$ countributing one-half times their expected multiplicity).

It is known that for large i, the $P_i(k)$ are problematic due to "tangles" [**Fri08**], which are certain low probability events in $\mathcal{G}_{n,d}$ that force a large second adjacency eigenvalue. Hene we might wish to modify the $P_i(k)$, as done in [**Fri08, FK14**], by introducing a variant of the Hashimoto trace. This leads us to later introduce the related functions $\widehat{P}_i(k)$, which we believe will be easier to study but contain almost the same information as the $P_i(k)$. We shall also explain a generalization of this calculation to models of covering graphs of degree n over a fixed, regular "base graph."

For reasons that we explain, this formal calculation takes a number of "leaps of faith" that we cannot justify at this point; on the other hand, it seems like a natural formal calculation to make, and it suggests an intriguing conjecture.

We have made a brief, preliminary experimental investigation of this conjecture with $d = 4$ and $d = 6$ for positive poles; these experiments are not particularly conclusive: according to [**MNS08**] the true trend may require very large values of n: this is based on the calculations that the "width of concentration" of the second largest adjacency eigenvalue will eventually overtake its mean's distance to $2(d-1)^{1/2}$. However, if this width of concentration is of the same order of magnitude as its distance to $2(d-1)^{1/2}$ as $n \to \infty$, then our conjecture does not contradict the other findings of [**MNS08**] (regarding a Tracy-Widom distribution over the width of concentration). Our experiments are made with graphs of only $n \leq 400,000$ vertices. For graphs of this size or smaller it does not lock like the expected number of real poles has stabilized; however, in all of our experiments this expected number is smaller than $1/4$ for all but very small values of n.

3. Graph Theoretic Preliminaries

In this subsection we specify our precise definitions for a number of concepts in graph and algebraic graph theory. We note that such definitions vary a bit in the literature. For example, in this paper graphs may have multiple edges and two types of self-loops—half-loops and whole-loops—in the terminology of [**Fri93**]; also see [**ST96, ST00, TS07**], for example, regarding half-loops.

3.1. Graphs and Morphisms.

DEFINITION 3.1. A *directed graph* (or *digraph*) is a tuple $G = (V, E^{\mathrm{dir}}, t, h)$ where V and E^{dir} are sets—the *vertex* and *directed edge* sets—and $t\colon E^{\mathrm{dir}} \to V$ is the *tail map* and $h\colon E^{\mathrm{dir}} \to V$ is the *head map*. A directed edge e is called *self-loop* if $t(e) = h(e)$, that is, if its tail is its head. Note that our definition also allows for *multiple edges*, that is directed edges with identical tails and heads. Unless specifically mentioned, we will only consider directed graphs which have finitely many vertices and directed edges.

A graph, roughly speaking, is a directed graph with an involution that pairs the edges.

DEFINITION 3.2. An *undirected graph* (or simply a *graph*) is a tuple $G = (V, E^{\mathrm{dir}}, t, h, \iota)$ where $(V, E^{\mathrm{dir}}, t, h)$ is a directed graph and where $\iota \colon E^{\mathrm{dir}} \to E^{\mathrm{dir}}$, called the *opposite map* or *involution* of the graph, is an involution on the set of directed edges (that is, $\iota^2 = \mathrm{id}_{E^{\mathrm{dir}}}$ is the identity) satisfying $t\iota = h$. The directed graph $G = (V, E^{\mathrm{dir}}, t, h)$ is called the *underlying directed graph* of the graph G. If e is an edge, we often write e^{-1} for $\iota(e)$ and call it the *opposite edge.* A self-loop e is called a *half-loop* if $\iota(e) = e$, and otherwise is called a *whole-loop.*

The opposite map induces an equivalence relation on the directed edges of the graph, with $e \in E^{\mathrm{dir}}$ equivalent to ιe; we call the quotient set, E, the *undirected edge* of the graph G (or simply its *edge*). Given an edge of a graph, an *orientation* of that edge is the choice of a representative directed edge in the equivalence relation (given by the opposite map).

NOTATION 3.3. For a graph, G, we use the notation $V_G, E_G, E_G^{\mathrm{dir}}, t_G, h_G, \iota_G$ to denote the vertex set, edge set, directed edge set, tail map, head map, and opposite map of G; similarly for directed graphs, G.

DEFINITION 3.4. Let G be a directed graph. The *adjacency matrix*, A_G, of G is the square matrix indexed on the vertices, V_G, whose (v_1, v_2) entry is the number of directed edges whose tail is the vertex v_1 and whose head is the vertex v_2. The *indegree* (respectively *outdegree*) of a vertex, v, of G is the number of edges whose head (respectively tail) is v.

The adjacency matrix of an undirected graph, G, is simply the adjacency matrix of its underlying directed graph. For an undirected graph, the indegree of any vertex equals its outdegree, and is just called its *degree.* The *degree matrix* of G is the diagonal matrix, D_G, indexed on V_G whose (v, v) entry is the degree of v. We say that G is *d-regular* if D_G is d times the identity matrix, i.e., if each vertex of G has degree d.

For any non-negative integer k, the number of closed walks of length k is a graph, G, is just the trace, $\mathrm{Tr}(A_G^k)$, of the k-th power of A_G.

NOTATION 3.5. Given a graph, G, the matrix A_G is symmetric, and hence the eigenvalues of A_G are real and can be ordered

$$\lambda_1(G) \geq \cdots \geq \lambda_n(G),$$

where $n = |V_G|$. We reserve the notation $\lambda_i(G)$ to denote the eigenvalues of A_G ordered as above.

If G is d-regular, then $\lambda_1(G) = d$.

DEFINITION 3.6. Let G be a graph. We define the *directed line graph* or *oriented line graph* of G, denoted $\mathrm{Line}(G)$, to be the directed graph $L = \mathrm{Line}(G) = (V_L, E_L^{\mathrm{dir}}, t_L, h_L)$ given as follows: its vertex set, V_L, is the set E_G^{dir} of directed edges of G; its set of directed edges is defined by

$$E_L^{\mathrm{dir}} = \left\{(e_1, e_2) \in E_G^{\mathrm{dir}} \times E_G^{\mathrm{dir}} \mid h_G(e_1) = t_G(e_2) \text{ and } \iota_G(e_1) \neq e_2\right\},$$

that is, E_L^{dir} corresponds to the non-backtracking walks of length two in G. The tail and head maps are simply defined to be the projections in each component, that is by $t_L(e_1, e_2) = e_1$ and $h_L(e_1, e_2) = e_2$.

The *Hashimoto matrix* of G is the adjacency matrix of its directed line graph, denoted H_G, which is, therefore, a square matrix indexed on E_G^{dir}. We use the symbol $\mu_1(G)$ to denote the Perron-Frobenius eigenvalue of H_G, and use $\mu_2(G),\ldots,\mu_m(G)$, where $m=|E_G^{\rm dir}|$, to denote the remaining eigenvalues, in no particular order (all concepts we discuss about the μ_i for $i\ge 2$ will not depend on their order).

If G is d-regular, then $\mu_1(G)=d-1$.

It is easy to see that for any positive integer k, the number of strictly non-backtracking closed walks of length k in a graph, G, equals the trace, $\mathrm{Tr}(H_G^k)$, of the k power of H_G; of course, the strictly non-backtracking walks begin and end in a vertex, whereas $\mathrm{Tr}(H_G^k)$ most naturally counts walks beginning and ending in an edge; the correspondence between the two notions can be seen by taking a walk of $\mathrm{Line}(G)$, beginning and ending an in a directed edge, $e\in E_G^{\rm dir}$, and mapping it to the strictly non-backtracking closed walk in G beginning at, say, the tail of e.

For graphs, G, that have half-loops, the Ihara determinantal formula takes the form (see [**Fri08, ST96, ST00, TS07**]):

$$\det(\mu I-H_G)=\det\bigl(\mu^2 I-\mu A_G+(D_G-I)\bigr)(\mu-1)^{|\mathrm{half}_G|}(\mu^2-1)^{|V_G|-|\mathrm{pair}_G|}, \tag{2}$$

where half_G is the set of half-loops of G, and pair_G is the set of undirected edges of G that are not half-loops, i.e., the collection of sets of the form, $\{e_1,e_2\}$ with $\iota e_1=e_2$ but $e_1\ne e_2$.

4. Variants of the Zeta Function

For any graph, G, recall that A_G denotes its adjacency matrix, and H_G denotes its Hashimoto matrix, i.e., the adjacency matrix of what is commonly called G's *oriented line graph*, $\mathrm{Line}(G)$. If $\zeta_G(u)$ is the Zeta function of G, then we have

$$\zeta_G(u)=\frac{1}{\det(I-uH_G)},$$

which, for d-regular G, we may alternatively write via the Ihara determinantal formula

$$\det(I-uH_G)=\det\Bigl(I-uA_G-u^2(d-1)\Bigr)(1-u^2)^{-\chi(G)} \tag{3}$$

(provided G has no half-loops, with a simple modification if G does).

DEFINITION 4.1. Let G be a d-regular graph. We call an eigenvalue of H_G *non-Ramanujan* if it is purely real and different from

$$\pm 1,\pm(d-1)^{1/2},\pm(d-1).$$

We say that an eigenvalue of A_G is *non-Ramanujan* if it is of absolute value strictly between $2(d-1)^{1/2}$ and d. It is known that the number of non-Ramanujan eigenvalues of H_G is precisely twice the number of eigenvalues of A_d. G is called *Ramanujan* if it has no non-Ramanujan H_G eigenvalues, or, equivalently, no non-Ramanujan A_G eigenvalues. Similary for positive non-Ramanujan eigenvalues of both H_G and A_G, and the same with "positive" replaced with "negative."

NOTATION 4.2. For any $\epsilon,\delta>0$, let $C^+_{\epsilon,\delta}$ be the boundary of the rectangle

$$\{x+iy\in\mathbb{C}\mid |1-x(d-1)^{1/2}|\le\epsilon,\ |y|\le\delta\}; \tag{4}$$

define $C^-_{\epsilon,\delta}$ similarly, with x replaced with $-x$.

DEFINITION 4.3. We say that a d-regular graph, G, is *ϵ-spectral* if G's real Hashimoto eigenvalues lie in set

$$\{-(d-1),-1,1,(d-1)\}\cup\{x\mid |1-x(d-1)^{-1/2}|<\epsilon\}.$$

We remark that for fixed $d\ge 3$ and $\epsilon>0$, it is known that a fraction $1-O(1/n)$ of random d-regular graphs on n vertices are ϵ-spectral [**Fri08**].

For ϵ-spectral G we have that the number of real, positive Hashimoto eigenvalues is given by

$$\frac{1}{2\pi i}\int_{C^+_{\epsilon,\delta}}\frac{-\zeta'_G(u)}{\zeta_G(u)}\,du$$

for $\delta>0$ sufficiently small, where $C^+_{\epsilon,\delta}$ is traversed in the counterclockwise direction; similarly for negative eigenvalues.

Observe that for each G, for large $|u|$ we have

$$-\zeta'_G(u)/\zeta_G(u)=\sum_{\mu\in\mathrm{Spec}(H_G)}-\mu\,(1-u\mu)^{-1},$$

where $\mathrm{Spec}(H_G)$ denotes the set of eigenvalues of H_G, counted with multiplicity, and hence

$$-\zeta'_G(u)/\zeta_G(u)=\sum_{\mu\in\mathrm{Spec}(H_G)}\sum_{k=0}^{\infty}u^{-1-k}\mu^{-k}.$$

By the Ihara determinantal formula, the set of eigenvalues of H_G consists of ± 1 with multiplicity $-\chi(G)$ plus, for each eigenvalue, λ, of A_G, the two roots μ_1,μ_2 of the equation

$$\mu^2-\mu\lambda+(d-1)=0;$$

note that any pair μ_1,μ_2 as above satisfy $\mu_1\mu_2=d-1$; hence to sum over μ^{-1} over the pairs μ_1,μ_2 as such is the same as summing over $\mu/(d-1)$ of all such eigenvalues. It easily follows that

$$\sum_{\mu\in\mathrm{Spec}(H_G)}\mu^{-k}=\bigl(1+(-1)^k\bigr)n(d-2)/2+\Bigl(\mathrm{Tr}(H_G^k)-\bigl(1+(-1)^k\bigr)n(d-2)/2\Bigr)(d-1)^{-k}$$

where Tr denotes the trace, and hence

$$-\zeta'_G(u)/\zeta_G(u)=\mathcal{L}_G(u)+e(u),$$

where

$$\mathcal{L}_G(u)=\sum_{k=0}^{\infty}u^{-1-k}\,\mathrm{Tr}(H_G^k)(d-1)^{-k},$$

and

$$\begin{aligned}e(u)&=\sum_{k=0}^{\infty}u^{-1-k}\bigl(1-(d-1)^{-k}\bigr)\bigl(1+(-1)^k\bigr)n(d-2)/2\\&=\frac{n(d-2)}{2u}\sum_{k\ge 0}\Bigl[u^{-k}+(-u)^{-k}-\bigl((d-1)u\bigr)^{-k}-\bigl(-(d-1)u\bigr)^{-k}\Bigr]\\&=\frac{n(d-2)}{2u}\left[\frac{1}{1-u}+\frac{1}{1+u}-\frac{1}{1-(d-1)u}-\frac{1}{1+(d-1)u}\right].\end{aligned}$$

It follows that $e(u)$ is a rational function with poles only at $u=\pm 1$ and $\pm 1/(d-1)$. Furthermore, the $e(u)$ poles at ± 1 have residue $-\chi(G)$.

DEFINITION 4.4. Let G be a d-regular graph without half-loops. We define the *essential logarithmic derivative* to be the meromorphic function complex function

$$\mathcal{L}_G(u) = \sum_{k=0}^{\infty} u^{-1-k}\,\mathrm{Tr}(H_G^k)(d-1)^{-k}.$$

We caution the reader that $\mathcal{L}_G(u)$ is the interesting part of *minus* the usual logarithmic derivative of $\zeta_G(u)$ (since we are interested in poles, not zeros). Clearly $\mathrm{Tr}(H_G^k)$ is bounded by the number of non-backtracking walks of length k in G, i.e., $|V_G|d(d-1)^{k-1}$, and hence the above expansion for $\mathcal{L}_G(u)$ converges for all $|u|>1$.

We summarize the above discussion in the follow proposition.

PROPOSITION 4.5. *Let G be a d-regular, ϵ-spectral graph. Then, with $C^+_{\epsilon,\delta}$ as in (4), we have that the number of positive non-Ramanujan Hashimoto eigenvalues is given by*

$$\frac{1}{2\pi i}\int_{C^+_{\epsilon,\delta}} \mathcal{L}(u)\,du$$

for $\delta>0$ sufficiently small; similarly for negative eigenvalues.

We remark that if G is ϵ-spectral for $\epsilon>0$ small, then $\mathrm{Tr}(H_G^k)$ is the sum of d^k plus $nd-1$ other eigenvalues, all of which are within the ball $|\mu|\le(d-1)^{1/2}+\epsilon'$ for some ϵ' that tends to zero as ϵ tends to zero; hence, for such G, the expression for $\mathcal{L}_G$ in Definition 4.4 has a simple pole of residue 1 at $u=1$, and the power series at infinity for

$$\mathcal{L}_G(u)-\frac{1}{u-1}$$

converges for all $|u|^{-1}<(d-1)^{1/2}+\epsilon'$.

5. The Expected Value of $\mathcal{L}_G$

For any even integer, d, and integer $n>0$, we define $\mathcal{G}_{n,d}$ to the probability space of d-regular random graphs formed by independently choosing $d/2$ permutations, $\pi_1,\ldots,\pi_{d/2}$ uniformly from the set of $n!$ permutations of $\{1,\ldots,\}$; to each such $\pi_1,\ldots,\pi_{d/2}$ we associate the random graph, $G=G(\{\pi_i\})$, whose vertex set is $V_G=\{1,\ldots,n\}$, and whose edge set, E_G, consists of all sets

$$E_G=\{\{i,\pi_j(i)\}\mid i=1,\ldots,n,\ j=1,\ldots,d/2\}.$$

It follows that G may have multiple edges and self-loops.

The following is a corollary of [**Fri08, FK14**].

THEOREM 5.1. *For a d-regular graph, G, define define $N_A^+(G)$ to be the number of positive non-Ramanujan adjacency eigenvalues of G plus the multiplicity of $2(d-1)^{-1/2}$ (if any) as an eigenvalue of G. Then for any even $d\ge 10$ and $\epsilon>0$ we have that*

$$\lim_{n\to\infty}\lim_{\delta\to 0}\mathbb{E}_{G\in\mathcal{G}_{n,d}}\left[N_{A,+}(G)-\frac{1}{4\pi i}\int_{C^+_{\epsilon,\delta}}\mathcal{L}_G(u)\,du\right]=0,$$

where we interpret the contour integral as its Cauchy principle for graphs, G, whose Zeta function has a pole on $C^+_{\epsilon,\delta}$.

PROOF. For $d \geq 10$ we know that the probability that a graph has any such eigenvalues is at most $O(1/n^2)$ (see [**Fri08**] for $d \geq 12$, and [**FK14**] for $d = 10$), and hence this expected number is at most $1/n$. □

For $d \geq 4$ one might conjecture the same theorem holds, although this does not seem to follow literally from [**Fri08, FK14**]; however, if one conditions on $G \in \mathcal{G}_{n,d}$ not having a (d, ϵ')-tangle (in the sense of [**FK14**]), which is an order $O(1/n)$ probability event, then we get equality. The problem is that one does not know where most of the eigenvalues lie in graphs that have tangles; we conjecture that graphs with tangles will not give more than an $O(1/n)$ expected number of eigenvalues strictly between $2(d-1)^{1/2} + \epsilon$ and d; hence we conjecture that the above above theorem remains true for all $d \geq 4$.

We remark that for any d-regular graph, we have that

$$\lim_{\delta \to 0} \frac{1}{4\pi i} \int_{C^+_{\epsilon,\delta}} \mathcal{L}_G(u)\, du = N_{A,+,\epsilon}(G),$$

where $N_{A,+,\epsilon}$ counts the number of positive, real Hashimoto eigenvalues, μ, such that

$$|1 - \mu(d-1)^{1/2}| \leq \epsilon.$$

Hence for any d, n, ϵ we have

$$\lim_{\delta \to 0} \mathbb{E}_{G \in \mathcal{G}_{n,d}} \left[\frac{1}{4\pi i} \int_{C^+_{\epsilon,\delta}} \mathcal{L}_G(u)\, du \right] = \mathbb{E}_{G \in \mathcal{G}_{n,d}} \left[N_{A,+,\epsilon}(G) \right].$$

Now we wish to conjecture a value for

$$\lim_{\delta \to 0} \mathbb{E}_{G \in \mathcal{G}_{n,d}} \left[\frac{1}{4\pi i} \int_{C^+_{\epsilon,\delta}} \mathcal{L}_G(u)\, du \right]. \tag{5}$$

We now seek to use trace methods to in order to conjecture what the value of (5) will be for fixed d and $n \to \infty$.

It is known that for d, r fixed and n large, we have that

$$\mathbb{E}_{G \in \mathcal{G}_{n,d}} \left[\mathrm{Tr}(H_G^k) \right] = P_0(k) + P_1(k) n^{-1} + \cdots + P_{r-1}(k) n^{1-r} + \mathrm{err}_r(n,k), \tag{6}$$

where $P_i(k)$ are functions of k alone, and

$$|\mathrm{err}_r(n,k)| \leq C_r k^{2r} (d-1)^k n^{-r}.$$

Furthermore, $P_0(k)$ is known (see [**Fri91, Fri08, FK14**], but essentially since [**BS87**]) to equal

$$P_0(k) = O(kd) + \sum_{k'|k} (d-1)^{k'},$$

where $k'|k$ means that k' is a positive integer dividing k; in particular, we have

$$P_0(k) = (d-1)^k + \mathbb{I}_{\mathrm{even}}(k)(d-1)^{k/2} + O(d-1)^{k/3}, \tag{7}$$

where $\mathbb{I}_{\mathrm{even}}(k)$ is the indicator function of k being even. Furthermore, we believe the methods of [**Fri91, FK14**] will show that for each i we have

$$\mathcal{P}_i(u) = \sum_{k=0}^{\infty} u^{-1-k} P_i(k)(d-1)^{-k} \tag{8}$$

is meromorphic with finitely many poles outside any disc about the origin of radius strictly greater than $1/(d-1)$; our idea is to use the "mod-S" function approach of [**FK14**] to show that each $P_i(k)$ is a polyexponential plus an error term (see [**FK14**]) and to argue for each "type" separately, although we have not written and checked this carefully as of the writing of this article; hence this belief may be regarded as a (plausible) conjecture at present. Let us give some conjectures based on the above assumption, and the (more speculative) assumption that we can evaluate the above asymptotic expansion by taking expected values term by term, and the (far more speculative) assumption that the $\mathcal{P}_i(u)$, formed by summing over arbitrarily large k, reflect the properties of $\mathcal{G}_{n,d}$ with n fixed.

DEFINITION 5.2. For any even $d \geq 4$, set

$$N_i = \frac{1}{2\pi i} \int_{C^+_{\epsilon,\delta}} \mathcal{P}_i(u)\, du,$$

where we assume the $\mathcal{P}_i(u)$, given in (8), based on functions $P_i(k)$ given in (6), are meromorphic functions, at least for $|u|$ near $(d-1)^{-1/2}$. We say that $\mathcal{G}_{\cdot,d}$ is *positively approximable to order* r if for some $\epsilon > 0$ we have

$$\lim_{\delta\to 0} \frac{1}{2\pi i} \int_{C^+_{\epsilon,\delta(n)}} \mathbb{E}_{G\in\mathcal{G}_{n,d}} \left[\mathcal{L}_G(u)\right] du = N_0 + N_1 n^{-1} + N_2 n^{-2} + \cdots + N_r n^{-r} + o(n^{-r})$$

as n tends to infinitly; we similarly define *negativly approximable* with negative real eigenvalues and $C^-_{\epsilon,\delta}$.

We now state a number of conjectures, which are successively weaker.

CONJECTURE 5.3. *For any even $d \geq 4$ we have $\mathcal{G}_{\cdot,d}$ is*

(1) *positively approximable to any order;*
(2) *positively approximable to order $r(d)$, where $r(d) \geq 0$, and $r(d) \to \infty$ as $d \to \infty$;*
(3) *positively approximable to order 0; and*
(4) *positively approximable to order 0 for d sufficiently large;*

and similarly with "postively" replaced with "negatively."

For the above conjecture, we note that $\mathcal{G}_{\cdot,d}$ exhibits (d,ϵ')-tangles of order 1 (see [**Fri08, FK14**]) for $\epsilon' > 0$ and $d \leq 8$; for such reasons, we believe that there may be a difference between small and large d.

Of course, the intriguing part of this conjecture is the calculation taking (7) to show that

$$\mathcal{P}_0(u) = \sum_{k=1}^{\infty} P_0(k) u^{-k-1} = \frac{u^{-1}}{1-u^{-1}} + \frac{u^{-1}}{1-(d-1)^{-1}u^{-2}} + h(u)$$

$$= \frac{1}{u-1} + \frac{u}{u^2-(d-1)} + h(u),$$

where $h(u)$ is holomorphic in $|u| > (d-1)^{-2/3}$. It follows that

$$N_0 = \frac{1}{2\pi i} \int_{C^+_{\epsilon,\delta}} \mathcal{P}_0(u)\, du = 1/2;$$

similarly with "positive" replaced with "negative." Hence the above conjecture would imply that the expected number of positive, non-Ramanujan adjacency eigenvalues for a graph in $\mathcal{G}n,d$, with d fixed and n large, would tend to 1/4. This establishes a main point of interest.

PROPOSITION 5.4. *Let $d \geq 4$ be an even integer for which $\mathcal{G}_{\cdot,d}$ is positively approximable to order* 0. *Then, as $n \to \infty$, the limit supremum of the expected of positive, non-Ramanujan Hashimoto eigenvalues of $G \in \mathcal{G}_{n,d}$ is at most $1/2$; similarly, the same for non-Ramanujan adjacency eigenvalues is at most $1/4$. The same holds with "positive(ly)" replaced everywhere with "negative(ly)."*

The reason we involve the limit supremum in the above is that it is conceivable (although quite unlikely in our opinion) that there is a positive expected multiplicity of the eigenvalue $(d-1)^{1/2}$ in H_G for $G \in \mathcal{G}_{n,d}$.

We finish this secion with a few remarks considering the above conjectures.

In the usual trace methods one estimates the expected value of A_G^k or H_G^k for k of size proportional to $\log n$; furthermore, the contributions to $P_0(k)$ consist of "single loops" (see [**Fri08**, **FK14**]), which cannot occur unless $k \leq n$. Hence, the idea of fixing n and formally summing in k cannot be regarded as anything but a formal summation.

We also note that for any positive integer, m, $\mathcal{P}_0(u)$ has poles at $(d-1)^{-(m-1)/m}\omega_m$, where ω_m is any m-th root of unity; hence this function does not resemble $\zeta_G(u)$ for a fixed d-regular graph G, whose poles are confined to the reals and the complex circle $|u| = (d-1)^{-1/2}$; hence if $\mathcal{P}_0(u)$ truly reflects some average property of $\zeta_G(u)$ for $G \in \mathcal{G}_{n,d}$ everwhere in $|u| > (d-1)^{-1}$, then there is some averaging effect that makes $\mathcal{P}_0(u)$ different that the typical $\zeta_G(u)$.

6. A Simpler Variant of the P_i and $\mathcal{P}_i$

Part of the problem in dealing with the P_i of (6) and $\mathcal{P}_i$ of (8) is that the $\mathcal{P}_i$ can, at least in principle (and we think likely), have roughly i^i real poles between 1 and $(d-1)^{1/2}$, where the i^i represents roughly the number of *types* [**Fri91**, **Fri08**, **FK14**] of order i (see also [**Pud15**] for similar problems).

However, the methods of [**FK14**] ("mod-S" functions, Section 3.5) show that for fixed d, and fixed i bounded by a constant times $(d-1)^{1/3}$, we have that $\mathcal{P}_i(u)$ has poles at only $u = \pm 1$ and $u = \pm(d-1)^{-1/2}$ for $|u| > (d-1)^{-2/3}$. Hence for fixed d and i sufficiently small, the $\mathcal{P}_i(u)$ are much simpler to analyze. However the calculations [**Fri91**, **Fri08**, **FK14**] show something a bit stronger: namely if we consider $\mathcal{P}_{i,d}(u)$ and $P_{i,d}(u)$ as depending both on i and d, then in fact

$$
\begin{aligned}
P_i(k) &= (d-1)^k Q_i(k, d-1) + \mathbb{I}_{\text{even}}(d-1)^{k/2} R_i(k, d-1) \\
&\quad + \mathbb{I}_{\text{odd}}(d-1)^{k/2} S_i(k, d-1) + O(d-1)^{k/3} C k^C
\end{aligned}
\tag{9}
$$

for some constant $C = C(i)$, and where $Q_i(k, d-1), R_i(k, d-1), S_i(k, d-1)$ are polynomial in k (whose degree is bounded by a function of i), whose coefficients are rational functions of $d-1$.

DEFINITION 6.1. The *large d polynomials of order i* are the functions $Q_i(k, d), R_i(k, d), S_i(k, d)$, determined uniquely in (9). The associated *approximate principle term* to Q_i, R_i, S_i is the function

$$
\widetilde{P}_i(k, d) = (d-1)^k Q_i(k, d-1) + \mathbb{I}_{k \text{ even}}(d-1)^{k/2} R_i(k, d-1) + \mathbb{I}_{k \text{ odd}}(d-1)^{k/2} S_i(k, d-1),
$$

and the associated *approximate generating function* is

$$
\widehat{\mathcal{P}}_i(u) = \sum_{k=0}^{\infty} u^{-1-k} \widehat{P}_i(k)(d-1)^{-k}.
$$

It follows that the $\widehat{\mathcal{P}}_i(u)$ have poles only at $u = 1$ and $u = \pm(d-1)^{-1/2}$ outside any disc about zero of radius strictly greater than $(d-1)^{-2/3}$.

The benefit of working with the $\widehat{P}_i(k,d)$ and $\widehat{\mathcal{P}}_i(u)$ is that there are various ways of trying to compute these functions. For example, one can fix k and n, and consider what happens as $d \to \infty$. In this case we are studying the expected number of fixed points of strictly reduced words of length k in the alphabet

$$\Pi = \{\pi_1, \pi_1^{-1}, \pi_2, \pi_2^{-1}, \ldots, \pi_{d/2}, \pi_{d/2}^{-1}\}$$

with d large. If such a word has exactly one occurrence of π_j and π_j^{-1} for some j, which is very likely since k is fixed and d is large, then the expected number of fixed points is exactly 1. Hence we are led to consider those words such that for every j, either π_j and π_j^{-1} do not occur, or they occur at least twice in total. The study of such words does not seem easy, although perhaps this can be understood, say using the recent works [**Pud15**, **PP15**].

This type of study also seems to ressemble more closely the standard (and much more studied) random matrix theory than the d-regular spectral graph theory with d fixed; perhaps methods from random matrix theory can be applied here.

7. Random Graph Covering Maps and Other Models

We remark that [**Fri08**] studies other models of random d-regular graphs on n vertices for d and n of arbitrary partity (necessarily having half-edges if d and n are odd). We remark that if d is odd and n is even, we can form a model of a d-regular graph based on d perfect matchings, called $\mathcal{I}_{n,d}$ in [**Fri08**]. A curious model of random regular graph is $\mathcal{H}_{n,d}$ (for d even), which is a variant of $\mathcal{G}_{n,d}$ where random permutations are chosen from the subset of permutations whose cyclic structure consists of a single cycle of length n; since such permutations cannot have self-loops (for $n > 1$), for most d, the probability of eigenvalues lying in the Alon region is larger for $\mathcal{H}_{n,d}$ than for $\mathcal{G}_{n,d}$.

For the rest of this section we give a natural extension of our main conjectures to the more general model of a random cover of a graphs. $\mathcal{G}_{n,d}$ and $\mathcal{I}_{n,d}$ are special cases of a "random degree n covering map of a base graph," where the base graphs are, respectively, a bouquet of $d/2$ whole-loops (requiring d to be even), and a bouquet of d half-loops (where d can be either even or odd). We shall be brief, and refer the reader to [**FK14**] for details.

Definition 7.1. A *morphism of directed graphs*, $\varphi\colon G \to H$ is a pair $\varphi = (\varphi_V, \varphi_E)$ for which $\varphi_V\colon V_G \to V_H$ is a map of vertices and $\varphi_E\colon E_G^{\rm dir} \to E_H^{\rm dir}$ is a map of directed edges satisfying $h_H(\varphi_E(e)) = \varphi_V(h_G(e))$ and $t_H(\varphi_E(e)) = \varphi_V(t_G(e))$ for all $e \in E_G^{\rm dir}$. We refer to the values of φ_V^{-1} as *vertex fibres* of φ, and similarly for edge fibres. We often more simply write φ instead of φ_V or φ_E.

Definition 7.2. A morphism of directed graphs $\nu\colon H \to G$ is a *covering map* if it is a local isomorphism, that is for any vertex $w \in V_H$, the edge morphism ν_E induces a bijection (respectively, injection) between $t_H^{-1}(w)$ and $t_G^{-1}(\nu(w))$ and a bijection (respectively, injection) between $h_H^{-1}(w)$ and $h_G^{-1}(\nu(w))$. We call G the *base graph* and H a *covering graph of* G.

If $\nu\colon H \to G$ is a covering map and G is connected, then the *degree* of ν, denoted $[H\colon G]$, is the number of preimages of a vertex or edge in G under ν (which does not depend on the vertex or edge). If G is not connected, we insist that the number of preimages of ν of a vertex or edge is the same, i.e., the degree is independent of the connected component, and we will write this number as $[H: G]$. In addition, we often refer to H, without ν mentioned explicitly, as a *covering graph* of G.

A morphism of graphs is a *covering map* if the morphism of the underlying directed graphs is a covering map.

DEFINITION 7.3. If $\pi: G \to B$ is a covering map of directed graphs, then an *old function (on V_G)* is a function on V_G arising via pullback from B, i.e., a function $f\pi$, where f is a function (usually real or complex valued), i.e., a function on V_G (usually real or complex valued) whose value depends only on the π vertex fibres. A *new function (on V_G)* is a function whose sum on each vertex fibre is zero. The space of all functions (real or complex) on V_G is a direct sum of the old and new functions, an orthogonal direct sum on the natural inner product on V_G, i.e.,

$$(f_1, f_2) = \sum_{v \in V_G} \overline{f_1(v)} f_2(v).$$

The adjacency matrix, A_G, viewed as an operator, takes old functions to old functions and new functions to new functions. The *new spectrum* of A_G, which we often denote $\mathrm{Spec}_B^{\mathrm{new}}(A_G)$, is the spectrum of A_G restricted to the new functions; we similarly define the *old spectrum*.

This discussion holds, of course, equally well if $\pi: G \to B$ is a covering morphism of graphs, by doing everything over the underlying directed graphs.

We can make similar definitions for the spectrum of the Hashimoto eigenvalues. First, we observe that covering maps induce covering maps on directed line graphs; let us state this formally (the proof is easy).

PROPOSITION 7.4. *Let $\pi: G \to B$ be a covering map. Then π induces a covering map π^{Line}:* $\mathrm{Line}(G) \to \mathrm{Line}(B)$.

Since $\mathrm{Line}(G)$ and $\mathrm{Line}(B)$ are directed graphs, the above discussion of new and old functions, etc., holds for $\pi^{\mathrm{Line}}: \mathrm{Line}(G) \to \mathrm{Line}(B)$; e.g., new and old functions are functions on the vertices of $\mathrm{Line}(G)$, or, equivalently, on E_G^{dir}.

DEFINITION 7.5. Let $\pi: G \to B$ be a covering map. We define the *new Hashimoto spectrum of G with respect to B*, denoted $\mathrm{Spec}_B^{\mathrm{new}}(H_G)$ to be the spectrum of the Hashimoto matrix restricted to the new functions on $\mathrm{Line}(G)$, and $\rho_B^{\mathrm{new}}(H_G)$ to be the supremum of the norms of $\mathrm{Spec}_B^{\mathrm{new}}(H_G)$.

We remark that

$$\sum_{\mu \in \mathrm{Spec}_B^{\mathrm{new}}(H_G)} \mu^k = \mathrm{Tr}(H_G^k) - \mathrm{Tr}(H_B^k),$$

and hence the new Hashimoto spectrum is independent of the covering map from G to B; similarly for the new adjacency spectrum.

To any base graph, B, one can describe various models of random covering maps of degree n. The simplest is to assign to each edge, $e \in E_B$, a permutation $\pi(e)$ on $1, \ldots, n$ with the stipulation that $\pi(\iota_B e)$ is the inverse permutation of $\pi(e)$; if $e = \{e_1, e_2\}$ is not a half-loop, and $\iota e_1 = e_2$, then we may assign an arbitrary random permutation to $\pi(e_1)$ and then set $\pi(e_2)$ to be the inverse permutation of $\pi(e_1)$; if e is a half-loop, then we can assign a random perfect matching to $\pi(e)$ if n is even, and otherwise choose $\pi(e)$ to consist of one fixed point (which is a half-loop in the covering graph) and a random perfect matching on the remaining $n-1$ elements. See [**FK14**] for a more detailed description of this model and other "algebraic" models of a random covering graph of degree n over a fixed based graph.

For $d \geq 4$ even, $\mathcal{G}_{n,d}$ is the above model over the base graph which is a bouquet of $d/2$ whole-loops, and $\mathcal{I}_{n,d}$ defined at the beginning of this section is the above model over the base graph consisting of a bouqet of d half-loops.

Again, one can make similar computations and conjectures as with $\mathcal{G}_{n,d}$. The analogue of $P_0(k)$ for random covers of degree n over general base graph, B, is easily seen to be

$$P_0(k) = O(kd) + \sum_{k'|k} \operatorname{Tr}(H_B^{k'}),$$

(see [**FK14**], but essentially done in [**Fri03**], a simple variant of [**BS87**]). As for regular graphs, the $\operatorname{Tr}(H_B^k)$ is essentially the old spectrum, and the new spectrum term is therefore

$$\mathbb{I}_{even}(k) \operatorname{Tr}(H_B^{k/2}) + O(k)(d-1)^{k/3}.$$

The analogue of $\mathcal{P}_0(u)$ is therefore determined by this

$$\operatorname{Tr}(H_B^{k/2})$$

term for k even. There are two cases of interest:

(1) B connected and not bipartite, in which case $d-1$ is an eigenvalue of H_B, and all other eigenvalues of H_B are of absolute value strictly less than $d-1$; then we get the similar formal expansions and conjectures as before; and
(2) B connected and bipartite, in which case $-(d-1)$ is also an eigenvalue, and we get a conjectured expectation of $1/2$ for the number of positive non-Ramanujan adjacency eigenvalues. In this case, of course, each positive eigenvalue has a corresponding negative eigenvalue. So, again, we would conjecture that there is a positive probability that a random degree n cover of B has two pairs (i.e., four) non-Ramanujan eigenvalues; and, again, these two conjectures imply that for fixed d-regular B and sufficiently large n, a strict majority of the degree n covers of B are relatively Ramanujan.

8. Numerical Experiments

Here we give some preliminary numerical experiments done to test our conjectures for random 4-regular graphs. As mentioned before, the results of [**MNS08**] indicate that one may need graphs of many more than the $n = 400,000$ vertices and fewer that we used in our experiments.

We also mention that our experiments are Bernoulli trials with probability of success that seems to be between .17 and .26. Hence for $n \geq 100,000$, where we test no more than 500 random sample graphs, the hundredths digit is not significant. We have tested $10,000$ examples only for $n \leq 10,000$.

For the model $\mathcal{G}_{n,4}$, of a $d = 4$-regular graph generated by two of the $n!$ random permutations of $\{1,\ldots,n\}$, we computed the total number of positive eigenvalues no smaller than $2(d-1)^{1/2}$. We sampled

(1) $10,000$ random graphs with $n = 100$, $n = 1000$, and $n = 10,000$, whose average number of such eigenvalues were 1.2681, 1.2258, and 1.1942, respectively;

(2) 500 random graphs with $n = 100,000$, with an average of 1.176;
(3) 250 random graphs with $n = 200,000$, with an average of 1.188;
(4) 79 random graphs with $n = 400,000$, with an average of 1.177.

Of course, there is always $\lambda_1(G) = d$, and there is a small chance, roughly $1/n + O(1/n^2)$ that G will be disconnected. So our formal calculations suggest that we should see 1.25 for very large n, or no more than this. However there is no particularly evident convergence of these average number of positive non-Ramanujan eigenvalues at these values of n.

We also tried the model $\mathcal{H}_{n,4}$, where the permutations are chosen among the $(n-1)!$ permutations whose cyclic structure is a single cycle of length n. Since $\mathcal{H}_{n,4}$ graphs are always connected, and generally have less tangles than $\mathcal{G}_{n,4}$ [**Fri08**], we felt this model may give more representative results for the same values of n. We sampled

(1) 10,000 random graphs with $n = 100$, $n = 1000$, and $n = 10,000$, whose average number of such eigenvalues were 1.1268, 1.161, and 1.1693, respectively;
(2) 500 random graphs with $n = 100,000$, with an average of 1.192;
(3) 55 random graphs with $n = 200,000$, with an average of 1.163;
(4) 87 random graphs with $n = 400,000$, with an average of 1.149.

Again, we see no evident convergence at this point.

References

[Alo86] N. Alon, *Eigenvalues and expanders*, Combinatorica **6** (1986), no. 2, 83–96, DOI 10.1007/BF02579166. Theory of computing (Singer Island, Fla., 1984). MR875835

[AM85] N. Alon and V. D. Milman, λ_1, *isoperimetric inequalities for graphs, and superconcentrators*, J. Combin. Theory Ser. B **38** (1985), no. 1, 73–88, DOI 10.1016/0095-8956(85)90092-9. MR782626

[BS87] Andrei Broder and Eli Shamir, *On the second eigenvalue of random regular graphs*, Proceedings 28th Annual Symposium on Foundations of Computer Science, 1987, pp. 286–294.

[dB46] N. G. de Bruijn, *A combinatorial problem*, Nederl. Akad. Wetensch., Proc. **49** (1946), 758–764 = Indagationes Math. 8, 461–467 (1946). MR0018142

[FK14] Joel Friedman and David-Emmanuel Kohler, *The relativized second eigenvalue conjecture of alon*, Available at `http://arxiv.org/abs/1403.3462`.

[FKS89] J. Friedman, J. Kahn, and E. Szemerédi, *On the second eigenvalue of random regular graphs*, 21st Annual ACM Symposium on Theory of Computing, 1989, pp. 587–598.

[Fri91] Joel Friedman, *On the second eigenvalue and random walks in random d-regular graphs*, Combinatorica **11** (1991), no. 4, 331–362, DOI 10.1007/BF01275669. MR1137767

[Fri93] Joel Friedman, *Some geometric aspects of graphs and their eigenfunctions*, Duke Math. J. **69** (1993), no. 3, 487–525, DOI 10.1215/S0012-7094-93-06921-9. MR1208809

[Fri03] Joel Friedman, *Relative expanders or weakly relatively Ramanujan graphs*, Duke Math. J. **118** (2003), no. 1, 19–35, DOI 10.1215/S0012-7094-03-11812-8. MR1978881

[Fri08] Joel Friedman, *A proof of Alon's second eigenvalue conjecture and related problems*, Mem. Amer. Math. Soc. **195** (2008), no. 910, viii+100, DOI 10.1090/memo/0910. MR2437174

[GG81] Ofer Gabber and Zvi Galil, *Explicit constructions of linear-sized superconcentrators*, J. Comput. System Sci. **22** (1981), no. 3, 407–420, DOI 10.1016/0022-0000(81)90040-4. Special issued dedicated to Michael Machtey. MR633542

[HN60] Frank Harary and Robert Z. Norman, *Some properties of line digraphs*, Rend. Circ. Mat. Palermo (2) **9** (1960), 161–168, DOI 10.1007/BF02854581. MR0130839

[Iha66] Yasutaka Ihara, *On discrete subgroups of the two by two projective linear group over $\mathfrak{p}$-adic fields*, J. Math. Soc. Japan **18** (1966), 219–235, DOI 10.2969/jmsj/01830219. MR0223463

[Knu67] Donald E. Knuth, *Oriented subtrees of an arc digraph*, J. Combinatorial Theory **3** (1967), 309–314. MR0214511

[LP10] Nati Linial and Doron Puder, *Word maps and spectra of random graph lifts*, Random Structures Algorithms **37** (2010), no. 1, 100–135, DOI 10.1002/rsa.20304. MR2674623

[LPS88] A. Lubotzky, R. Phillips, and P. Sarnak, *Ramanujan graphs*, Combinatorica **8** (1988), no. 3, 261–277, DOI 10.1007/BF02126799. MR963118

[Mar88] G. A. Margulis, *Explicit group-theoretic constructions of combinatorial schemes and their applications in the construction of expanders and concentrators* (Russian), Problemy Peredachi Informatsii **24** (1988), no. 1, 51–60; English transl., Problems Inform. Transmission **24** (1988), no. 1, 39–46. MR939574

[McK81] Brendan D. McKay, *The expected eigenvalue distribution of a large regular graph*, Linear Algebra Appl. **40** (1981), 203–216, DOI 10.1016/0024-3795(81)90150-6. MR629617

[MNS08] Steven J. Miller and Tim Novikoff, *The distribution of the largest nontrivial eigenvalues in families of random regular graphs*, Experiment. Math. **17** (2008), no. 2, 231–244. MR2433888

[Mor94] Moshe Morgenstern, *Existence and explicit constructions of $q+1$ regular Ramanujan graphs for every prime power q*, J. Combin. Theory Ser. B **62** (1994), no. 1, 44–62, DOI 10.1006/jctb.1994.1054. MR1290630

[MSS13] Adam W. Marcus, Daniel A. Spielman, and Nikhil Srivastava, *Interlacing families I: Bipartite Ramanujan graphs of all degrees*, Ann. of Math. (2) **182** (2015), no. 1, 307–325, DOI 10.4007/annals.2015.182.1.7. MR3374962

[PP15] Doron Puder and Ori Parzanchevski, *Measure preserving words are primitive*, J. Amer. Math. Soc. **28** (2015), no. 1, 63–97, DOI 10.1090/S0894-0347-2014-00796-7. MR3264763

[Pud15] Doron Puder, *Expansion of random graphs: new proofs, new results*, Invent. Math. **201** (2015), no. 3, 845–908, DOI 10.1007/s00222-014-0560-x. MR3385636

[Ser03] Jean-Pierre Serre, *Trees*, Springer Monographs in Mathematics, Springer-Verlag, Berlin, 2003. Translated from the French original by John Stillwell; Corrected 2nd printing of the 1980 English translation. MR1954121

[ST96] H. M. Stark and A. A. Terras, *Zeta functions of finite graphs and coverings*, Adv. Math. **121** (1996), no. 1, 124–165, DOI 10.1006/aima.1996.0050. MR1399606

[ST00] H. M. Stark and A. A. Terras, *Zeta functions of finite graphs and coverings. II*, Adv. Math. **154** (2000), no. 1, 132–195, DOI 10.1006/aima.2000.1917. MR1780097

[Tan84] R. Michael Tanner, *Explicit concentrators from generalized N-gons*, SIAM J. Algebraic Discrete Methods **5** (1984), no. 3, 287–293, DOI 10.1137/0605030. MR752035

[Ter11] Audrey Terras, *Zeta functions of graphs*, Cambridge Studies in Advanced Mathematics, vol. 128, Cambridge University Press, Cambridge, 2011. A stroll through the garden. MR2768284

[TS07] A. A. Terras and H. M. Stark, *Zeta functions of finite graphs and coverings. III*, Adv. Math. **208** (2007), no. 1, 467–489, DOI 10.1016/j.aim.2006.03.002. MR2304325

Department of Computer Science, University of British Columbia, Vancouver, BC V6T 1Z4, Canada –and– Department of Mathematics, University of British Columbia, Vancouver, BC V6T 1Z2, Canada.

Email address: jf@cs.ubc.ca, jf@math.ubc.ca

URL: http://www.math.ubc.ca/~jf

Contemporary Mathematics
Volume **739**, 2019
https://doi.org/10.1090/conm/739/14894

Rank and Bollobás-Riordan polynomials: Coefficient measures and zeros

Dmitry Jakobson, Tomas Langsetmo, Igor Rivin, and Lise Turner

ABSTRACT. We discuss some (numerical and theoretical) results about the coefficients and zeros of Tutte (dichromatic) polynomial of graphs of bounded degree whose size increases. We also discuss related results for Bollobás-Riordan polynomials.

1. Introduction

In this paper we discuss some numerical and theoretical results on the coefficients and zeros of Tutte and Bollobás-Riordan polynomials. The numerical results on the coefficients of Tutte polynomials inspired the paper [**JMNT**], where weak convergence of certain natural coefficient measures was investigated for sequences of bounded degree graphs that converge in the sense of Benjamini-Schramm; we generalize those results for Bollobás-Riordan polynomials in this paper. The MSc thesis [**Tur**] of one of the authors also included a similar result for the coefficient measure of the Tutte polynomial in the case of Benjamini-Schramm convergent sequences of planar graphs with bounded face degree. We also establish some *a priori* results for the coefficient measures of Tutte polynomials. Finally, we study (numerically) zeros of the Tutte and Bollobás-Riordan polynomials.

Below, we summarize the results in our paper.

1.1. Tutte polynomials.

The *Tutte (or dichromatic)* polynomial was introduced by Tutte; it is the most general graph invariant satisfying a deletion-contraction recurrence formula. After a change of variables, it can be transformed into a *rank polynomial* (or Whitney rank generating function). It contains important information about G, in particular about its connectivity properties, and about nowhere-zero flows on G. In Statistical Physics, it describes the partition function for the Potts model on G. When restricted to certain curves (or points), the dichromatic polynomial specializes to some well-known graph invariants, including chromatic polynomial, the number of spanning trees, the number of acyclic

2010 *Mathematics Subject Classification.* Primary 05C31,05C63,05C80,57M20.
Key words and phrases. Graphs, ribbon graphs, Benjamini-Schramm convergence, Tutte polynomial, Bollobás-Riordan polynomial, coefficients, zeros.
The first author was supported by NSERC and FQRNT.
The second and fourth authors were supported by NSERC.

orientations etc. It is closely related to important invariants in knot theory, including the Jones polynomial. See [**Wel**] for an excellent survey on the properties of the Tutte polynomial.

The asymptotic behaviour of many graph invariants, including Laplace spectrum, cycle distribution, colouring properties, non-concentration of eigenvectors etc. has been studied extensively before. However, several asymptotic properties of the Tutte polynomial have not been considered before, to our knowledge. In our paper, we focus on the *coefficients* of the Tutte polynomial T_G (defined in (2.1)) as well as its zeroes. In our numerical experiments, we focus on random d-regular graphs G. In particular, we define a probability measure describing the concentration of the (normalized) coefficients of this polynomial in 2.4. We study these probability measures numerically, averaged over random regular graphs, in section 4.

In Section 3, we use the spanning tree expansion of T_G and results about overlap of spanning trees in G to establish Theorem 3.3, bounding the degree of T_G in terms of the size of overlap between two spanning trees of G. This result is generalized to more than two spanning trees in Lemma 3.5.

In Section 3.1, the previous results are specialized to the case of d-regular graphs G on n vertices. First, we note the following result of Catlin: a bound on the edge connectivity of G implies existence of disjoint spanning trees, Proposition 3.7. Next, we combine results due to Wormald about the edge connectivity of random regular graphs, with Proposition 3.7 and Lemma 3.5 to establish Theorem 3.8 which gives an a.s. bound on the degree of T_G for d-regular G on n vertices, as $n \to \infty$. The case of cubic graphs (regular graphs of degree 3) is considered in Theorem 3.11 about the overlap of spanning trees on such graphs. All graphs in this paper will be simple unless loops or multiple edges are explicitly allowed.

Section 4 collects the results of numerical experiments on the coefficient measure of the Tutte polynomial of random regular graphs of varying size and degree. Some comments are made regarding the shape of these distributions.

Section 5 contains various numerical representations of the zero sets of the Tutte polynomial of random regular graphs. Both *real* and *complex* zeros are considered. The real zeros lie in $\mathbb{R}^2$; the complex zeros lie in $\mathbb{C}^2$ and so we picture three-dimensional "slices" of the zero sets. Various cross sections are presented in an effort to better understand their structure. Some theoretical results are also presented.

1.2. Bollobás-Riordan polynomials. The Bollobás-Riordan polynomial is a 3-variable polynomial defined for *ribbon graphs* (graphs with a cyclic edge orientation at every vertex) which provides an extension of the rank polynomial to such graphs. Those polynomials were introduced in [**BR01, BR02**].

In the second part of our paper, we investigate asymptotic properties of the coefficients and zeros of Bollobás-Riordan polynomials for bounded degree ribbon graphs whose size increases. We define ribbon graphs in §6. Bollobás-Riordan polynomials are defined in §8. Random ribbon graphs were studied in several papers, including [**Gam, Fl-P, Ch-Pit**]. We summarize some of the relevant results in §9, and provide a natural extension of Benjamini-Schramm convergence for ribbon graphs. In section 10 we show that natural analogues of the coefficient measures for those graphs converge to a delta function (as the graph size increases), extending one of the main results in [**JMNT**]. In section 11 we include some numerical investigations of the coefficients and the zero sets of Bollobás-Riordan polynomials.

Finally, in section 12 we list some natural questions that provide further study directions.

2. Tutte polynomial

The form of the dichromatic polynomial considered in this section is the Tutte polynomial which we will denote $T_G(x,y)$. It is defined as follows: let G have n vertices and m edges. Choose an ordering of the edges of G (the result will not depend on a particular choice); for every spanning tree T of G, this ordering will define its *internal activity* $\mathrm{int}(T)$ and *external* activity $\mathrm{ext}(T)$; we refer to [**Big1**, Ch. 13] for details.[1] We note that $\mathrm{int}(T) \leq n-1$, and $\mathrm{ext}(T) \leq m-n+1$. Then the Tutte polynomial is defined by

$$T_G(x,y) = \sum_{i,j} t_G(i,j)x^i y^j, \tag{2.1}$$

where $t_G(i,j)$ is the number of spanning trees of G with $\mathrm{int}(T) = i$ and $\mathrm{ext}(T) = j$. We note that

$$T_G(1,1) = \sum_{i,j} t_G(i,j) = \tau(G), \tag{2.2}$$

the number of spanning trees of G.

An equivalent definition of the Tutte polynomial for connected graphs is given by ([**CsFr**, §2])

$$T_G(x,y) = \sum_{A\subseteq E} (x-1)^{k(A)-1}(y-1)^{k(A)+|A|-n}$$

where $k(A)$ is the number of connected components of A. This is $(x-1)^{-n+1}R_G(x-1, y-1)$ where R_G is the rank polynomial of G [**Big1**, Ch. 13].

2.1. Coefficient measures for the Tutte polynomial.

Below, we focus on the distribution of the coefficients of the Tutte polynomials $T_G(x,y)$.

We first discuss the question of normalization. It follows from (2.2) that $T_G(1,1) = \tau(G)$, and its behaviour for random k-regular graphs $G_{n,k}$ with n vertices was studied in [**McK**], where it was shown that for such graphs (assuming $k \geq 3$) we have almost surely

$$\lim_{n\to\infty} \tau(G_{n,k})^{1/n} \to \frac{(k-1)^{k-1}}{(k^2-2k)^{k/2-1}}. \tag{2.3}$$

Given a connected graph G with n vertices and m edges, we associate to each graph a probability measure μ_G in the unit square $[0,1]\times[0,1]$ associated to the coefficients of the Tutte polynomial $T_G(x,y)$ as follows:

$$\mu(G) := \frac{1}{\tau(G)} \sum_{i,j} \tau_G(i,j)\cdot\delta(i/m, j/m), \tag{2.4}$$

---

[1] Order the edges of G in an arbitrary way. An edge $e \notin T$ is *externally active* if it is minimal in its fundamental cycle, which is formed by the union of e and the path connecting the endpoints of e inside T; $\mathrm{ext}(T)$ is equal to the number of externally active edges in $E(G)\setminus T$. An edge $e \in T$ is called *internally active* if it is minimal in its fundamental cocycle, which is formed by the union of e and all the edges in $E(G)\setminus T$ having exactly one endpoint in each of the two subtrees obtained from T by removing e; $\mathrm{int}(T)$ is equal to the number of internally active edges in T.

where δ denotes the Dirac delta-function. It follows from (2.2) that $\mu(G)$ is indeed a probability measure. Also, it follows from earlier remarks that for any spanning tree T of G we have

$$\operatorname{int}(T) + \operatorname{ext}(T) \leq (n-1) + (m-n+1) = m.$$

It follows that in (2.1), for any monomial we have $(i+j)/m \leq 1$, and hence in (2.4) $\mu(G)$ is actually supported in the triangle Δ with vertices at $(0,0), (1,0)$ and $(0,1)$.

2.2. Sequences of graphs. Consider a sequence G_i of graphs with n_i vertices and m_i edges. By weak compactness of the space of probability measures on Δ, we know that the sequence of probability measures $\mu(G_i)$ will have convergent subsequences. We would like to understand the limit measures of those sequences.

We concentrate on sequences of regular graphs $G_{n,k}$, but remark that in principle this question can be studied for arbitrary sequences.

2.3. Previous work. Limits of Tutte polynomials for recursive families of graphs ([**BDS**]) have been studied in [**CS**]; several other families were considered in [**Man**] and [**LFH**]. Weak convergence of measures on the roots of Tutte polynomials with one fixed variable was found by [**CsFr**]

3. Distribution of the coefficients: a priori results

Below, we establish some *a priori* results about the coefficients of the Tutte polynomial T_G that are of independent interest.

We first remark that the constant term of the Tutte polynomial is always 0. Indeed, since there is a finite number of edges, there must be a first one in the arbitrary ordering. Call it e. Let T be any spanning tree. If $e \in T$, then clearly e is the smallest bridge between the two subsets disconnected by removing it from T. Similarly, if $e \notin T$, then e must be the smallest edge in $\operatorname{Cyc}(T, e)$ (a cycle formed by the union of e and the path connecting the endpoints of e in T). Thus e must always be either internally or externally active and so either x or y must have exponent at least 1 in any spanning tree.

We next define the concept of maximal and minimal spanning trees. We number the edges by their ordering and assign each a length equal to its number. Then construct the spanning trees that maximize and minimize the total lengths and denote them $T_{\min}$ and $T_{\max}$ respectively. This leads to the following two Lemmas; they are probably standard, but we include the proofs for completeness.

LEMMA 3.1. *Let T be a spanning tree and let e be internally active with respect to T. Then $e \in T \cap T_{\min}$.*

PROOF. Removing e from T creates two disconnected trees spanning sets of vertices A and B. Since e is internally active, it is the smallest bridge between A and B (an edge with one endpoint in A and another endpoint in B). Suppose $e \notin T_{\min}$. Adding it to $T_{\min}$ will create a cycle $\operatorname{Cyc}(T_{\min}, e)$. This cycle includes vertices in both A and B and therefore contains at least two bridges between the two sets. Let f be the smallest bridge other than e. Now, since e is the smallest bridge between A and B, $e < f$. Thus replacing f by e in $T_{\min}$ reduces the total length, contradicting the definition. Hence, e must have been in $T_{\min}$ and so in $T \cap T_{\min}$. □

Let T be a spanning tree of G; we denote its complement by by T^{C}.

LEMMA 3.2. *Let T be any spanning tree and let e be externally active with respect to T. Then $e \in T^{\complement} \cap T_{\max}^{\complement}$.*

PROOF. Let e be externally active and suppose $e \in T_{\max}$. Remove e from $T_{\max}$ to obtain two trees spanning sets A and B. Now consider $\mathrm{Cyc}(T, e)$. It must contain at least one other bridge f between A and B. Now, by definition, e is the smallest edge in $\mathrm{Cyc}(T, e)$ so $e < f$. Hence replacing e with f in $T_{\max}$ increases its total length. This contradicts the definition of $T_{\max}$ and so $e \notin T_{\max}$, proving our theorem. □

It is important to note that the previous two lemmas are not if and only if statements. They only specify sets that must contain all the internally or externally active edges, not the set of such edges. They lead us to the following theorem.

THEOREM 3.3. *Let G be a graph with n vertices and m edges. Suppose also it has two spanning trees T_1 and T_2 that overlap in only k edges. Suppose also that this overlap is minimal. Then the Tutte polynomial of G cannot have terms of degree more than $m - (n - k - 1)$.*

PROOF. Let T_1 and T_2 be the two trees overlapping on only k edges. The Tutte polynomial is independent of the ordering so we may select a specific ordering. Set T_1 to be $T_{\min}$ by numbering its edges first. Let the k overlapping edges be $n - k$ to $n - 1$. We now want T_2 to be $T_{\max}$. Assign the largest $m - 1 - k$ numbers to the remaining edges of T_2, and the remaining numbers to the edges of $(T_1 \cup T_2)^{\complement}$ in any way. We claim that T_2 is now $T_{\max}$.

Let T be any spanning tree other than T_2. We need to show that its total length is less than that of T_2. Since, by hypothesis, any two spanning trees have overlap at least k, $T \cap T_{\min}$ must contain at least k edges. Since both T and T_2 have $n - 1$ edges, we define a bijection f from T_2 to T as follows.

(1) f is the identity on $T \cap T_2$.
(2) f maps $T_2 \cap T_{\min}$ into $T_{\min} \cap T$. This is possible since $T_2 \cap T_{\min}$ has k elements and $T_{\min} \cap T$ contains at least that many.

We now claim that f is a non-increasing function when the edges are ordered by length. Denote the length function by L. Let e be an edge in T_2. If $e \in T$, then $f(e) = e$ and so $L(f(e)) = L(e)$. If not and $e \in T_{\min} \cap T_2$, then $L(e) \geq n - k$ given how the edges were numbered in $T_{\min}$. Furthermore, since $e \notin T \cap T_2$, $f(e) \notin T \cap T_2$. Now apply condition (2) and we have that $f(e) \in T_{\min}$ but not in T_2. Hence, by our numbering, $L(f(e)) < n - k$ and so $L(f(e)) < L(e)$. The last case is if $e \notin T_{\min}$ and $e \notin T$. By the first statement and our numbering, $L(e) \geq m - k$. By the second statement, $f(e) \notin T_2$ meaning that $L(f(e)) < m - k$. Hence once again we have $L(f(e)) < L(e)$.

Thus, since f is constant on the overlap and strictly decreasing on $T_2 \setminus (T \cap T_2)$, the total length of T is less than that of T_2 for any spanning tree T which is not identically T_2. Hence $T_2 = T_{\max}$.

Now, by the previous two lemmas, an edge can only be active (externally or internally) if it is in $S = T_{\min} \cup T_{\max}^{\complement}$. Since $T_{\max}$ contains $n - 1 - k$ elements not in $T_{\min}$, S has size $m - (n - 1 - k)$ and so the maximum degree of a term in the Tutte polynomial is $m - (n - 1 - k)$. □

REMARK 3.4. In the special case where two non-overlapping spanning trees exist, the total activity cannot exceed $m + 1 - n$.

Some graphs have many entirely non-overlapping spanning trees. The following theorem gives some bounds on activity of certain spanning subtrees of these graphs.

LEMMA 3.5. *Let G be a graph with n nodes and m edges having k non-overlapping spanning trees numbered $T_1, T_2, ..., T_k$. Let T be a spanning tree that does not overlap with the last ℓ trees of this list for some $\ell < k-1$. Then the contribution of T to the Tutte polynomial of G cannot have degree more than $m-(\ell+1)(n-1)$.*

PROOF. Number the edges of T_1 with the numbers 1 through $n-1$. Number the edges in the trees T_2 to T_k with the numbers $m-k(n-1)+1$ through m, with the edges in T_i having smaller numbers than the edges in T_j if $i < j$. Number the remaining edges with the numbers in between.

Now define $G^* = G \setminus T_{\max,G}$. Recall that T_{min}, T_{max} were defined just before Lemma 3.1. Let G^{p*} denote G after p applications of the $*$ operator. By our choice of numbering $T_1 = T_{\min,G}$, $T_k = T_{\max,G}$, $T_{k-1} = T_{\max,G^*}$ and so on. Furthermore, since edges in $T_{\max}$ cannot be externally active, the activity of any spanning subtree T contained in G^* will be the same in G as in G^* provided the maximal and minimal trees have no overlap.

Now choose a spanning tree T satisfying the conditions of the theorem for some $\ell < k-1$. Then $T \subseteq G^{\ell*}$. Each of these applications of $*$ removes $n-1$ edges from G meaning that $G^{\ell*}$ has $m-\ell(n-1)$ edges. Since $\ell < k-1$, there are at least two non-overlapping spanning trees so, by theorem 3.4, the total activity of T is at most $m-\ell(n-1)-(n-1) = m-(\ell+1)(n-1)$. □

The previous results concerned the bounds on the external and internal activity of specific trees. The following theorem give bounds on the average activity over all spanning trees of a graph. For what follows let any order be given on the edges and denote it by $\leq$. Internal and external activity will be defined with respect to $\leq$. Reverse internal activity and reverse external activity will be defined to be internal and external activity with respect to the reverse order $\geq$.

The following result may be standard, but we include the proof for completeness.

THEOREM 3.6. *Let G be a simple graph on n vertices and m edges with b bridges.*

(1) *The average external activity over all spanning trees of G is at most $(m-n+1)/2$.*
(2) *The average internal activity over all spanning trees of G is at most $(n-1+b)/2$.*

PROOF. For (1), let T be a spanning tree of G and consider an edge e not in T. Let C denote the unique cycle in $T \cup e$. C must contain at least 3 edges and so e cannot be both the largest and the smallest edge in C. Thus e cannot be both externally active and reverse externally active. Hence the sum of the external and reverse external activity of any spanning tree cannot exceed $m-n+1$. Thus the average of this quantity over all spanning trees is at most $m-n+1$. However, since the Tutte polynomial does not depend on the ordering of the edges, the average external activity and the average reverse external activity are equal. Thus twice the average external activity is at most $m-n+1$ and the average external activity is at most $(m-n+1)/2$.

The proof of (2) is similar. Let T be a spanning tree of G and let e be an edge in T. Let C be the set of edges in G joining the two components of $T - e$. If e is a bridge, it is the only element of C and thus it is both internally and reverse internally active. If e is not a bridge, C contains at least two elements and so e cannot be both internally and reverse internally active. Thus the sum of the internal and reverse internal activity of T is at most $n - 1 + b$. Proceeding as before, the average internal activity is at most $(n - 1 + b)/2$. □

3.1. Applications to regular graphs. We next apply the results from Section 3 to families of regular graphs. Theorem 3.3 implies that in a graph G with n vertices, $m \geq 2n - 2$ edges, and at least two edge-disjoint spanning trees, the degree of the Tutte polynomial is at most $m - n + 1$. So, finding edge-disjoint spanning trees in G would imply a bound on the degree of $T_G(x, y)$.

In the graph theory literature, the maximal number of edge-disjoint spanning trees in G is called a *tree packing number* of G and is often denoted by $\sigma(G)$. Many interesting results about $\sigma(G)$ can be found in a survey [**Pal**] and references therein. The following basic observation is due to Catlin, see [**Cat**] or [**Pal**, Cor. 4]:

PROPOSITION 3.7. *If the edge connectivity $\lambda(G)$ of the graph G satisfies $\lambda(G) \geq 2k$, then $\sigma(G) \geq k$, i.e. G has k edge-disjoint spanning trees.*

It is known (see [**Wor**]) that for random d-regular graphs $G_{n,d}$ on n vertices, $\lambda(G_{n,d}) = d$ asymptotically almost surely, as $n \to \infty$. It follows that $\sigma(G_{n,d}) = \lfloor d/2 \rfloor$. If $d \geq 4$, then we have $\lfloor d/2 \rfloor \geq 2$, and so Theorem 3.3 applies.

We remark that an d-regular graph on n vertices has $m = dn/2$ edges.

THEOREM 3.8. *Let $d \geq 4$. Then the degree of the Tutte polynomial of a random d-regular graph $G \in G_{n,d}$ on n vertices satisfies $\deg T_G \leq dn/2 - n + 1$ almost surely as $n \to \infty$.*

This explains the "coefficient-free strips" in the figures in Section 4

For cubic graphs (regular graphs of degree 3), $\lfloor d/2 \rfloor = 1$, so this case requires a separate consideration. We first consider the case of simple graphs. The depth-first search algorithm and the construction of the tree are first discussed.

ALGORITHM 1. Let G be a finite graph. Let S be an empty stack. Let T, the depth-first search tree begin empty.

(1) Choose any vertex v of G and add it to T.
(2) Push each of the neighbours of v onto S, together with the edge connecting it to v.
(3) Pop S and call the vertex v. Call its associated edge e.
(4) If v is not already in T, add v and e to T.
(5) Repeat steps 2 through 4 until T is a spanning tree.

The tree described above T is called the depth-first search tree. The method for choosing the first vertex and the order in which neighbours are pushed to the stack are not described. Some authors add vertices to the tree as they are pushed to the stack, rather than after they have been popped. This is called the preorder depth first search tree and behaves quite differently. What is described in the above algorithm and will be used in what follows is sometimes referred to as the postorder depth first search tree.

LEMMA 3.9. *The complement of a depth-first search tree in a cubic graph G without loops or multiple edges is acyclic.*

PROOF. Choose a vertex v_0 and perform a depth-first search starting at v_0, building the tree T as we go. Arguing for a contradiction, suppose that C is a cycle contained in $G \setminus E(T)$. Since G has no loops or multiple edges, C must be incident on at least three vertices. Call this set of vertices V_C. Each of these vertices has two edges in C and one edge in the complement of C. Hence $V_C \neq V$.

Suppose $v_0 \notin V_C$. Let v be the first vertex in V_C discovered by the search. It must have been reached via its edge not in C. Now consider the two remaining edges incident on v. Both are in C and lead to vertices that have not been discovered since v is the first. Thus, following the next edge incident on v does not create a cycle and so that edge is added to T hence $C \nsubseteq G \setminus T$.

Now suppose $v_0 \in V_C$. Then let v be the second vertex in V_C to be discovered by the search. If v was reached via an edge in C, then we are done since this edge is in T. Otherwise, consider either of the other two edges incident to v. It is in C. If it does not lead to v_0, it does not create a cycle and is added to T. If it leads to v_0, it is not added to T and we move to the third edge incident on v. This edge is also in C and leads to an undiscovered vertex. So, either way, we find an edge in C to add to T. □

The following lemma is a standard result in matroid theory. We include the proof for completeness.

LEMMA 3.10. *Any acyclic subset of edges S of a connected graph G can be extended to a spanning tree.*

PROOF. This can be done with a greedy algorithm. The subgraph S has k connected components. Since G is connected, there must be an edge in $G \setminus S$ connecting two of these components. Add it to S. It cannot create a cycle. There are now $k-1$ connected components in S. Repeat the algorithm until there is only one. The set S is then a spanning tree. □

THEOREM 3.11. *Any cubic graph G on n vertices without loops or multiple edges has two spanning trees T_1 and T_2 such that $T_1 \cup T_2 = G$. In particular, $\#(T_1 \cap T_2) = n/2 - 2$. Thus the degree of the Tutte polynomial for cubic simple graphs is at most $3n/2 - n + 1 + (2n - 2 - 3n/2) = n - 1$.*

PROOF. Choose a vertex v_0 and let T_1 be a depth-first search tree starting there. Let T_2 be a spanning tree obtained by extending $G \setminus T_1$. Then $T_1 \cup T_2 = G$ since $G \setminus T_1 \subseteq T_2$. Thus, by a counting argument, the spanning trees are minimally overlapping. □

Next consider the case where we allow multiple edges, but not loops. Lemma 3.9 can be extended to cover these cases provided there is at least one vertex without multiple edges. In this case, set v_0 to be such a vertex. For any cycle C, which can be incident on as few as 2 vertices, let v be the first vertex in V_C reached by the depth-first search. It must be reached by an edge not in C and so the remaining two edges incident on v must be in C and must therefore lead to undiscovered edges. Thus one of them is added to T and so C is not in $G \setminus T$.

Triple edges are impossible in connected cubic graphs of more than 2 vertices. If every vertex has a double edge, then we must have one large ring with double edges every second link. A depth first search tree on this kind of graph will have an acyclic complement provided its first edge added in the depth first search is one of a double edge.

Thus, even allowing multiple edges, the degree of Tutte polynomial of a cubic graph on n vertices without loops is at most $n-1$.

In the case where there are loops in the graph, the loops cannot be part of any spanning tree. By a similar reasoning to that above, the complement of a depth first search is acyclic once we remove the loops. Hence, in a cubic graph with ℓ loops, n vertices and no multiple edges, the size of the minimal overlap is $2n-2-(3n/2-\ell)$ and so the degree of the Tutte polynomial is at most $n-1+\ell$.

3.2. Another argument in the case of cubic graphs. In this section, we use the second definition of the Tutte polynomial to give an upper bound on the degree of the Tutte polynomial of a simple random cubic graph. Essentially, we want to bound

$$2k(A)+|A|+n-1$$

First of all, we rewrite this as

$$2k(A)-\left|A^{\mathsf{C}}\right|-n+m-1 \le \sum_{C \text{ component of } A} (2-out(C)-|C|)+m-1$$

where $out(C)$ is the number of edges joining C and C^{C} in E. Now we claim that $2-out(C)-|C| \le -|C|/2$ for all components C.

First of all, if $|C| \ge 4$, $2-|C| \le |C|/2$ so the claim holds. For $|C| \le 3$, we check the cases one at a time. If a C contains only one vertex, $out(C)=3$ since the graph is cubic. Thus $2-out(C)-|C| = 2-4 = -2 \le -1/2$. If $|C|=2$ then we must have one edge joining the two vertices leaving $out(C)=4$. Thus $2-out(C)-|C| = 2-4-2 = -4 \le -1$. If $|C|=3$, then there are at most three edges inside C. Thus $out(C) \ge 3$. Hence $2-out(C)-|C| \le 2-3-3 = -4 \le -3/2$. Thus in all cases the claim is satisfied.

Thus, since every vertex is in a component, $\sum_{C \text{ component of } A}(2-out(C)-|C|)$ $\le -n/2$. Finally, since in a cubic graph on n vertices, $m=3n/2$, this gives us that

$$2k(A)-\left|A^{\mathsf{C}}\right|-n+m-1 \le -\frac{n}{2}+\frac{3n}{2}-1 = n-1$$

Thus the maximum degree of the Tutte polynomial of a simple cubic graph on n vertices is $n-1$.

4. Numerical experiments on the coefficient measure

In this section, we present some numerical experiments on the coefficient measure. These were obtained by uniformly sampling random regular graphs, computing the coefficient measure of each and then averaging over 100 graphs.

It was very convenient for us to use MATHEMATICA, since it conveniently has built-in command `RandomGraph[DegreeDistribution]`, sampling random graphs with a given degree distribution.

In addition, MATHEMATICA has a built-in command `TuttePolynomial[G, {x, y}]` which produces the Tutte polynomial of the graph G. The standard MATHEMATICA commands were then used to plot the zeros and the coefficients.

In Figure 1 below, the distribution of the coefficients of T_G was plotted, averaged over 100 graphs as follows: (A) 3-regular graphs with 22 vertices; (B) 3-regular graphs with 24 vertices.

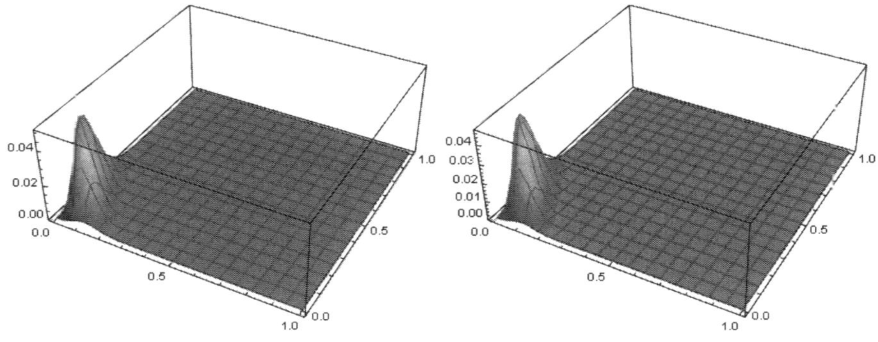

FIGURE 1. Coefficients of the Tutte polynomial: 3-regular graphs

In Figure 2 below, the distribution of the coefficients of T_G was plotted, averaged over 100 graphs as follows: (A) 4-regular graphs with 19 vertices; (B) 4-regular graphs with 20 vertices; (C) 5-regular graphs with 16 vertices; (D) 6-regular graphs with 17 vertices.

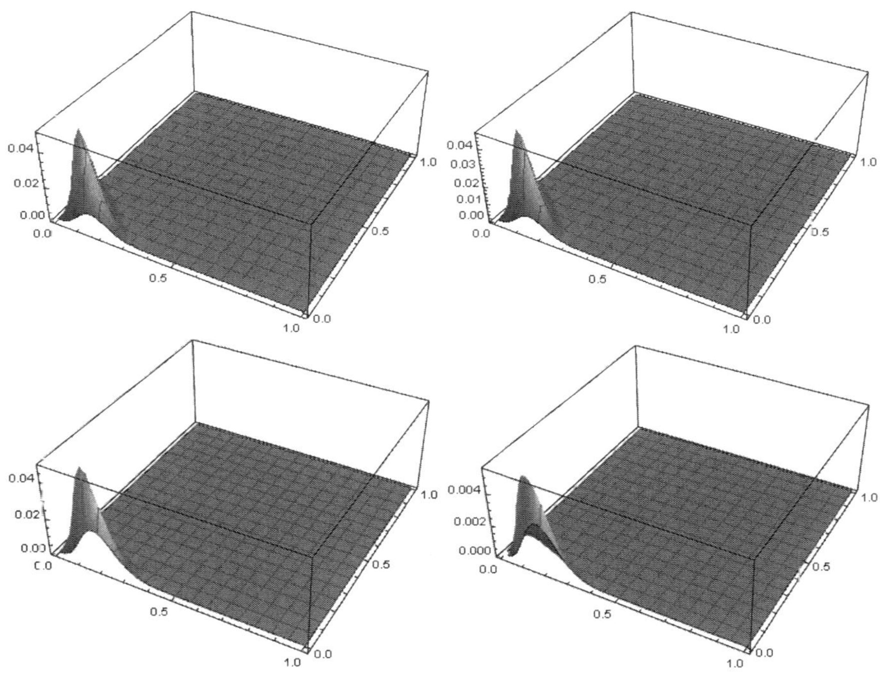

FIGURE 2. Coefficients of the Tutte polynomial: 4- 5- and 6-regular graphs

To generate random planar graphs, we used `RandomGraph` command inMATHEMATICA to generate random regular graphs, then used `PlanarGraphQ[G]` command to check for planarity, and discarded the graphs that were not planar. The following coefficient distributions are obtained by sampling only planar graphs.

In Figure 3 below, the distribution of the coefficients of T_G was plotted, averaged over 100 planar graphs as follows: (A) 3-regular planar graphs with 20 vertices; (B) 3-regular planar graphs with 22 vertices; (C) 3-regular planar graphs with 24 vertices; (D) 4-regular planar graphs with 19 vertices.

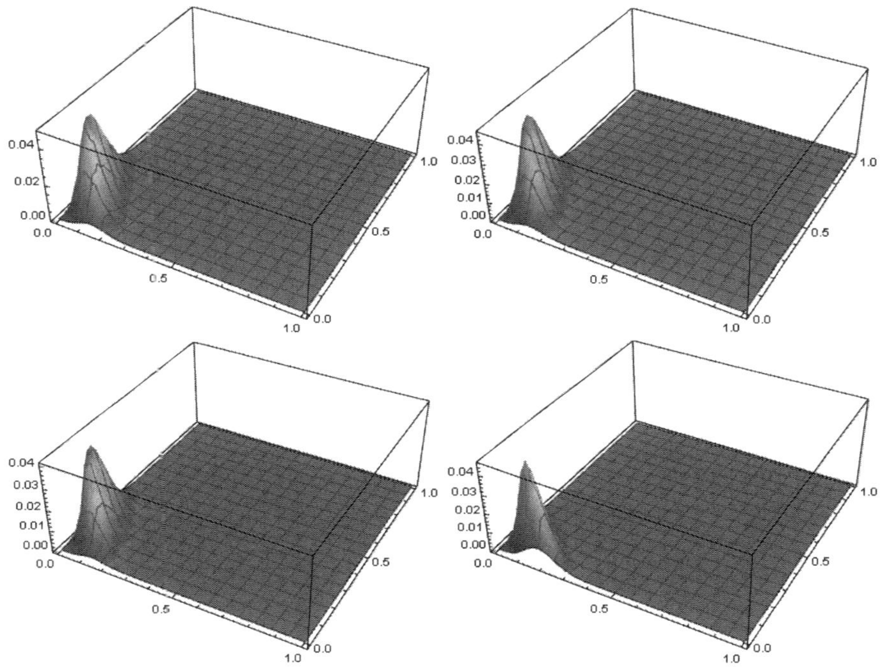

FIGURE 3. Coefficients of the Tutte polynomial: planar graphs

We end this section by plotting the distribution of the Tutte coefficients coefficients of the Petersen graph in Figure 4:

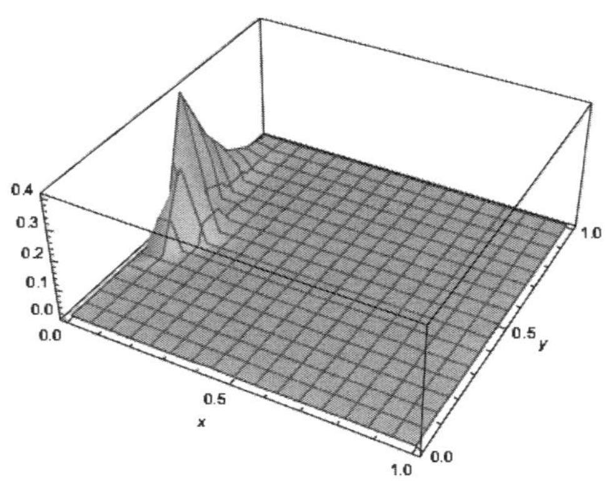

FIGURE 4. Tutte coefficients: the Petersen graph

4.1. Asymptotic behaviour of the coefficients. It appears, based on the numerical experiments described above, that the coefficient measures μ_G, averaged over regular graphs on n vertices, converge to a delta function as $n \to \infty$. The coefficient measures for planar graphs seem to have a similar behaviour. A related result for the coefficient measures of *rank polynomials* were established in the paper [**JMNT**], and in the MSc thesis [**Tur**] of one of the authors. The authors plan to further study questions about the coefficients and other properties of the graph polynomials in the near future. Some of those questions are discussed in Section 12.

5. Zeros of Tutte polynomials

Although the main focus of this paper is the coefficient measure of the Tutte polynomial, the *zero sets* $\mathcal{N}(T_G)$ of T_G are also of interest. This section presents a brief overview of results in this area. It concludes with some numerical results illustrating various cross sections of the zero sets. The convergence of roots of $T_G(x, y)$ *for fixed* y was addressed in several papers, including [**Sok**], [**CsFr**] and many others. Here we would like to study the behaviour of $\mathcal{N}(T_G)$ as a subset of $\mathbb{R}^2$ (or of $\mathbb{C}^2$).

Often the following form of the Tutte polynomial is considered in the literature:

$$Z_G(q, v) = \sum_{A \subseteq E} q^{k(A)} v^{|A|}, \tag{5.1}$$

where the sum is taken over all subsets A of the edge set E of G, and $k(A)$ denotes the number of connected components of the graph (V, A). The two forms are related (see e.g. the Introduction in [**Sok**]) by

$$T_G(x, y) = (x-1)^{-k(E)}(y-1)^{-|V|} Z_G((x-1)(y-1), y-1);$$
$$Z_G(q, v) = (q/v)^{k(E)} v^{|V|} T_G(1 + q/v, 1 + v),$$

so their zero sets are equivalent. In statistical physics, limits of zeros of $Z_G(q, v)$ for fixed q correspond to phase transitions in the q-state Potts model (cf. the Introduction in [**Sok**] and references therein). Many result surveyed below were established for the zero set $\mathcal{N}(Z_G)$.

Below, we shall restrict ourselves to connected graphs $G = (V, E)$. We first discuss some heuristics about the behaviour of the nodal set $\mathcal{N}(Z_G)$ as $|q|, |v| \to \infty$. That behaviour of $\mathcal{N}(Z_G)$ is determined by the highest powers of q and v in Z_G. Clearly, for any $A \subseteq G$, we have $k(A) \leq |V|$, with equality iff $A = \emptyset$. Accordingly, $Z_G(q, v) = q^{|V|} + \mathit{terms\ of\ lower\ degree\ in}\ q$. Similarly, for any $A \subseteq G$, we have $|A| \leq |E|$, with equality iff $A = E$. Accordingly, $Z_G(q, v) = v^{|E|}q + \mathit{terms\ of\ lower\ degree\ in}\ v$. Keeping just those highest degree terms of Z_G , we get $q^{|V|} + qv^{|E|} + \mathit{lower\ order}$. The zero set of the highest degree terms (after division by q) is $v^{|E|} + q^{|V|-1} = 0$. For d-regular graphs, we have $|V| = n$ and $|E| = dn/2$, so the previous equation becomes $q^{n-1} + v^{nd/2} = 0$. It seems interesting to see how accurately this very naive expression approximates the real nodal set of Z_G in the limit $|q|, |v| \to \infty$; we hope to address that question in a future project.

An important question is to find explicit bounds for $\mathcal{N}(Z_G)$ and to see how they depend on a graph G. In the paper [**JPS**, Theorems 1.2 and 1.3], the authors proved that for simple graphs, if v is fixed, then all zeros q of $Z_G(q, v)$ lie in the disc

$$|q| < K_\mu^* \Delta^*(G, v),$$

where $K_\mu^* \leq 5 + 2\mu, \mu = \widehat{\Delta}(G, v)/\Delta^*(G, v)$ and where $\Delta^*(G, v)$ and $\widehat{\Delta}(G, v)$ are given by the following expressions:

$$\Delta^*(G, v) = \max_{x \in V} \sum_{x \in e, e=(xy)} \min\{|v|, \frac{|v|}{\sqrt{|1+v|}}\} \prod_{y \in f} \max\{1, |1+v|\}^{1/2};$$

and

$$\widehat{\Delta}(G, v) = \max_{x \in V} \sum_{x \in e, e=(xy)} \min\{|v|, \frac{|v|}{|1+v|}\} \prod_{y \in f} \max\{1, |1+v|\}^{1/2}.$$

We remark that we have specialized the formulas in [**JPS**] (valid for multivariate Tutte polynomials) to the case where the edge weights $w_e = v$ for every edge $e \in E$. The results in [**JPS**] generalized earlier results of Sokal [**Sok**].

In the special case of d-regular graphs $G_{n,d}$, we have

$$\Delta^*(G, v) = d \min\{|v|, \frac{|v|}{\sqrt{|1+v|}}\} \max\{1, |1+v|\}^{d/2},$$

and

$$\widehat{\Delta}(G, v) = d \min\{|v|, \frac{|v|}{|1+v|}\} \max\{1, |1+v|\}^{d/2};$$

therefore

$$\mu = \frac{\min\{1, |1+v|^{-1}\}}{\min\{1, |1+v|^{-1/2}\}}.$$

5.1. Zeros: experimental results. In this section, we looked at the zeros of $T_G(x, y)$ for one random regular graph at a time (without averaging). We first considered real zeros. As $|x|, |y| \to \infty$, the zero set seems to be asymptotic to an algebraic curve; we do prove any rigorous results in that direction, but hope to address this question in future work.

We used standard `Mathematica` commands to plot zeros of Tutte polynomials.

In Figure 5 below, real zeros of T_G were plotted, for the following random regular graphs: (A) a 3-regular graph with 16 vertices; (B) a 4-regular graph with 16 vertices; (C) a 6-regular graph with 16 vertices; (D) a 5-regular graph with 10 vertices.

We next plot the real zeros of the Tutte polynomial of the Petersen graph in Figure 6.

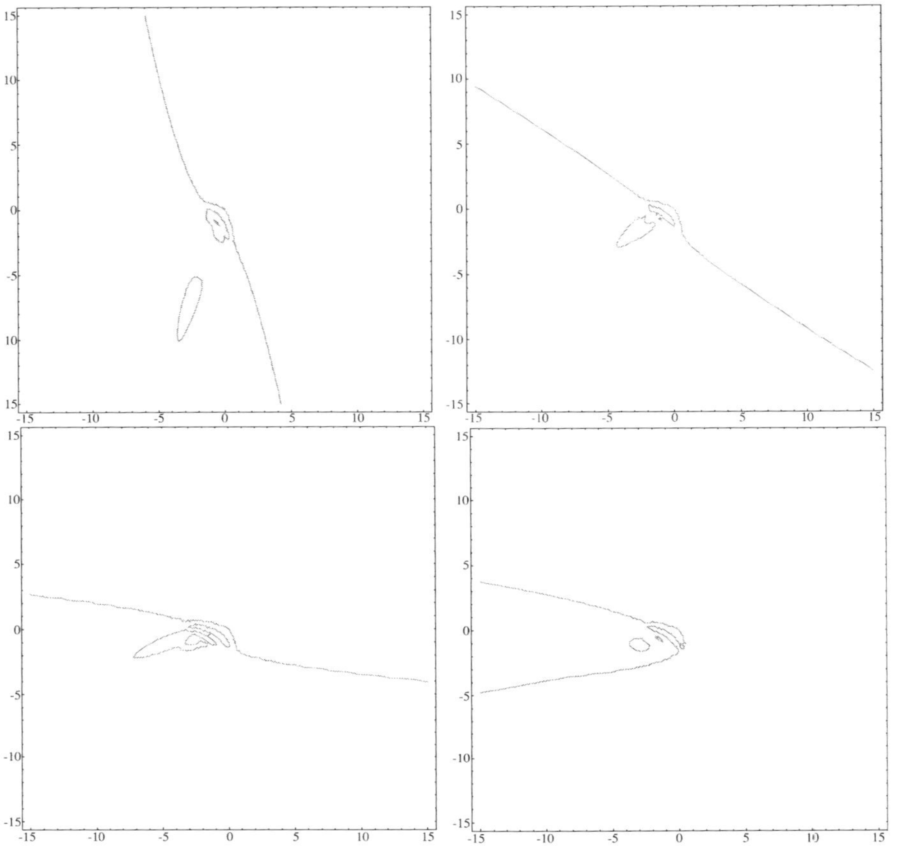

FIGURE 5. Real Tutte zeros: random regular graphs

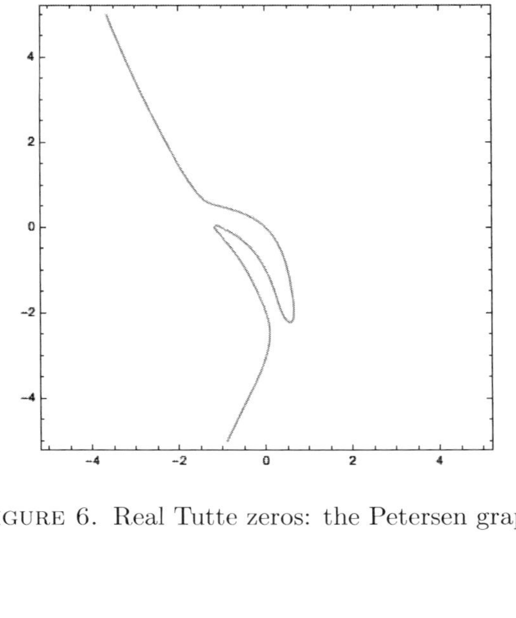

FIGURE 6. Real Tutte zeros: the Petersen graph

It also seems interesting to complexify x and y variables, and to consider the null variety of $T_G(x, y)$ in $\mathbb{C}^2$. In two dimensional complex space, representing the zero sets is more difficult. The following images show the zero sets for the real (blue) and imaginary parts (red) of the Tutte polynomial of various random regular graphs. The zero sets of the functions are the intersections of these curves.

In Figure 7 below, complex zeros of T_G were plotted, for the following random regular graphs: (A) a 3-regular graph with 8 vertices in the space with x complex and y real; (B) a 3-regular graph with 8 vertices in the space with y complex and x real; (C) a 3-regular graph with 10 vertices in the space with x complex and y real; (D) a 3-regular graph with 10 vertices in the space with y complex and x real.

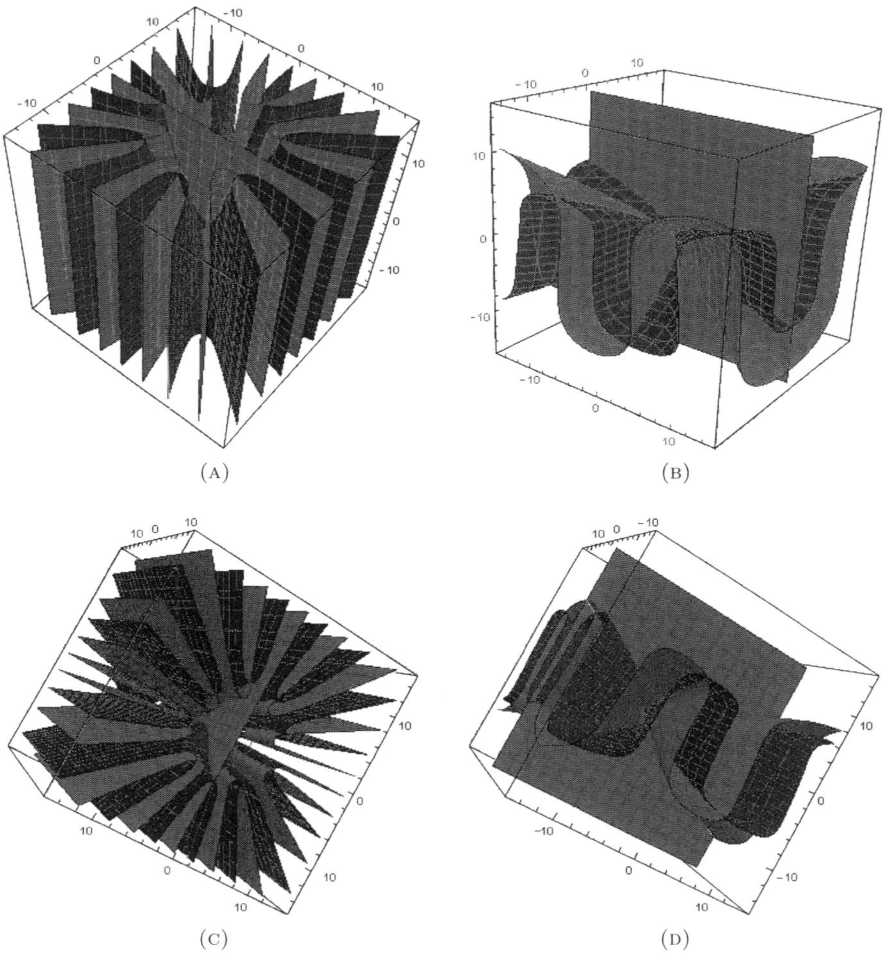

FIGURE 7. Complex zeros of the Tutte polynomial: 3-regular graphs

In Figure 8 below, complex zeros of T_G were plotted, for the following random regular graphs: (A) a 4-regular graph with 8 vertices in the space with x complex and y real; (B) a 4-regular graph with 8 vertices in the space with y complex and x

real; (C) a 4-regular graph with 9 vertices in the space with x complex and y real; (D) a 4-regular graph with 9 vertices in the space with y complex and x real.

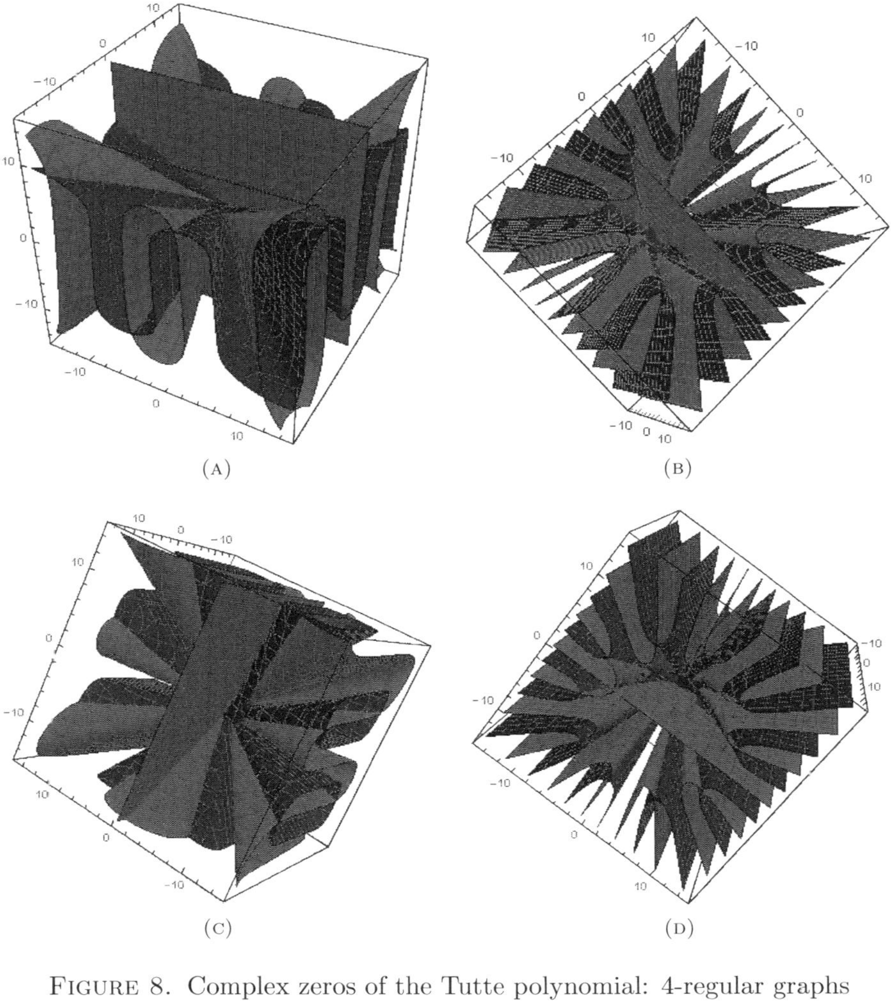

FIGURE 8. Complex zeros of the Tutte polynomial: 4-regular graphs

In Figure 9 below, complex zeros of T_G were plotted, for the following random regular graphs: (A) a 4-regular graph with 10 vertices in the space with x complex and y real; (B) a 4-regular graph with 10 vertices in the space with y complex and x real; (C) a 5-regular graph with 8 vertices in the space with x complex and y real; (D) a 5-regular graph with 8 vertices in the space with y complex and x real.

6. Ribbon graphs: summary

Below we summarize our results about Bollobás-Riordan polynomials associated to ribbon graphs. In the paper [**JMNT**], the authors studied limiting distribution of coefficients of rank polynomials for random sparse graphs. In the current paper, we would like to extend those results to the case of oriented *ribbon graphs*,

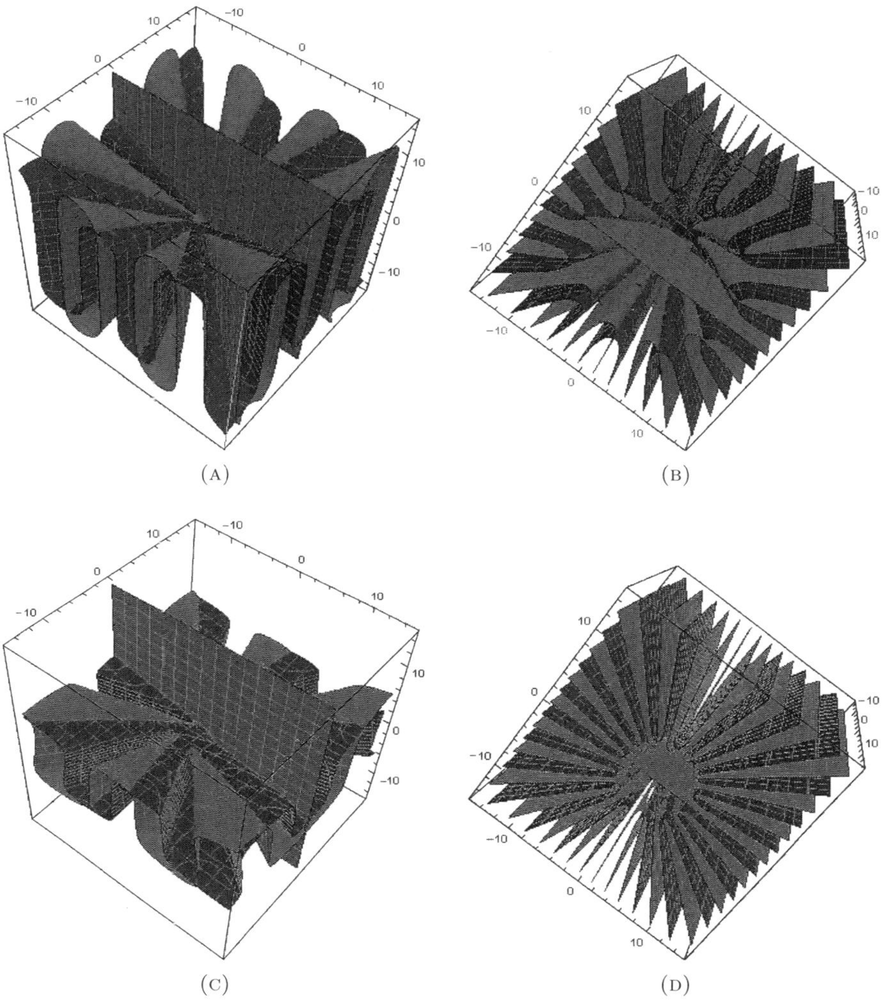

FIGURE 9. Complex zeros of the Tutte polynomial: 4- and 5-regular graphs

which are graphs with a cyclic orientation of edges at each vertex. Such graphs arise in the study of knot invariants, in quantum field theory and in other areas.

A natural extension of the rank polynomial to ribbon graphs is the *Bollobás-Riordan* polynomial ([**BR01**, **BR02**]), which is a polynomial in 3 variables. First, we define the normalized coefficient measures for such polynomials. Next, we extend the definition of Benjamini-Schramm convergence from graphs to ribbon graphs. Finally, we extend some of the results in [**JMNT**] to the coefficients of Bollobás-Riordan polynomials for sequences of ribbon graphs that converge in the Benjamini-Schramm sense. Next, we compute the limit measure for sequences of random ribbon graphs arising from random edge orientations of regular graphs, see e.g. [**Gam**] and [**Ch-Pit**].

7. Ribbon graphs and "left-hand turn" surfaces

Below, we discuss ribbon graphs the associated left hand turn (LHT) surfaces.

An *orientable* ribbon graph is a graph embedded on an orientable surface such that every face such that every face of the resulting polyhedral surface is contractible. In this paper, we restrict ourselves to *orientable* ribbon graphs; this simplifies our exposition. We refer to [**Chm**] and [**MY**] for more detailed description of ribbon graphs, including discussion about *non-orientable* ribbon graphs, sometimes called *Möbius* graphs.

Let G be a graph embedded on an orientable surface S. A choice of orientation on S defines a cyclic ordering of edges at every vertex; we denote such an ordering by O. Conversely, given a graph G with a cyclic ordering O of edges at every vertex, one can reconstruct in a canonical way an oriented surface S (considered in the papers of Brooks, Makover, Monastyrski [**BM01, BM04, BrMon**] and Gamburd [**Gam**]) as follows.

First, consider the *left-hand turn paths* defined by (G, O): start along an oriented edge, then "turn left" according to the cyclic orientation at the head of the oriented edge, and continue until you get a closed cycle. The set of directed edges of G decomposes into a disjoint union of such cycles. Those cycles correspond to *boundary components* of a polyhedral surface $S(G, O)$, obtained by filling in every cycle with a disk. We call the corresponding surface the *left hand turn* (LHT) surface.[2]

The surface $S(G, O)$ can be defined equivalently as follows: cut every edge of G in the middle, and consider the corresponding "half-edges." The cyclic ordering of half-edges at every vertex determines a permutation β of the set of $2E(G)$ half-edges; a cycles in the cyclic decomposition of β consists of the half-edges incident to a given vertex, with the cyclic ordering determined by O. The cycle structure of β coincides with the degree sequence of G. Another permutation α is an involution: it interchanges the half-edges that form a given edge. The cycle structure of α is $(2, 2, \ldots, 2)$. One can show ([**Chm, Gam, BM01, BM04, BrMon**]) that the left-hand turn paths defined by (G, O) correspond to cycles in the cyclic decomposition of $\beta\alpha$. We discuss this further in Section 9.

We remark that for any subgraph F of G, a cyclic orientation $O(G)$ of edges of G incident to a given vertex u induces a cyclic orientation $O(F)$ of the edges of F incident to u. Accordingly, given an oriented ribbon graph (G, O), one can canonically define a ribbon subgraph $(F, O(F))$.

An orientation O is called *prime* if (G, O) has precisely *one* LHT path (equivalently, the surface $S(G, O)$ has one boundary component). The following result was shown in [**BM01, Xu**]; see also [**BrMon**]. Suppose that d is odd, and $n \equiv 2 (\text{mod } 4)$; or that n odd, and $d \equiv 2 (\text{mod } 4)$. Let G be a d-regular graph, and T a spanning tree of G such that $G \setminus T$ is connected (or more generally, so that each component of $G \setminus T$ has an even number of edges). Then G admits a prime orientation. Moreover, this last condition holds with asymptotic probability one as $n \to \infty$.

[2]Given a combinatorial graph G, there are clearly $\prod_{v \in V(G)} (deg(v) - 1)!$ choices of the orientation O, and of the corresponding orientable ribbon graphs.

8. Bollobás-Riordan polynomials

We refer to the papers by [**Ch-Pak**] and [**Mof**] for the definition of Bollobás-Riordan polynomial. Let (G, O) be a ribbon graph, as described in Section 7. Given a subset F of edges of G, we define a subgraph of G as follows: we keep all the vertices of G, but only the edges from F; we shall call the corresponding subgraph F as well.

We denote by $|V(G)|$ the number of vertices of G; by $|E(G)|$ the number of its edges; and by $k(G)$ the number of its connected components. Also, we denote by $r(G) = |V(G)| - k(G)$ the rank of G; by $\text{null}(G) = |E(G)| - r(G)$ the nullity (or co-rank) of G.

As discussed in Section 7, to every ribbon graph G we can canonically associate a surface with boundary $S(G, O)$ (whose boundary components correspond to the left hand turn paths defined by (G, O)). This coincides with the definition given in [**Ch-Pak**, §2]. We denote the number of its boundary components by $bc(G, O)$. We also remark that for any subgraph F of G, a cyclic orientation $O(G)$ of ribbon "half-edges" of G incident to a given vertex u induces a cyclic orientation $O(F)$ of ribbon "half-edges" of F incident to u. Accordingly, we can canonically define a ribbon graph $(F, O(F))$.

The *Bollobás-Riordan polynomial* is defined as follows. Denote by $\mathcal{F}(G)$ the set of all spanning subgraphs of G.

$$BR_{(G,O)}(x, y, z) = \sum_{F \in \mathcal{F}(G)} x^{r(G)-r(F)} y^{\text{null}(F)} z^{k(F)+\text{null}(F)-bc(F)}. \tag{8.1}$$

We remark that if we set $z = 1$, we get the following identity:

$$BR_{(G,O)}(x, y, 1) = x^{r(G)} R_G(1/x, y),$$

where $r(G) = |V(G)| - 1$ and R_G denotes the rank polynomial of G as defined in [**JMNT**]. We remark that $\deg(x) \leq |V(G)| = n, \deg(y) \leq |E(G)| = m$. We now make some elementary remarks about $\deg(z)$.

Let F be a subgraph of G with k connected components, say $C_1, \ldots, C_k$. Let the component C_i have n_i vertices, m_i edges and b_i boundary components. The power of z in the term corresponding to F is given by $2k(F) + |E(F)| - |V(F)| - bc(F)$, where $2k(F) = 2k$ is twice the number of connected components of F; $|E(F)| = \sum_{i=1}^{k} m_i \leq m = |E(G)|$; $|V(F)| = |V(G)| = n$; and $bc(F) = \sum_{i=1}^{k} b_i \geq k$. We remark that $k \leq n$. Accordingly,

$$\deg(z) \leq 2k + m - n - k = m + k - n \leq m = |E(G)|. \tag{8.2}$$

For orientable ribbon graphs (G, O) considered in this paper, the number of boundary components $bc(F)$ of a subgraph F coincides with the number of LHT paths of F with an induced orientation (an isolated vertex of F is defined to correspond to one LHT path).

We note that the sum of all the coefficients of $BR_{(G,O)}(x, y, z) = 2^{|E(G)|} = 2^m$. Accordingly, if

$$BR_{(G,O)}(x, y, z) = \sum_{(i,j,k)} \rho(i, j, k) x^i y^j z^k,$$

we define the *normalized coefficient measure* of $BR(G,O)$ by

$$\mu_{BR}(G,O) = \frac{1}{2^m} \sum_{(i,j,k)} \rho(i,j,k)\delta\left(\frac{i}{n}, \frac{j}{m}, \frac{k}{m}\right). \tag{8.3}$$

By the previous remarks, $\mu_{BR}(G,O)$ is a probability measure supported in the unit cube $[0,1]^3 \subset \mathbb{R}^3$.

We would like to study limit points of the measures $\mu_{BR}(G_\ell, O_\ell)$ for sequences (G_ℓ, O_ℓ) of sparse graphs with increasing number of vertices; we remark that by compactness, limit points will exist for arbitrary sequences of graphs, but we restrict ourselves to sparse graphs in this paper.

We finally remark that for an oriented ribbon subgraph F of an oriented ribbon graph G, the power of z in (8.1) is twice the genus $2\gamma(F,O)$ of the surface $S(F,O)$, see e.g. [**Chm**, p. 4]. It follows that the largest degree of z is attained by $F = G$.

$$\deg_z(BR_{(G,O)}(x,y,z)) = 2\gamma(G,O). \tag{8.4}$$

9. Random graphs with orientations

In this section, we apply some of the results obtained in [**Gam**] and [**Ch-Pit**] to the study of Bollobás-Riordan polynomials.

Random d-regular graphs with orientation considered in [**Gam**, §3,4] were described in Section 7 using permutations β and α. For such graphs, the cycle structure of β is $(d,d,\dots,d)$. The cycle structure of α is $(2,2,\dots,2)$. The LHT paths (corresponding to the faces of $S(G,O)$) correspond to the cycles of $\beta\alpha$. To state the next result, we introduce the following notation: A_N denotes the alternating subgroup (of the permutation group S_N); C_d denotes the conjugacy class in A_N consisting of permutations whose cycle decomposition is the product of (N/d) disjoint d-cycles; the convolution of probability measures on S_N is denoted by $*$. Finally, the total variation distance between probability measures μ,ν on $G = A_N$ is defined by $||\mu - \nu|| = \max_{A\subset G} |\mu(A) - \nu(A)|$.

The main result in [**Gam**] is the following theorem ([**Gam**, Thm. 4.1]):

THEOREM 9.1. *Let $N = dn$ and let P_d denote the probability measure on A_N supported on C_d. Let U denote the uniform distribution on A_N. Then for $d \geq 3$,*

$$\lim_{n\to\infty} ||P_d * P_2 - U|| = 0.$$

This allowed to determine the asymptotic behaviour of the number of LHT paths for random cubic graphs G with a random orientation O at every vertex, cf. [**Gam**, Corollary 5.1].

THEOREM 9.2. *Let $L(n)$ denote the number of LHT paths in a random cubic graph on n vertices with random orientation. Then, as $n \to \infty$, $\mathbb{E}(L(n)) = \log(3n) + \gamma + o(1)$ and $\mathrm{Var}(L(n)) = \log(3n) + \gamma - \pi^2/6 + o(1)$, where $\gamma = 0.5772\dots$ is Euler's constant. Further, $(L(n) - \log n)/\sqrt{\log n}$ converges to standard normal distribution $N(0,1)$.*

The corresponding result for random regular graphs is [**Gam**, Corollary 4.1].

THEOREM 9.3. *The distribution of LHT paths for random regular graphs with random orientation converges to Poisson-Dirichlet distribution.*

These results were extended in [**Fl-P**, **Ch-Pit**]. Below we formulate the results in [**Ch-Pit**] in a form that is convenient for applications to BS convergent sequences of graphs. We assume all vertices of our graphs will have degree d satisfying $3 \le \delta \le d \le D$ (minimum degree $\delta \ge 3$; maximum degree D). It follows from the definition of BS convergence that there exist *limiting proportions* $b_\delta, \dots, b_D$ of vertices of degree $\delta, \dots, D$ respectively.

9.1. Random oriented graphs with a given degree sequence. The results in [**Ch-Pit**] can be reformulated to provide a model for random graphs with orientations, generalizing the model considered in [**Gam**].

Let G_j have n_j vertices. Let $(n_{j,\delta}, \dots, n_{j,D})$ be a sequence that is realizable as a degree sequence of a graph with n_j vertices; we have $\sum_{k=\delta}^{D} n_{j,k} = n_j$. We assume that

$$\lim_{j\to\infty} n_{j,k}/n_j = b_k, \qquad \delta \le k \le D.$$

Let H_j denote the half-edges of G_j; we have $N_j := |H_j| = \sum_{k=\delta}^{D} k n_{j,k} = 2E(G_j)$. We remark that

$$N_j \asymp 2n_j \sum_{k=\delta}^{D} k b_k \tag{9.1}$$

Consider now a conjugacy class β in the alternating group S_{N_j} whose cycle structure is

$$C_{\mathbf{m}} := (n_{j,\delta} C_\delta, \dots, n_{j,D} C_D);$$

it determines the cyclic orientation O_j at the vertices.

Let α be a permutation whose cycle structure is $C_2 := (2, 2, \dots, 2)$ (it determines the edges of G_j). We are interested in the cycle structure of the product $\beta\alpha$. Denote by $P(\mathbf{m}, j) := P(n_{j,\delta}, \dots, n_{j,D})$ the probability measure supported on the conjugacy class of $C_{\mathbf{m}}$. Let $P(2, j)$ denote the measure P_2 on S_{N_j} supported on the conjugacy class $(2, \dots, 2)$. An analogue of Theorem 9.1 in this setting was proved in [**Ch-Pit**, Theorem 2.2], who determined the asymptotic distribution of $P(\mathbf{m}, j) * P_2$. We state their result below.

THEOREM 9.4. *Let $P(\mathbf{m}, j)$ and $P(2, j)$ be as above. Let U_j denote the uniform distribution on A_{N_j} if $C_{\mathbf{m}}$ and C_2 are of the same parity; and the uniform distribution on $S_{N_j} \setminus A_{N_j}$ if $C_{\mathbf{m}}$ and C_2 are of different parity. Then as $N_j \to \infty$, we have*

$$||P(\mathbf{m}, j) * P(2, j) - U_j|| = O(N_j^{-1}).$$

The following results follow as corollaries (see [**Ch-Pit**, §3]). We note that the construction in [**Ch-Pit**] is *dual* to the construction in the current paper: polygons glued along pairs of sides to form a surface in [**Ch-Pit**] correspond to cyclically oriented edges around a vertex in our paper. Accordingly, the number of vertices in [**Ch-Pit**, Thm. 3.1] is equal to the number of faces in our paper.

In the next statement, we keep the notation of Theorem 9.4. Below, $\mathbb{P}$ denotes the uniform measure on the set of LHT surfaces (G_j, O) considered in 9.4; recall that $\mathbb{P} = P(\mathbf{m}, j) * P(2, j)$. Also, let F_j denote the number of faces in (G_j, O); those faces are in bijection with cycles in the product of two permutations. Let C_N^e (resp. C_N^o) denote the total number of cycles of the permutation chosen uniformly at random from all even permutations (respectively all odd permutations) in S_N. The next result is a restatement of [**Ch-Pit**, Thm. 3.1].

PROPOSITION 9.5. *If $C_{\mathbf{m}}$ and C_2 are of the same parity, then $||\mathbb{P}(F_j - C^e_{N_j})|| = O(N_j^{-1})$; if $C_{\mathbf{m}}$ and C_2 are of the opposite parity, then $||\mathbb{P}(F_j - C^o_{N_j})|| = O(N_j^{-1})$.*

Denote by C_N the number of cycles of a random permutation in S_N. It is known that

$$\mathbb{E}[C_N] = \sum_{j=1}^{N} 1/j = \log N + O(1); \qquad \mathrm{Var}[C_N] = \sum_{j=1}^{N} (1/j)(1 - 1/j) = \log N + O(1).$$

We continue the summary of results from [**Ch-Pit**, §3,4]. The first result concerns the number X_j of components of the surface $S(G_j, O_j)$, [**Ch-Pit**, Thm. 4.1]:

$$\mathbb{P}(X_j = 1) = 1 - O(N_j^{-1}).$$

Fix $a \in \mathbb{N}$. The genus γ_j of $S(G_j, O_j)$ was determined in [**Ch-Pit**, Thm. 4.2].

THEOREM 9.6. *With notation as above, for all admissible ℓ, the genus $\gamma_j = \gamma$ of $S(G_j, O_j)$ satisfies (here we denote $N_j = N, m_j = m$)*

$$\mathbb{P}(\gamma_j = 1 + N/4 - m/2 - \ell/2) = \frac{(2 + O(\log^{-1/2} N))}{\sqrt{2\pi \mathrm{Var}[C_N]}} \exp\left(- \frac{(\ell - \mathbb{E}[C_N])^2}{2\mathrm{Var}[C_N]} \right),$$

where ℓ is admissible provided $(\ell - \mathbb{E}[C_N])/\sqrt{\mathrm{Var}[C_N]} \in [-a, a]$.

Using (9.1), find that

$$\mathbb{E}[\gamma(G_j, O_j)] \asymp 1 + \frac{n_j}{2} \left(\sum_{k=\delta}^{D} k b_k - 1 \right) - \log \left(2n_j (\sum_{k=\delta}^{D} k b_k) \right) \tag{9.2}$$

By (8.4), we find that

$$\mathbb{E}[\deg_z(BR_{G_j}(x, y, z))] = 2\mathbb{E}[\gamma(G_j, O_j)]$$

10. Convergence of the coefficient measures of Bollobás-Riordan polynomials

In this section, we extend the results of [**JMNT**] showing that the normalized coefficient measures of rank polynomials converge to a delta function, provided a sequence of the corresponding (bounded degree) graphs converges in the sense of Benjamini-Schramm convergent sequences. We extend that result to the normalized coefficient measures (8.3), provided a sequence of (bounded degree) oriented ribbon graphs converges as in (10.1) defined below.

10.1. BS Convergence for oriented ribbon graphs and LHT surfaces. In the section 9, we considered random oriented graphs with a given degree sequence. In this section, we would like to study more general sequences of oriented graphs. We concentrate on two cases:

(a) Take a sequence of graphs G_j (of minimal degree 3) that converges BS, and put a random cyclic orientation of the edges at every vertex.
(b) Generalize the notion of BS convergence to oriented graphs.

Motivated by our approach in [**JMNT**], we extend the definition of Benjamini-Schramm convergence "left-hand turn" surfaces $S(G, O)$; this coincides with the BS convergence for oriented ribbon graphs.

In what follows, we will consider an extension of Benjamini-Schramm convergence to ribbon graphs and oriented graphs. Everything remains the same, except that every instance of rooted graph is replaced by rooted oriented or rooted ribbon graph. Isomorphisms of such graphs are required to also preserve the additional structure.

Let $\{(G_j, O_j)\}$ be a sequence of ribbon graphs; we restrict ourselves to graphs of bounded degree, and assume that $|V(G_j)| \to \infty$ as $j \to \infty$. Denote by $S(G_j, O_j)$ the corresponding LHT surfaces. By analogy with the usual Benjamini-Schramm convergence, we say that (G_j, O_j) converges to an infinite ribbon graph (G_∞, O_∞) with a given probability measure on its vertices, provided the following holds. Given $R \in \mathbb{N}$ and a finite ribbon graph α, denote by $\mathbb{P}_{(G_j,O_j)}(\alpha, R)$ the probability that a ball of radius R centered at a random vertex $u \in (G_j, O_j)$ is isomorphic to α.

DEFINITION 10.1. The sequence (G_j, O_j) converges to (G_∞, O_∞) provided that for any $R \in \mathbb{N}$, and for any α, there exists $0 \leq P_\infty(\alpha, R) \leq 1$ such that $\mathbb{P}_{(G_j,O_j)}(\alpha, R) \to P_\infty(\alpha, R)$ as $j \to \infty$.

This is not the original definition given by Benjamini and Schramm, but it is equivalent in the case of graphs of uniformly bounded degree.

For what follows, instead or choosing the root from the uniform distribution, we choose the root from the stationary distribution, essentially choosing a uniformly random root edge. For sequences of graphs of uniformly bounded degree (and no isolated points), Benjamini-Schramm convergence according to the uniform distribution is equivalent to that under the stationary distribution. Indeed the probability measures in the two cases differ only by multiplication by a uniformly bounded, continuous function. Since the graph is oriented, we will more specifically choose a specific direction of traversal of the root edge. In the case of ribbon graphs, this will mean the side of the ribbon to the left as the edge is traversed in that direction. Let $\vec{uv}$ denote the edge uv traversed from u to v. Benjamini-Schramm convergence is stated in terms of functions of graphs rooted at vertices, however for this proof it is more convenient to consider graphs rooted at edges. The two are equivalent for locally finite graphs since a function f on the directed edges can be considered a function of the vertices by letting $f(v)$ be the average of the $f(\vec{vu})$, for arcs pointing out of v.

Let $L_{\vec{uv}}$ denote the length of the boundary component containing $\vec{uv}$. For an edge subgraph A, let $L^A_{\vec{uv}}$ denote the length of the boundary component in A containing $\vec{uv}$.

THEOREM 10.2. *The sequence of coefficient measures of the Bollobás-Riordan polynomials of a Benjamini-Schramm convergent sequence of oriented ribbon graphs with uniformly bounded degree converges to a δ function.*

PROOF. Consider a sequence (G_j, O_j) of oriented ribbon graphs of bounded degree, that converges in the sense of Definition 10.1; we denote the number of vertices of G_j by n_j, and the number of edges of G_j by m_j.

For each j, let A_j be a subset of the edge set of G_j chosen by including each edge uniformly and independently at random with probability $\frac{1}{2}$.

It is shown in Theorem 18 of [**JMNT**] that

- The measure associated to $|A_j|/n_j$ converges weakly to a δ function.
- The measure associated to $k(A_j)/n_j$ (where $k(A_j)$ is the number of connected components of A_j) converges weakly to a δ function.

- The density m_j/n_j converges to a non-zero real number.

This implies that the measures associated to $r(A_j)/m_j$ and $s(A_j)/m_j$ where r is the rank and s the co-rank of A_j converge weakly to δ functions. In Lemma 17 of [**JMNT**], it is shown that the subgraphs A_j themselves form a Benjamin-Schramm convergent sequence of un-oriented graphs. This result extends naturally to ribbon graphs. Similarly to Lemma 17, one could also formulate the convergence result in the language of probability measures on random subgraphs of (G_j, O_j); we leave it as an exercise.

It suffices to show that $\mathbb{E}\left[bc(A_j)/m_j\right]$ converges and $\operatorname{Var}\left(bc(A_j)/m_j\right) \to 0$. The expectations and variance here are taken with respect to the choice of a uniformly random edge subgraph A_j by independently removing each edge with probability $\frac{1}{2}$. The proof will be very similar to the treatment of the number of connected components in [**JMNT**]. Let us first note that

$$bc(A_j) = \sum_{(u,v):uv\in A_j} \left(L_{\vec{uv}}^{A_j}\right)^{-1}$$

and let

$$\ell^{A_j}(\vec{uv}) = \begin{cases} \left(L_{\vec{uv}}^{A_j}\right)^{-1}, & \text{if } uv \in A_j \\ 0, & \text{otherwise} \end{cases}.$$

Then $bc(A_j) = \sum_{(u,v):uv\in E(G_j)} \ell^{A_j}(\vec{uv})$. Define further

$$\ell_R^{A_j}(\vec{uv}) = \begin{cases} \ell^{A_j}(\vec{uv}), & \text{if } \ell^{A_j}(\vec{uv}) > \frac{1}{R} \\ 0, & \text{otherwise} \end{cases}.$$

The quantity $\ell_R^{A_j}(\vec{uv})$ depends only on what happens within the ball of radius R about uv and is within $1/R$ of $\ell^{A_j}(\vec{uv})$.

To show convergence of the expectation, we use linearity of expectation.

$$\begin{aligned} \mathbb{E}\left[\frac{bc(A_j)}{m_j}\right] &= \mathbb{E}\left[\sum_{\vec{uv}:uv\in E} \frac{1}{m_j}\ell^{A_j}(\vec{uv})\right] \\ &= \frac{1}{m_j}\sum_{\vec{uv}:uv\in E(G_j)} \mathbb{E}\left[\ell^{A_j}(\vec{uv})\right] \end{aligned}$$

We may now approximate $\mathbb{E}\left[\ell^{A_j}(\vec{uv})\right]$ using $\ell_R^{A_j}(\vec{uv})$. $\mathbb{E}\left[\ell^{A_j}(\vec{uv})\right]$ and $\mathbb{E}\left[\ell_R^{A_j}(\vec{uv})\right]$ differ by at most $\frac{1}{R}$. However, $\ell_R^{A_j}(\vec{uv})$ is entirely determined by the behaviour of A_j in the ball of radius R about uv. Hence $\mathbb{E}\left[\ell_R^{A_j}(\vec{uv})\right]$ is entirely determined by the ball of radius R about uv. Since the G_j have uniformly bounded degree, such a ball can possibly be isomorphic to only a finite number of graphs. Since the G_j are Benjamini-Schramm convergent, the probability that the ball of radius R about a randomly selected $\vec{uv}$ is isomorphic to any given graph converges. Thus $D_{R,j} = (1/m_j)\sum_{\vec{uv}} \mathbb{E}\left[\ell_R^{A_j}(\vec{uv})\right]$ converges to a limit D_R as $j \to \infty$. The sequence of D_R (indexed by R) is Cauchy, and thus converges to a limit D as $R \to \infty$. Let $\epsilon > 0$ be given and choose $R > 3/\epsilon$. Then, for all j, $\frac{1}{m_j}\sum_{\vec{uv}} \mathbb{E}\left[\ell^{A_j}(\vec{uv})\right]$ is within $\epsilon/3$ of $D_{R,j}$. Choose J large enough that for all $j \geq J$, $D_{R,j}$ is within $\epsilon/3$

of D_R, which is within $1/R$ of D. Then, summing up all the errors for all $j \geq J$, $(1/m_j)\sum_{\vec{uv}} \mathbb{E}\left[\ell^{A_j}(\vec{uv})\right]$ is within ϵ of D. Thus $(1/m_j)\sum_{\vec{uv}} \mathbb{E}\left[\ell^{A_j}(\vec{uv})\right]$ converges to D.

For the variance, let $\epsilon > 0$ be given. Let $R > \frac{16}{\epsilon}$ and consider again $\ell_R^{A_j}(\vec{uv})$. Since the graph has bounded degree, there is a bound on the number of edges at distance at most R from any point. Hence, for all but $O(m_j)$ pairs (uv, wx), $\ell_R^{A_j}(\vec{uv})$ and $\ell_R^{A_j}(\vec{wx})$ are independent. Hence $\operatorname{Var}\left(\sum_{(u,v):uv\in E(G_j)} \ell_R^{A_j}(\vec{uv})\right) = O(m_j)$. Now let $f_R^{A_j}(\vec{uv}) = \ell^{A_j}(\vec{uv}) - \ell_R^{A_j}(\vec{uv})$ and note that this is always bounded above by $1/R < \epsilon/16$. We note that

$$bc(A_j) = \sum_{(u,v):uv\in E(G_j)} \ell_R^{A_j}(\vec{uv}) + \sum_{(u,v):uv\in E(G_j)} f_R^{A_j}(\vec{uv})$$

and so

$$\begin{aligned}\operatorname{Var}(bc(A_j)) &= \operatorname{Var}\left(\sum_{(u,v),uv\in E(G_j)} \ell_R^{A_j}(\vec{uv}) + \sum_{(u,v),uv\in E(G_j)} f_R^{A_j}(\vec{uv})\right)\\ &= Var\left(\sum_{(u,v):uv\in E(G_j)} \ell_R^{A_j}(\vec{uv})\right) + \operatorname{Var}\left(\sum_{(u,v):uv\in E(G_j)} f_R^{A_j}(\vec{uv})\right) +\\ &\quad Cov\left(\sum_{(u,v):uv\in E(G_j)} f_R^{A_j}(\vec{uv}), \sum_{(u,v):uv\in E(G_j)} \ell_R^{A_j}(\vec{uv})\right)\end{aligned}$$

We know that the first term is $O(m_j)$. The other two are each less than $(\epsilon/8)4m_j^2$ and so their sum is less than ϵm_j^2. Choose a graph sufficiently far in the sequence G_j that for all subsequent graphs, the first term is smaller than $(\epsilon/2)m_j^2$. Hence $\operatorname{Var}(bc(A_j)/m_j) < 3\epsilon/2$ past a certain point in the sequence. This holds for all $\epsilon > 0$ and so $\operatorname{Var}(bc(A_j)/m_j) \to 0$. This finishes the proof. □

This theorem deals with the case of deterministic Benjamini-Schramm convergent sequences of oriented ribbon graphs. This corresponds to point (b) stated at the start of this section. There can also be Benjamini-Schramm convergent sequences of random ribbon graphs. In these, each $\mathcal{G}_j$ is a distribution on the graphs of a given size. The random vertex is selected by first choosing a graph G_j from $\mathcal{G}_j$ and then selecting a random vertex from it. Point (a) given at the start of the section is such a case. There is an underlying Benjamini-Schramm convergent sequence of graphs G_j. It is made into an oriented ribbon graph by assigning independently and uniformly at random a cyclic ordering to the edges at every vertex. This gives us a sequence $(G_j, \mathcal{O}_j)$ of random ribbon graphs. They are Benjamini-Schramm convergent since, for all oriented ribbon graphs α with underlying graph α' and radii R, $\mathbb{P}_{(G_j,\mathcal{O}_j)}(\alpha, R) = \mathbb{P}_{G_j}(\alpha', R)\mathbb{P}(\alpha|\alpha')$ where $\mathbb{P}(\alpha|\alpha')$ denotes the probability of obtaining α from α' by assigning a cyclic ordering of edges to every vertex. The first factor converges as $j \to \infty$ since the G_j were Benjamini-Schramm convergent. The second factor does not depend on j.

The theorem and proof as stated do not apply to Benjamini-Schramm convergent sequences of random oriented ribbon graphs. However, the proof can be modified to account for the r andomness that arises from giving a cyclic orientation O_j to the edges around each vertex in a Benjamini-Schramm convergent sequence of random graphs. Replace each instance of $L_{\vec{uv}}^{A_j}$, $\ell^{A_j}(\vec{uv})$ and $\ell_R^{A_j}(\vec{uv})$ by $L_{\vec{uv}}^{A_j,O_j}$, $\ell^{A_j,O_j}(\vec{uv})$ and $\ell_R^{A_j,O_j}(\vec{uv})$ respectively to represent the fact that the length of boundary components depends also on the cyclic ordering which is now also random. The crucial properties that $\ell^{A_j,O_j}(\vec{uv})$ and $\ell_R^{A_j,O_j}(\vec{uv})$ are within $1/R$ of each other and that $\ell_R^{A_j,O_j}(\vec{uv})$ depends only on the behaviour of A_j and O_j in the ball of radius R about uv remain unchanged. Take the expectation and variance over both A_j and O_j. The proof proceeds exactly as before. Thus Theorem 10.2 applies to the sequences of ribbon graphs described in point (a) at the start of this section.

In Section 6 of [**JMNT**], the precise coordinates of the limiting δ function of the coefficient measure of the rank polynomial were found in the case of random regular graphs. We now find the z-axis coordinate for the limit of the coefficient measure of the Bollobás-Riordan polynomial for oriented ribbon graphs obtained from random regular graphs by adding a cyclic ordering of the edges at every vertex. Random regular graphs are not a sequence of Benjamini-Schramm convergent deterministic graphs. However, as discussed in section 6 of [**JMNT**], they converge to a fixed graph and thus are almost uniformly Benjamini-Schramm convergent as defined on page 17 of the same paper, allowing them to be treated similarly to a deterministic sequence. By 8.4, the degree $\deg_z(BR_{(G,O)}(x,y,z)) = 2\gamma(G,O)$, where $\gamma(G,O)$ denotes the genus of the orientable LHT surface $S(G,O)$. By Euler's formula, it is equal to $2+|E(G)|-|V(G)|-L(G,O)$, where $L(G,O)$ denotes the number of LHT paths of (G,O); it is also equal to the number of faces of $S(G,O)$. By Proposition 9.5, for d-regular graphs on n vertices with random orientation, $L(G,O)$ grows logarithmically in n. We have $|V(G) = n, |E(G)| = dn/2$. The normalized z-coordinate of the δ function is equal to $\mathbb{E}[2\gamma(G,O)]/|E(G)|$. To leading order in n, it is asymptotic to $[n(d/2-1)]/[dn/2]$. After cancellations, we find that the normalized z-coordinate of the limiting δ function is equal to $(d-2)/d$.

11. Numerical investigations: coefficients and zeros of Bollobás-Riordan polynomials

To compute Bollobás-Riordan on random graphs, we generate random cyclic orientations of half-edges around each vertex, and then use the SAGE ribbon graph library to count the number of boundary components for each induced subgraph we must sum over. The SAGE code is included in the section 12. Since the evaluation of the Tutte polynomial can yield the number of 3-colourings, which is a $\#P$-complete problem, the computation of the Tutte polynomial is $\#P$-hard (the class $\#P$ is a complexity class that contains function problems of counting solutions that correspond to underlying NP decision problems). Therefore, computing the Tutte and Bollobás-Riordan polynomials is not feasible for large graphs in reasonable amounts of time. Note that the computation employed below is exponential in the number of edges, so the problem quickly becomes intractable as the the graphs grow. However, the computation is highly parallelizable (we can sum over induced graphs separately on different processors and then sum up the polynomials at the end), which can yield notable increases in speed and potentially allow us to study slightly larger graphs.

We first provide pictures of the coefficient measures for Bollobás-Riordan polynomials.

Figure 10 shows the coefficient measure for a ribbon graph obtained by choosing a random orientation on a 3-regular graph on 12 vertices; the graph itself is shown as well.

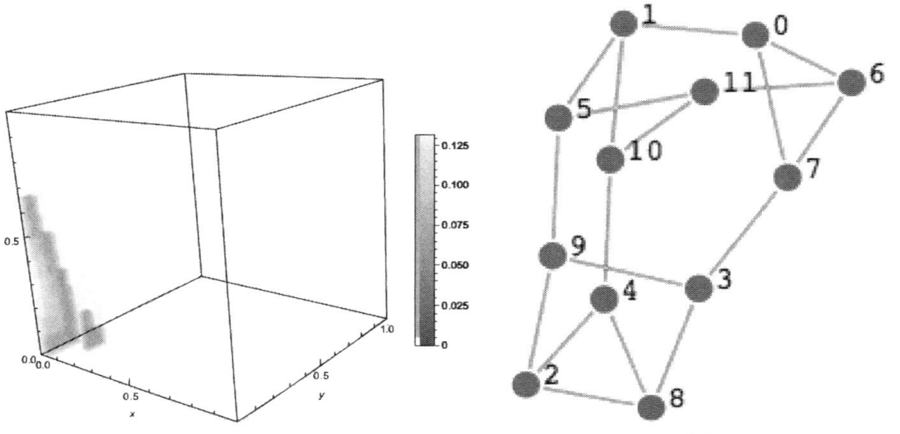

(A) (B)

FIGURE 10. BR Coefficients: 3-regular graph on 12 vertices

Next, Figure 11 shows the coefficient measure for a ribbon graph obtained by choosing a random orientation on a 4-regular graph on 10 vertices; the graph itself is shown as well.

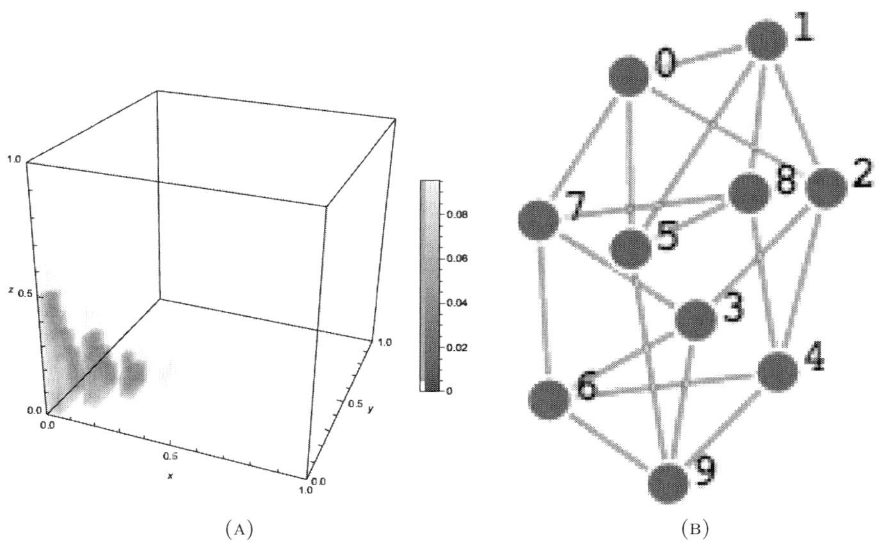

(A) (B)

FIGURE 11. BR Coefficients: 4-regular graph on 10 vertices

Next, we include the pictures of the zero sets in $\mathbb{R}^3$ of Bollobás-Riordan polynomials. Figure 12 shows the zero set for a 3-regular graph on 12 vertices; it is the same graph as in Figure 10.

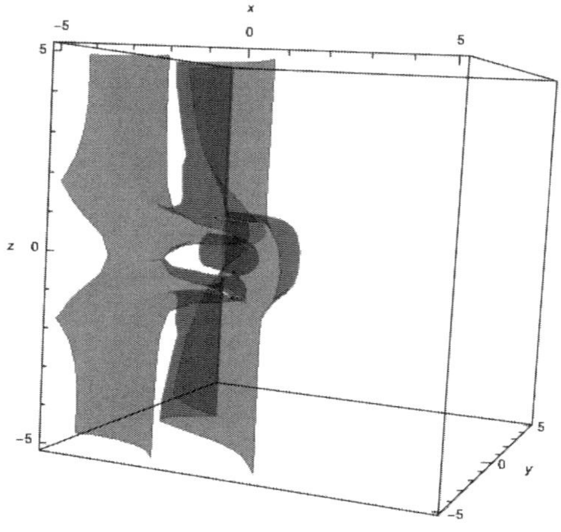

FIGURE 12. Bollobás-Riordan zeros: 3-regular graph on 12 vertices

Figure 13 shows the zero set for a 4-regular graph on 10 vertices; it is the same graph as in Figure 11.

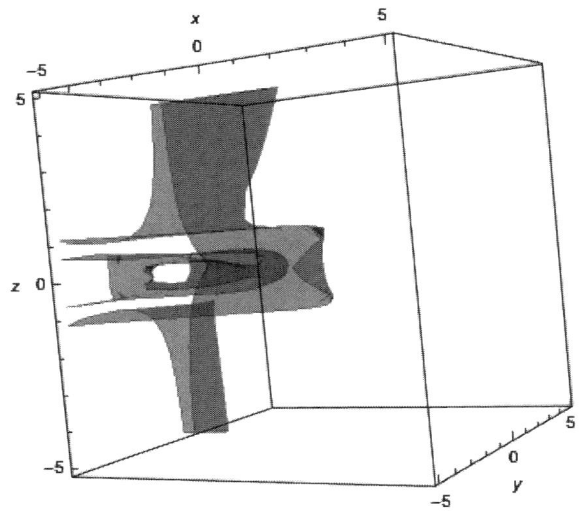

FIGURE 13. Bollobás-Riordan zeros: 4-regular graph on 10 vertices

12. Conclusion

In conclusion, we would like to discuss some natural questions that arise from the results in [**JMNT**] and the present paper.

PROBLEM 1. It seems natural to generalize results in [**JMNT**] and the present paper to higher-dimensional simplicial complexes; a natural generalization of rank and Bollobás-Riordan polynomials are the Kruskal-Renardi polynomials, see e.g. [**KR**], [**BBC**]. We expect that an analogue of Theorem 10.2 holds for (suitably defined) sequences of converging CW complexes, at least under suitable bounded degree restrictions.

PROBLEM 2. We hope to establish large deviation results for the coefficient measures of rank polynomials (for random regular graphs, and possibly for more general models of random graphs with degrees). This is work in progress with O. Angel, C. MacRury and L. Silberman. We also hope to establish similar results for the coefficient measures Bollobás-Riordan polynomials (e.g. for random regular graphs with random orientations, and possibly for more general models considered e.g. in [**Ch-Pit**]).

PROBLEM 3. It seems natural to study asymptotic behaviour of the zero sets of rank, BR and related polynomials; in particular beyond the asymptotic behaviour "at infinity" discussed briefly in Section 5. Since the degree of the polynomials is growing, their nodal sets should be rescaled appropriately. It could be interesting to compare their behaviour with the asymptotic behaviour of nodal sets of high energy Laplace eigenfunctions, cf. e.g. [**Zel**] for a recent survey.

PROBLEM 4. It seems very interesting to explore in more detail possible applications to Statistical Physics, e.g. in connection to the q-state Potts model on graphs.

PROBLEM 5. It seems interesting to explore possible connections to knot polynomials and asymptotic properties of knot invariants.

PROBLEM 6. A natural question is to study the asymptotic behaviour of the coefficient measures and zero sets for sequences of *dense* graphs, removing the bounded degree restriction of [**JMNT**] and the present paper; one could consider various notions of convergence for such graphs (e.g. graphon convergence etc).

Appendix: Computer code

In this section we include the computer code used in the numerical investigations described in Section 11. The MATHEMATICA commands we used to generate random regular graphs and their Tutte polynomials were described in section 4 and 5 .

For Bollobás-Riordan polynomials, we used SAGE to generate random graphs with random orientations, and standard MATHEMATICA commands to plot the coefficient densities and zeros of the corresponding polynomials. We include the SAGE code below.

```
from copy import deepcopy

G = graphs.RandomRegular(3,20) #uniform random 3-regular graph on 20
vertices
V = G.vertices()
dartlist = [[] for _in range(len(V))]
#We label each edge with a positive number k, and each half-edge
with either 2k or 2k-1. This allows us to pass between edges and
```

```
half-edges to make use of the ribbon graph library while we are
cycling over spanning edge subsets during the computation of the
Bollobas-Riordan polynomial.
r = [] #involution of darts (sends each dart to its other half)
for i, e in enumerate(G.edges()):
    dartlist[e[0]].append(2*i+1)
    dartlist[e[1]].append(2*i+2)
    G.set_edge_label(e[0], e[1], i+1)
    r.append([2*i+1,2*i+2])
E = G.edges()}
rank_G = len(V) - G.connected_components_number()
nullity_G = len(E) - rank_G
s = [] #cyclic permutations of darts in mutable list form,
generated uniformly at random
for dl in dartlist:
    shuffle(dl)
    s.append(dl)
edge_subsets = []
#generates a list of edge subsets
for i in range(1,1 << len(E)):
    subset = [E[bit] for bit in range(len(E))
    if i & (1 << bit) > 0]
    edge_subsets.append(subset)
#we now loop over all spanning subgraphs
x,y,z = var('x,y,z')
brpoly(x,y,z) = x^(rank_G);
for E_F in edge_subsets:
    F = Graph([V,E_F])
    removededges = [l[2] for l in
    list(Set(E).difference(Set(E_F)))]
    removeddarts = []
    r_F = []
    for e in removededges:
        removeddarts.append(2*e-1)
        removeddarts.append(2*e)
    for d in r:
        if d[0] not in removeddarts and d[1] not in removed darts:
            r_F.append(d);
    s_F = deepcopy(s)
    for i, dl in enumerate(s):
        for d in dl:
            if d in removeddarts:
                s_F[i].remove(d)
    s_F = [a for a in s_F if (a != [] and len(a) != 1)]
    sigma_F = PermutationGroupElement(''.join([str(tuple(a))
    for a in s_F]))
    rho_F = PermutationGroupElement(''.join([str(tuple(a))
    for a in r_F]))
```

```
    isolated = F.degree_histogram()[0]
    bc = RibbonGraph(sigma_F,rho_F).number_boundaries()
    bc += isolated
   components_F = F.connected_components_number()
    rank_F = len(V) - components_F
    nullity_F = len(E_F) - rank_F
brpoly = brpoly + x^(rank_G - rank_F) * y^(nullity_F) *
z^(components_F - bc + nullity_F)
 print(brpoly)
```


Acknowledgements

This work is partly based on a 2015 summer research project by Lise Turner and Michaël Medeiros Charbonneau. The authors would like to acknowledge Michaël's contributions to Sections 2, 3, 4 and 5. The authors would like to thank L. Addario-Berry, O. Angel, M. Aizenman, C. MacRury, S. Norin, I. Pak, P. Sarnak and L. Silberman for inspiring conversations. The authors would like to thank anonymous referees for very interesting and useful comments.

References

[BBC] Carlos Bajo, Bradley Burdick, and Sergei Chmutov, *On the Tutte-Krushkal-Renardy polynomial for cell complexes*, J. Combin. Theory Ser. A **123** (2014), 186–201, DOI 10.1016/j.jcta.2013.12.006. MR3157807

[Big1] Norman Biggs, *Algebraic graph theory*, 2nd ed., Cambridge Mathematical Library, Cambridge University Press, Cambridge, 1993. MR1271140

[BDS] N. L. Biggs, R. M. Damerell, and D. A. Sands, *Recursive families of graphs*, J. Combinatorial Theory Ser. B **12** (1972), 123–131, DOI 10.1016/0095-8956(72)90016-0. MR0294172

[B-D] Joseph Blitzstein and Persi Diaconis, *A sequential importance sampling algorithm for generating random graphs with prescribed degrees*, Internet Math. **6** (2010), no. 4, 489–522, DOI 10.1080/15427951.2010.557277. MR2809836

[Bol] Béla Bollobás, *Random graphs*, 2nd ed., Cambridge Studies in Advanced Mathematics, vol. 73, Cambridge University Press, Cambridge, 2001. MR1864966

[BR01] Béla Bollobás and Oliver Riordan, *A polynomial invariant of graphs on orientable surfaces*, Proc. London Math. Soc. (3) **83** (2001), no. 3, 513–531, DOI 10.1112/plms/83.3.513. MR1851080

[BR02] Béla Bollobás and Oliver Riordan, *A polynomial of graphs on surfaces*, Math. Ann. **323** (2002), no. 1, 81–96, DOI 10.1007/s002080100297. MR1906909

[BM01] Robert Brooks and Eran Makover, *Belyi surfaces*, Entire functions in modern analysis (Tel-Aviv, 1997), Israel Math. Conf. Proc., vol. 15, Bar-Ilan Univ., Ramat Gan, 2001, pp. 37–46. MR1890529

[BM04] Robert Brooks and Eran Makover, *Random construction of Riemann surfaces*, J. Differential Geom. **68** (2004), no. 1, 121–157. MR2152911

[BrMon] R. Brooks and M. Monastyrsky, *K-regular graphs and Hecke surfaces*, Geometry, spectral theory, groups, and dynamics, Contemp. Math., vol. 387, Amer. Math. Soc., Providence, RI, 2005, pp. 65–74, DOI 10.1090/conm/387/07234. MR2179786

[Cat] Paul A. Catlin, *Super-Eulerian graphs: a survey*, J. Graph Theory **16** (1992), no. 2, 177–196, DOI 10.1002/jgt.3190160209. MR1163250

[CS] Shu-Chiuan Chang and Robert Shrock, *Tutte polynomials and related asymptotic limiting functions for recursive families of graphs*, Adv. in Appl. Math. **32** (2004), no. 1-2, 44–87, DOI 10.1016/S0196-8858(03)00077-0. Special issue on the Tutte polynomial. MR2037143

[Chm] S. Chmutov. *Topological Tutte polynomial*. arxiv:1708.08132

[Ch-Pak] Sergei Chmutov and Igor Pak, *The Kauffman bracket of virtual links and the Bollobás-Riordan polynomial* (English, with English and Russian summaries), Mosc. Math. J. **7** (2007), no. 3, 409–418, 573, DOI 10.17323/1609-4514-2007-7-3-409-418. MR2343139

[Ch-Pit] Sergei Chmutov and Boris Pittel, *On a surface formed by randomly gluing together polygonal discs*, Adv. in Appl. Math. **73** (2016), 23–42, DOI 10.1016/j.aam.2015.09.016. MR3433499

[CsFr] Péter Csikvári and Péter E. Frenkel, *Benjamini-Schramm continuity of root moments of graph polynomials*. part B, European J. Combin. **52** (2016), no. part B, 302–320, DOI 10.1016/j.ejc.2015.07.009. MR3425982

[DFKLS] Oliver T. Dasbach, David Futer, Efstratia Kalfagianni, Xiao-Song Lin, and Neal W. Stoltzfus, *The Jones polynomial and graphs on surfaces*, J. Combin. Theory Ser. B **98** (2008), no. 2, 384–399, DOI 10.1016/j.jctb.2007.08.003. MR2389605

[Flem] K. Fleming. *Boundary cycles in random triangulated surfaces.* PhD Thesis, Department of Mathematics, Harvey Mudd College, 2008.

[Fl-P] Kevin Fleming and Nicholas Pippenger, *Large deviations and moments for the Euler characteristic of a random surface*, Random Structures Algorithms **37** (2010), no. 4, 465–476, DOI 10.1002/rsa.20331. MR2760359

[Gam] Alex Gamburd, *Poisson-Dirichlet distribution for random Belyi surfaces*, Ann. Probab. **34** (2006), no. 5, 1827–1848, DOI 10.1214/009117906000000223. MR2271484

[JPS] Bill Jackson, Aldo Procacci, and Alan D. Sokal, *Complex zero-free regions at large $|q|$ for multivariate Tutte polynomials (alias Potts-model partition functions) with general complex edge weights*, J. Combin. Theory Ser. B **103** (2013), no. 1, 21–45, DOI 10.1016/j.jctb.2012.08.002. MR2995717

[JMNT] Dmitry Jakobson, Calum MacRury, Sergey Norin, and Lise Turner, *Distribution of coefficients of rank polynomials for random sparse graphs*, Electron. J. Combin. **25** (2018), no. 4, Paper 4.50, 29. MR3907781

[KR] Vyacheslav Krushkal and David Renardy, *A polynomial invariant and duality for triangulations*, Electron. J. Combin. **21** (2014), no. 3, Paper 3.42, 22. MR3262279

[LFH] Yunhua Liao, Aixiang Fang, and Yaoping Hou, *The Tutte polynomial of an infinite family of outerplanar, small-world and self-similar graphs*, Phys. A **392** (2013), no. 19, 4584–4593, DOI 10.1016/j.physa.2013.05.021. MR3083114

[Man] Arun P. Mani, *On some Tutte polynomial sequences in the square lattice*, J. Combin. Theory Ser. B **102** (2012), no. 2, 436–453, DOI 10.1016/j.jctb.2011.08.003. MR2885429

[McK] Brendan D. McKay, *Spanning trees in regular graphs*, European J. Combin. **4** (1983), no. 2, 149–160, DOI 10.1016/S0195-6698(83)80045-6. MR705968

[Mof] Iain Moffatt, *Knot invariants and the Bollobás-Riordan polynomial of embedded graphs*, European J. Combin. **29** (2008), no. 1, 95–107, DOI 10.1016/j.ejc.2006.12.004. MR2368618

[MY] Motohico Mulase and Josephine T. Yu, *Non-commutative matrix integrals and representation varieties of surface groups in a finite group* (English, with English and French summaries), Ann. Inst. Fourier (Grenoble) **55** (2005), no. 6, 2161–2196. MR2187951

[Pal] E. M. Palmer, *On the spanning tree packing number of a graph: a survey*, Discrete Math. **230** (2001), no. 1-3, 13–21, DOI 10.1016/S0012-365X(00)00066-2. Paul Catlin memorial collection (Kalamazoo, MI, 1996). MR1812338

[Sok] Alan D. Sokal, *Bounds on the complex zeros of (di)chromatic polynomials and Potts-model partition functions*, Combin. Probab. Comput. **10** (2001), no. 1, 41–77. MR1827809

[Tur] L. Turner. *Asymptotic behaviour of the distribution of coefficients of dichromatic polynomials of convergent sequences of sparse graphs.* Master's thesis. 2018.

[Tut] W. T. Tutte, *On dichromatic polynominals*, J. Combinatorial Theory **2** (1967), 301–320. MR0223272

[Wel] Dominic Welsh, *The Tutte polynomial*, Random Structures Algorithms **15** (1999), no. 3-4, 210–228, DOI 10.1002/(SICI)1098-2418(199910/12)15:3/4⟨210::AID-RSA2⟩3.0.CO;2-R. Statistical physics methods in discrete probability, combinatorics, and theoretical computer science (Princeton, NJ, 1997). MR1716762

[Wor] Nicholas C. Wormald, *The asymptotic connectivity of labelled regular graphs*, J. Combin. Theory Ser. B **31** (1981), no. 2, 156–167, DOI 10.1016/S0095-8956(81)80021-4. MR630979

[Xu] Nguyen Huy Xuong, *Sur quelques classes de graphes possédant des propriétés topologiques remarquables* (French, with English summary), C. R. Acad. Sci. Paris Sér. A-B **283** (1976), no. 11, Ai, A813–A816. MR0424601

[Zel] Steve Zelditch, *Local and global analysis of nodal sets*, Surveys in differential geometry 2017. Celebrating the 50th anniversary of the Journal of Differential Geometry, Surv. Differ. Geom., vol. 22, Int. Press, Somerville, MA, 2018, pp. 365–406. MR3838125

Department of Mathematics and Statistics, McGill University, 805 Sherbrooke Str. West, Montréal QC H3A 0B9, Canada. IR: Department of Mathematics, Temple University, Wachman Hall, 1805 North Broad Street, Philadelphia, PA 19122
Email address: dmitry.jakobson@mcgill.ca

Department of Mathematics and Statistics, McGill University, 805 Sherbrooke Str. West, Montréal QC H3A 0B9, Canada. IR: Department of Mathematics, Temple University, Wachman Hall, 1805 North Broad Street, Philadelphia, PA 19122
Email address: tomaslangsetmo@gmail.com

Department of Mathematics and Statistics, McGill University, 805 Sherbrooke Str. West, Montréal QC H3A 0B9, Canada. IR: Department of Mathematics, Temple University, Wachman Hall, 1805 North Broad Street, Philadelphia, PA 19122
Email address: igor.rivin@temple.edu

Department of Mathematics and Statistics, McGill University, 805 Sherbrooke Str. West, Montréal QC H3A 0B9, Canada. IR: Department of Mathematics, Temple University, Wachman Hall, 1805 North Broad Street, Philadelphia, PA 19122
Email address: lise.turner@uwaterloo.ca

Contemporary Mathematics
Volume **739**, 2019
https://doi.org/10.1090/conm/739/14895

The Brownian motion on $\text{Aff}(\mathbb{R})$ and quasi-local theorems

V. Konakov, S. Menozzi, and S. Molchanov

ABSTRACT. This paper is concerned with Random walk approximations of the Brownian motion on the Affine group $\text{Aff}(\mathbb{R})$. We are in particular interested in the case where the innovations are discrete. In this framework, the return probabilities of the walk have fractional exponential decay in large time, as opposed to the polynomial one of the continuous object. We prove that integrating those return probabilities on a suitable neighborhood of the origin, the expected polynomial decay is restored. This is what we call a *Quasi-local theorem.*

1. Introduction

In his seminal paper [**Yor92**] (see also the related survey work [**MY05**]), M. Yor studied the distribution density and the moments for the following particular exponential functional of the Brownian motion $(B_s)_{s\geq 0}$:

$$A_t^{\nu} = \int_0^t \exp(2B_s + \nu s)ds, \tag{1.1}$$

which corresponds, up to a normalization in t^{-1}, to the quantity appearing in the Asian options in the Black and Scholes setting (see again [**Yor92**]). The general case $\nu \neq 0$ can be reduced to $\nu = 0$ using the Girsanov theorem and the central object will be from now on the functional $A_t = \int_0^t \exp(2B_s)ds$.

This functional appears in several different situations, including the study of the Brownian motion on the group $\text{Aff}(\mathbb{R})$ of the affine transformations of $\mathbb{R} : x \mapsto ax + b,\ a, b \in \mathbb{R},\ a > 0$. This group can be isomorphically represented in the upper triangular 2×2 matrices setting $g = \begin{bmatrix} a & b \\ 0 & 1 \end{bmatrix}$, $a > 0$. The affine group provides the simplest example of solvable Lie group. We announced several results on the Brownian motion $x_t := \big(a_t, b_t\big)$ on $\text{Aff}(\mathbb{R})$ in the short communication [**KMM11**] which partly rely on the results by Yor [**Yor92**].

The central result of [**KMM11**] is the following Theorem.

Key words and phrases. Affine group, random walks and Brownian motion, local and quasi-local limit theorems, heat-kernel bounds.

The study has been funded by the Russian Science Foundation (project no. 17-11-01098) and the Russian Academic Excellence Project 5-100 for the second author.

THEOREM 1.1. *Let $p(t,\cdot,\cdot)$ be the transition density of the Brownian motion $x_t = \big(a_t, b_t\big)$ on* Aff(ℝ) *w.r.t. the corresponding Riemannian volume. Then, for all $g \in$* Aff(ℝ)*:*

$$p(t,g,g) = p(t,e,e) \sim_{t\to+\infty} \sqrt{\frac{\pi}{2}}\frac{1}{t^{\frac{3}{2}}}, \tag{1.2}$$

where $e = I_2$ is the neutral element of Aff(ℝ).

The note [**KMM11**] contains similar results for other solvable Lie groups. We also refer to [**KMM17**] for related topics. We will prove Theorem 1.1 in Section 2.

The most interesting fact in Theorem 1.1 is the slow decay of $p(t,g,g)$, $t \to +\infty$, which looks contradictory to the exponential growth of Aff(ℝ). Observe that such an exponential growth occurs for all finitely generated solvable groups that have no nilpotent subgroups of finite index, i.e. that are not virtually nilpotent, by Milnor's theorem [**Mil68**] (see also [**Wol68**]). We will establish that the random walks on the subgroups of Aff(ℝ) cannot directly give a *good* approximation of the Brownian motion $(x_t)_{t\geq 0}$ on Aff(ℝ). There are several reasons for that. First, Aff(ℝ) is non-unimodular, i.e. it cannot be approximated with increasing accuracy by discrete subgroups (lattices). Secondly, any such subgroup is typically dense and chaotically distributed in Aff(ℝ), see the results below, in particular Proposition 3.2 and Theorem 4.2. These arguments potentially apply to much more general situations.

More precisely, if $(x_n^\varepsilon)_{n\in\mathbb{N}}$ is the Markov chain corresponding to a symmetric random walk on the subgroup $G^\varepsilon \subset G$ generated by the matrices $\begin{bmatrix} \exp(\pm\varepsilon) & 0 \\ 0 & 1 \end{bmatrix} = g_1^{\pm\varepsilon}$; $\begin{bmatrix} 1 & \pm\varepsilon \\ 0 & 1 \end{bmatrix} = g_2^{\pm\varepsilon}$ with step ε^2 in time, $\varepsilon \in \mathbb{Q}$, then for $t = n\varepsilon^2 \in \mathbb{R}^+$ we have that:

$$P^\varepsilon(t,g,g) := P^\varepsilon(n,g,g) = \mathbb{P}_g(x_n^\varepsilon = g) \leq \exp(-cn^{\frac{1}{3}}\ln(n)^{\frac{2}{3}}),\ g \in G^\varepsilon. \tag{1.3}$$

Let us stress that the exponential estimation $P^\varepsilon(n,g,g) \leq \exp(-cn), c > 0$, which one could expect due to the exponential growth of the group cannot hold. Indeed, the solvable groups are amenable and it therefore follows from Kesten [**Kes59**] that $P^\varepsilon(n,g,g)$ decays at a sub-exponential rate. We will establish in Section 3 by direct elementary arguments this estimate which is a particular case of the typical asymptotics obtained for the return probabilities of random walks on general solvable groups studied e.g. by Pittet and Saloff-Coste [**PSC02**] and Tessera [**Tes13**]. The striking point is here that the return probability has fractional exponential decay and does not behave as $\frac{c}{t^{\frac{3}{2}}}$ as one could have expected from Theorem 1.1. The cause of this phenomenon is the special nature of the subgroup G^ε (which is dense but again highly chaotically distributed).

Note that for the nilpotent groups, like e.g. the Heisenberg one $\mathbf{H}^3$, the corresponding local limit theorems hold, see e.g. Breuillard [**Bre05**] (like in the case of the random walk on $\mathbb{Z}^d$ see e.g. [**IL71**], [**Pet05**], [**BR76**] or [**LL10**]). In the nilpotent setting we refer as well to the work of Alexopoulos [**Ale02**], where the most general local limit theorem on finitely generated groups of polynomial growth (i.e. virtually nilpotent by Gromov [**Gro81**]) is given.

We also mention that for absolutely continuous innovations, a local Theorem on Aff(ℝ), with the expected rate of order $n^{-3/2}$, matching the diagonal decay of the heat-kernel in (1.2) for large times, was proved by Bougerol [**Bou83**].

In this work, we will establish what we call *quasi-local* theorems for the previously described random walk on the discrete subgroup. Our first quasi-local theorem gives the estimation of the probability that x_n^ε belongs to a small neighborhood of the unit element $e = I$ which shrinks to e when $n \to +\infty$. We establish that the corresponding limit theorem holds with the expected convergence rate (see Section 4). It will be specified as well in Section 4.1 how this phenomenon, i.e. the dramatic difference between the fractional exponential decay of return probabilities stated in (1.3) and the polynomial one appearing when taking into account an associated neighborhood (which precisely corresponds to the large time behavior in (1.2)), already appears for a specific simple random walk on the dense locally uniformly distributed subgroup of ℝ generated by finitely many rationally independent numbers $\pm\alpha_i$, $i \in \{1, \cdots, N\}$. Roughly speaking, this dichotomy emphasizes that the paths of the random walk on the subgroup are quite *dense*. We will then eventually show that introducing (partially) an absolutely continuous component in the Markov chain x_n^ε on Aff(ℝ), one can check that the densities of the finite dimensional distributions of x_n^ε converge uniformly to the corresponding densities of the diffusion on Aff(ℝ).

2. Diffusion on Aff(ℝ) and similar groups

We briefly recall the construction of the Brownian motion on Aff(ℝ), see e.g. McKean [**McK69**], Ibéro [**Ibé76**] or Rogers and Williams [**RW85**]. The Lie algebra $\mathfrak{A}(\mathrm{Aff}(\mathbb{R}))$ consists of the matrices of the form $\begin{bmatrix} x & y \\ 0 & 0 \end{bmatrix}$, $x, y \in \mathbb{R}$. The metric on this algebra (i.e. in each plane of the tangent bundle of Aff(ℝ)) has the form $ds^2 = dx^2 + dy^2$. The exponential mapping Exp from the algebra $\mathfrak{A}(\mathrm{Aff}(\mathbb{R}))$ to the group Aff(ℝ) then writes:

$$g = \mathrm{Exp}\left(\begin{bmatrix} x & y \\ 0 & 0 \end{bmatrix}\right) = \begin{bmatrix} a & b \\ 0 & 1 \end{bmatrix} = \sum_{k \geq 0} \frac{1}{k!} \begin{bmatrix} x & y \\ 0 & 0 \end{bmatrix}^k = \begin{bmatrix} e^x & e^x y \\ 0 & 1 \end{bmatrix}, \tag{2.1}$$

i.e. $x = \ln(a), y = be^{-x} = \frac{b}{a}$ so that

$$ds^2 = dx^2 + dy^2 = \frac{da^2 + db^2}{a^2}, \tag{2.2}$$

i.e. the Riemannian metric on Aff(ℝ) is given by the same formula as the hyperbolic metric on the Poincaré model of the Lobachevskii plane (i.e. upper half plane of ℂ):

$$\mathbb{C}_+ = \{b + ia, a > 0\}.$$

The ball of radius R in this metric has an exponentially growing volume, i.e. $Vol(B(R)) = 2\pi\big(\cosh(R) - 1\big)$ (see e.g. Gruet [**Gru96**]).

In Section 3 we will consider the symmetric random walk on the finitely generated subgroups $G^\varepsilon \subset G$. We consider the simplest subgroups with two generators:

$$g_1^\varepsilon = \begin{bmatrix} \exp(\varepsilon) & 0 \\ 0 & 1 \end{bmatrix}, \quad g_2^\varepsilon = \begin{bmatrix} 1 & \varepsilon \\ 0 & 1 \end{bmatrix}. \tag{2.3}$$

The number of different words of length n with the alphabet $\{g_1^{\varepsilon}, g_1^{-\varepsilon}, g_2^{\varepsilon}, g_2^{-\varepsilon}\}$ again grows exponentially with n from Milnor [**Mil68**] (non-niplotent or non-abelian solvable groups with finite number of generators have exponential growth).

The symmetric Brownian motion on G can be constructed as the exponential mapping in the Stratonovich sense of the Brownian motion (B_t^1, B_t^2), i.e. B^1, B^2 are two independent scalar Brownian motions, on $\mathfrak{A}(\mathrm{Aff}(\mathbb{R}))$:
(2.4)

$$g_t = \begin{bmatrix} a_t & b_t \\ 0 & 1 \end{bmatrix} = (\circ) \bigcap_{s=0}^{t} \begin{bmatrix} (1+dB_s^1) & dB_s^2 \\ 0 & 0 \end{bmatrix} = \begin{bmatrix} \exp(B_t^1) & \int_0^t \exp(B_s^1) dB_s^2 \\ 0 & 1 \end{bmatrix}.$$

The generator of $(a_t, b_t)_{t\geq 0}$ writes for all $\varphi \in C^2(\mathbb{R}_+\backslash\{0\} \times \mathbb{R}, \mathbb{R})$:

$$L\varphi(a,b) = \frac{1}{2}\Big(a^2(\partial_a^2 + \partial_b^2)\varphi + a\partial_a\varphi\Big)(a,b) =: \Delta_{\mathrm{Aff}(\mathbb{R})}\varphi(a,b), \tag{2.5}$$

where $\Delta_{\mathrm{Aff}(\mathbb{R})}$ stands for the Laplace-Beltrami operator on $\mathrm{Aff}(\mathbb{R})$. Observe that the diffusion matrix $a^2 I_2$ is indeed the inverse of the Riemannian metric tensor $a^{-2}I_2$.

To find the fundamental solution of the parabolic equation $\partial_t p = Lp$, i.e. the transition density of the Brownian motion on $\mathrm{Aff}(\mathbb{R})$, we will apply the Doob transform to the well known density of the Brownian diffusion on the hyperbolic space, see Karpelevich *et al.* [**KTS59**] and Gruet [**Gru96**] for multi-dimensional generalizations. We also refer to Bougerol [**Bou15**] for other applications of Doob transforms on algebraic structures.

PROPOSITION 2.1 (Transition Density of the Brownian Motion on the hyperbolic plane $\mathbb{H}^2$). *The density of the diffusion with generator*

$$\mathcal{L}\varphi(a,b) = \frac{1}{2}a^2\Delta\varphi(a,b)$$

w.r.t. the corresponding Riemann volume $\frac{dadb}{a^2}$ *is given by:*

$$p_{\mathbb{H}^2}(t,x,y) = \frac{\sqrt{2}\exp(-\frac{t}{8})}{(2\pi t)^{3/2}} \int_r^{+\infty} \frac{u\exp(-\frac{u^2}{2t})}{\sqrt{\cosh(u) - \cosh(r)}} du, \tag{2.6}$$

where $r = d_{\mathbb{H}^2}(x,y)$ *is the hyperbolic distance between* $x = (a_1, b_1), y = (a_2, b_2) \in \mathbb{H}^2$*, namely:*

$$d_{\mathbb{H}^2}(x,y) = \mathrm{arcosh}\left(1 + \frac{|x-y|^2}{2a_1 a_2}\right),$$

where $|x-y|^2 = |a_1 - a_2|^2 + |b_1 - b_2|^2$ *is the usual squared Euclidean distance in* $\mathbb{R}^2$.

Now we want to use the Doob transform. The following Proposition holds, see e.g. Pinsky [**Pin95**].

PROPOSITION 2.2 (Doob transform). *Let M be a Riemannian manifold with metric $ds^2 = g_{ij}dx^i dx^j$ and corresponding Laplace-Beltrami operator*

$$\Delta_M f(x) = \frac{1}{\sqrt{\det(g)}}\partial_{x_i}\Big(g^{ij}\sqrt{\det(g)}\partial_{x_j} f\Big)(x).$$

Let $p(t,x,y)$ be the fundamental solution of the heat equation $\partial_t p = \frac{1}{2}\Delta_M p = -\frac{1}{2}\Delta_M^* p$ and $\psi(x) > 0$ be the positive λ-harmonic function, i.e. it solves $\frac{1}{2}\Delta_M \psi = \lambda\psi$. Put

$$p_\lambda(t,x,y) = \exp(-\lambda t)\frac{p(t,x,y)}{\psi(x)}\psi(y).$$

Then, $p_\lambda(t,x,y)$ is the transition density of a new diffusion on M with generator:

$$L_\lambda f(x) = \frac{1}{2}\frac{\Delta_M(f\psi)(x)}{\psi(x)} - \lambda f(x) = \frac{1}{2}\Delta_M f(x) + \nabla_M f(x)\cdot\nabla\ln(\psi(x)).$$

Here ∇_M stands for the Riemannian gradient, and the densities are always intended to be w.r.t. the corresponding Riemannian volume $\sqrt{\det g}dy$.

Observe now that for $\psi(a,b) = a^{\frac{1}{2}}$, simple computations give that

$$\frac{1}{2}\Delta_{\mathbb{H}^2}\psi(a,b) = \frac{a^2}{2}\left(a^{\frac{1}{2}}\right)'' = -\frac{1}{8}\psi(a,b),\ \lambda = -\frac{1}{8}.$$

Combining Propositions 2.1 and 2.2 and the above expression for ψ, we derive that the density of the Brownian motion on Aff($\mathbb{R}$) can be expressed as the Doob-transform of the density of the Brownian motion on $\mathbb{H}^2$.

THEOREM 2.3 (Density of the Brownian motion in Aff($\mathbb{R}$) and Diagonal behavior in long time). *The density $p_{\mathrm{Aff}}(t,e,\cdot)$ of the Brownian motion in* Aff($\mathbb{R}$) *writes for all $(t,g,h) \in \mathbb{R}_+^* \times \mathrm{Aff}(\mathbb{R})^2$:*
(2.7)

$$p_{\mathrm{Aff}(\mathbb{R})}(t,g,h) = \exp\left(\frac{t}{8}\right)\frac{p_{\mathbb{H}^2}(t,g,h)}{\psi(g)}\psi(h) = \exp\left(\frac{t}{8}\right)\frac{p_{\mathbb{H}^2}(t,g,h)}{a^{\frac{1}{2}}}c^{\frac{1}{2}},\ g=(a,b), h=(c,d).$$

with $p_{\mathbb{H}^2}$ as in (2.6). *For $t\to+\infty$ one has for all $g \in$* Aff($\mathbb{R}$)*:*

$$p_{\mathrm{Aff}(\mathbb{R})}(t,g,g) \sim \frac{1}{(2\pi t)^{\frac{3}{2}}}\int_0^{+\infty}\frac{u}{\sinh(\frac{u}{2})}du = \sqrt{\frac{\pi}{2}}\frac{1}{t^{\frac{3}{2}}}.$$

The previous theorem has an important application in spectral theory (together with the remark that $p_{\mathrm{Aff}(\mathbb{R})}(t,g,g) \sim \frac{C}{t}$ as $t \to 0$, since $\dim(\mathrm{Aff}(\mathbb{R})) = 2$, see e.g. [**Mol75**]).

THEOREM 2.4. *Consider on* Aff($\mathbb{R}$) *the Schrödinger operator with non-positive fast decreasing potential $W(g)$:*

$$H = -\Delta_{\mathrm{Aff}(\mathbb{R})} + W(g), \Delta_{\mathrm{Aff}(\mathbb{R})} = \frac{1}{2}a^2\left(\partial_a^2 + \partial_b^2\right) + \frac{1}{2}a\partial_a,$$

and the spectral problem $H\psi = \lambda\psi$. Then, since operator H has at most a finite negative spectrum $\{\lambda_j \leq 0\}$, one has:

$$N_0(W) := \sharp\{j : \lambda_j \leq 0\} \leq$$
$$C_1\int_{g\in\mathrm{Aff}(\mathbb{R}):0\leq|W(g)|\leq 1}|W(g)|^{\frac{3}{4}}\sigma(dg) + C_2\int_{g\in\mathrm{Aff}(\mathbb{R}):|W(g)|>1}|W(g)|\sigma(dg),$$

where for $g = (a,b)$, $\sigma(dg) = \frac{dadb}{a^2}$ is the Riemannian volume element on Aff($\mathbb{R}$). *Also, the constants C_1, C_2 here are independent of the considered potential W and can be computed directly.*

The previous Theorem is a direct consequence of the work by Molchanov and Vainberg [**MV08**].

Eventually, we can also refer to Melzi [**Mel02**] for a global upper bound of the density of the Brownian motion on $\mathrm{Aff}(\mathbb{R})$. This work provides a tractable control for the diagonal and off-diagonal behavior of the heat-kernel in large time.

3. Approximation of diffusion by random walks and associated return probability estimates

In this section, we are interested in the approximation of the Brownian motion on $\mathrm{Aff}(\mathbb{R})$ by a discrete random walk. Let now ε be a given small parameter. The time step of our random walk $(x_n^\varepsilon)_{n\geq 0}$ will be ε^2 (with the usual parabolic scaling). In particular for a given time $t > 0$, it makes

$$n^\varepsilon(t) = \lfloor \frac{t}{\varepsilon^2} \rfloor \tag{3.1}$$

steps on the interval $[0, t]$. Set $x_0^\varepsilon = \begin{bmatrix} 1 & 0 \\ 0 & 1 \end{bmatrix}$, and for all $n \geq 1$:

$$x_{n+1}^\varepsilon = x_n^\varepsilon A_{\varepsilon,n+1},\ A_{\varepsilon,n+1} = \begin{bmatrix} \exp(\varepsilon X_{n+1}) & \varepsilon Y_{n+1} \\ 0 & 1 \end{bmatrix},$$

where the $(X_i)_{i\in\mathbb{N}^*}, (Y_i)_{i\in\mathbb{N}^*}$ are independent symmetric random variables, defined on some given probability space $(\Omega, \mathcal{A}, \mathbb{P})$, sharing the moment of the standard Gaussian law up to order two. Hence, the above dynamics rewrites at time n:

$$\begin{aligned} x_n^\varepsilon &:= \begin{bmatrix} a_n^\varepsilon & b_n^\varepsilon \\ 0 & 1 \end{bmatrix} = \begin{bmatrix} e^{\varepsilon \sum_{i=1}^n X_i} & \varepsilon\big(\sum_{i=1}^n Y_i \exp(\varepsilon \sum_{j=1}^{i-1} X_i)\big) \\ 0 & 1 \end{bmatrix} \\ &=: \begin{bmatrix} e^{\varepsilon S_n} & \varepsilon\big(\sum_{i=1}^n Y_i \exp(\varepsilon S_{i-1})\big) \\ 0 & 1 \end{bmatrix}, \end{aligned} \tag{3.2}$$

where we use the usual convention $\sum_{j=1}^0 = 0$. We will consider here mainly two cases.

- The Bernoulli Case: both $(X_i)_{i\in\mathbb{N}^*}, (Y_i)_{i\in\mathbb{N}^*}$ are independent sequences of independent Bernoulli random variables, i.e. $\mathbb{P}[X_1 = 1] = \mathbb{P}[X_1 = -1] = \mathbb{P}[Y_1 = 1] = \mathbb{P}[Y_1 = -1] = \frac{1}{2}$. In such case, it is easy to see that the random walk stays on the subgroup G^ε.*
- The mixed case: $(X_i)_{i\in\mathbb{N}^*}, (Y_i)_{i\in\mathbb{N}^*}$ are independent sequences. The $(X_i)_{i\in\mathbb{N}^*}$ are still Bernoulli random variables whereas the $(Y_i)_{i\in\mathbb{N}^*}$ have an absolutely continuous law.

For the rest of the section we focus on the Bernoulli case and the associated return probability estimates (see (1.3) and Theorem 3.1). The mixed case is developed in Section 4.3, since it can be handled rather directly from the proof of our main results in Section 4.2. In particular, we emphasize that for the mixed case, the density assumption for the $(Y_i)_{i\in\mathbb{N}^*}$ is sufficient to restore the LLT (see Theorem 4.8).

---

*Observe that this would as well be the case for any integer valued independent sequences $(X_i, Y_i)_{i\in\mathbb{N}^*}$ of independent random variables sharing the two first moments of the Gaussian law.

In the Bernoulli case, the idea is to express the non-diagonal element b_n^ε in (3.2) in terms of the *local times* $L(a,n)$ of the random walk $(S_k)_{k\geq 0}$ at level $a \in [\![M_n^-, M_n^+]\!]$[†], where

$$M_n^- := \min_{k\leq n} S_k \leq 0, \ M_n^+ := \max_{k\leq n} S_k \geq 0.$$

We also precisely define:

$$L(n,a) := \sharp\{k : S_k = a, \ 0 < k \leq n\}.$$

With these notations, we readily derive from the definition in (3.2) the following discrete *occupation time formula*:

$$b_n^\varepsilon = \varepsilon \sum_{a=M_n^-}^{M_n^+} \Big(\sum_{k\in[\![1,n]\!]:S_{k-1}=a} Y_k \Big) \exp(\varepsilon a). \tag{3.3}$$

The simplest (and yet very important) local theorem for x_n^ε concerns the asymptotic behaviour of the return probability $\pi_{2n} = \mathbb{P}_e[x_{2n}^\varepsilon = e] = \mathbb{P}[S_{2n} = 0, \sum_{k=1}^{2n} Y_k e^{\varepsilon S_{k-1}} = 0]$.

The exact asymptotic convergence rates of π_{2n} can be found in [**PSC02**] (see Theorem 3.11, i) therein). Precisely, the following result holds.

THEOREM 3.1 (Asymptotics of the return probabilities on the subgroup). *Assume that e^ε is transcendental. Then, there exists $c \geq 1$ s.t. for n large enough:*

$$c^{-1} n^{\frac{1}{3}} (\ln(n))^{\frac{2}{3}} \leq -\ln(\pi_{2n}) \leq c n^{\frac{1}{3}} (\ln(n))^{\frac{2}{3}}.$$

In the quoted article, the authors actually consider $\varepsilon = 1$, which readily gives the transcendence property. Actually, when e^ε is transcendental, the group generated by g_1^ε and g_2^ε defined in (2.3) is isomorphic to $\mathbb{Z}\wr\mathbb{Z}$, where g_1^ε corresponds to the *walk* generator on the base $\mathbb{Z}$ and g_2^ε to the *switch* generator in the lamp group $\mathbb{Z}$. Hence, Theorem 3.1 is again a direct consequence of Theorem 3.11, i) in [**PSC02**]. The bounds follow from some properties of the local time of the simple random walk on $\mathbb{Z}$. For the sake of completeness, we prefer to give below a slightly different proof of the lower bound of Theorem 3.1 which directly uses the transcendentality of e^ε. We also hope that our approach might extend to higher order solvable matrix groups for which the reduction to the random walk on the wreath product is less clear.

In our work, we are indeed interested in Donsker-Prokhorov type results (see Proposition 4.1 below), which will require the previous scaling of (3.1). This leads us to consider the previous transcendence condition. Namely, if e^ε is transcendent, and since $(S_i)_{i\in\mathbb{N}}$ is $\mathbb{Z}$ valued, we will have that:

$$\begin{aligned}\sum_{i=1}^{2n} Y_i \exp(\varepsilon S_{i-1}) &= \sum_{a\in[\![M_{2n}^-, M_{2n}^+]\!]} \sum_{k\in[\![1,2n]\!], S_{k-1}=a} Y_k \exp(\varepsilon a) \\ &= 0 \iff \forall a \in [\![M_{2n}^-, M_{2n}^+]\!], \quad \sum_{k\in[\![1,2n]\!], S_{k-1}=a} Y_k = 0.\end{aligned}$$

We now mention that, from the Lindemann-Weierstrass theorem, a sufficient condition for e^ε to be transcendental is that ε is algebraic, which for instance happens if $\varepsilon \in \mathbb{Q}$.

[†] from now on we denote by $[\![\cdot,\cdot]\!]$ intervals of integers.

We now provide a proof for this lower bound, which relies on stochastic analysis arguments associated with some controls for the local time of the simple random walk, see e.g. [**Rev05**].

PROPOSITION 3.2. *If e^{ε} is transcendental then there exists $c \geq 1$ s.t. for n large enough:*

$$\pi_{2n} \leq \exp(-c^{-1} n^{\frac{1}{3}} \ln(n)^{\frac{2}{3}}).$$

In particular, the proof emphasizes that the upper bound of the return probability does not depend on ε as soon as it is algebraic.

PROOF. The numbers $e^{k\varepsilon}$, $k \in \mathbb{Z}$ being rationally independent the probability π_{2n} rewrites:

$$\begin{aligned} \pi_{2n} = \mathbb{P}\Big[\cap_{a \in [\![M_{2n-1}^-, M_{2n-1}^+]\!]} & L(2n-1, a) \\ &= 0 \text{ Mod } 2, S_{2n} \\ &= 0, \forall a \in [\![M_{2n-1}^-, M_{2n-1}^+]\!] \sum_{k \in [\![1,n]\!] : S_{k-1} = a} Y_k = 0 \Big]. \end{aligned} \tag{3.4}$$

Set now, $A := \{\cap_{a \in [\![M_{2n-1}^-, M_{2n-1}^+]\!]} L(2n-1, a) = 0 \text{ Mod } 2, S_{2n} = 0\}$. We can thus write:

$$\pi_{2n} = \mathbb{E}\left[\prod_{a \in [\![M_{2n-1}^-, M_{2n-1}^+]\!]} \frac{\binom{L(2n-1,a)}{\frac{L(2n-1,a)}{2}}}{2^{L(2n-1,a)}} \mathbb{I}_A \right]. \tag{3.5}$$

Observe that, on the considered event A, for $a \in [\![M_{2n-1}^-, M_{2n-1}^+]\!]$, the local time $L(2n-1, a)$ is even. The contribution $\frac{\binom{L(2n-1,a)}{\frac{L(2n-1,a)}{2}}}{2^{L(2n-1,a)}}$ then corresponds to the probability that a symmetric Binomial law with parameter $L(2n-1, a)$ is equal to 0. This exactly describes the event $\sum_{k \in [\![1,n]\!] : S_{k-1} = a} Y_k = 0$.

Observe importantly that on A:

$$\frac{\binom{L(2n-1,a)}{\frac{L(2n-1,a)}{2}}}{2^{L(2n-1,a)}} \leq \frac{1}{2}.$$

Let us now localize w.r.t. the position of the minimum M_{2n-1}^- and maximum M_{2n-1}^+. Namely, we want to get rid of the *large deviations* for our current problem. Introduce the set $D_\alpha := \{M_{2n-1}^- \leq -\alpha\} \bigcup \{M_{2n-1}^+ \geq \alpha\}$. Observe that

$$\begin{aligned} T_{2n}^{D_\alpha} &:= \mathbb{E}\left[\prod_{a \in [\![M_{2n-1}^-, M_{2n-1}^+]\!]} \frac{\binom{L(2n-1,a)}{\frac{L(2n-1,a)}{2}}}{2^{L(2n-1,a)}} \mathbb{I}_{D_\alpha \cap A} \right] \leq \left(\frac{1}{2}\right)^{\alpha} 2\mathbb{P}[M_{2n-1}^+ \geq \alpha] \\ &\leq 4 \exp(-\alpha \ln 2) \exp(-\frac{\alpha^2}{4n}), \end{aligned}$$

using the Bernstein inequality for the last control. Now in order to equilibrate the contributions of these *large deviations* w.r.t the stated bound in Proposition 3.2 we want to solve the equation $\frac{\alpha^2}{n} + \alpha \ln 2 = n^{\frac{1}{3}} \ln(n)^{\frac{2}{3}}$. It is then easily checked that the positive root α_n of the equation is s.t. $\alpha_n \sim \frac{n^{\frac{1}{3}} \ln(n)^{\frac{2}{3}}}{\ln(2)} =: m_n$. It thus follows that there exists C_0^1 s.t. for n large enough:

$$T_{2n}^{D_{m_n}} \leq \exp(-C_0^1 m_n).$$

On the other hand, we can as well derive the required control provided the extremas are *small* with the previously emphasized threshold. Namely, introducing:

$$\begin{aligned}
T_{2n}^S &:= \mathbb{E}\left[\prod_{a \in [\![M_{2n-1}^-, M_{2n-1}^+]\!]} \frac{\binom{L(2n-1,a)}{\frac{L(2n-1,a)}{2}}}{2^{L(2n-1,a)}} \mathbb{I}_{|M_{2n-1}^-| \leq \frac{m_n}{\ln(n)}, |M_{2n-1}^+| \leq \frac{m_n}{\ln(n)}} \mathbb{I}_A\right] \\
&\leq \mathbb{P}\Big[|M_{2n-1}^-| \leq \frac{m_n}{\ln(n)}, M_{2n-1}^+ \leq \frac{m_n}{\ln(n)}, S_{2n=0}\Big] \\
&\leq \mathbb{P}\Big[\forall k \in [\![0, 2n]\!], \frac{S_k}{\sqrt{n}} \in [-\frac{m_n}{\sqrt{n}\ln(n)}, \frac{m_n}{\sqrt{n}\ln(n)}]\Big].
\end{aligned} \tag{3.6}$$

To control the last inequality we use the following important Lemma concerning *tube estimates* for the random walk:

Lemma 3.3 (Tube Estimates for the Random Walk). *There exists constants $c \leq 1, C \geq 1$ s.t.:*

$$\mathbb{P}[\forall k \in [\![1, 2n]\!], |S_k| \leq \frac{m_n}{\ln(n)}] \leq C \exp(-cn^{\frac{1}{3}} \ln(n)^{\frac{2}{3}}),$$

$$\sum_{a=-m_n}^{m_n} \mathbb{P}[L(2n-1, a) > c^{-1} n^{\frac{2}{3}} \ln(n)^{\frac{1}{3}}] \leq C \exp(-cn^{\frac{1}{3}} \ln(n)^{\frac{2}{3}}).$$

The above result can be viewed as a discrete analogue of the tube estimates for the Brownian motion that can be found in [**IW80**]. The proof is postponed to the end of the Section for the sake of clarity.

From Lemma 3.3 and (3.6) we get $T_{2n}^S \leq C \exp(-cn^{\frac{1}{3}} \ln(n)^{\frac{2}{3}})$. Thus, it suffices to restrict to the study of:

$$\begin{aligned}
T_{2n}^M \quad := \quad \mathbb{E}\Bigg[& \prod_{a \in [\![M_{2n-1}^-, M_{2n-1}^+]\!]} \frac{\binom{L(2n-1,a)}{\frac{L(2n-1,a)}{2}}}{2^{L(2n-1,a)}} \mathbb{I}_{|M_{2n-1}^-| \leq m_n, M_{2n-1}^+ \leq m_n} \\
& \times \Big(\mathbb{I}_{M_{2n-1}^+ > \frac{m_n}{\ln(n)}} + \mathbb{I}_{|M_{2n-1}^-| > \frac{m_n}{\ln(n)}}\Big) \mathbb{I}_A \Bigg].
\end{aligned}$$

Fix now a $\delta \in (0, 1)$ and introduce the random set:

$$A_\delta := \{a \in [\![M_{2n-1}^-, M_{2n-1}^+]\!] : L(2n-1, a) > n^\delta\}.$$

Let us now fix $c \in (0,1)$. If $\sharp A_\delta \geq c\frac{m_n}{\ln(n)}$, then:

$$\begin{aligned}
T_{2n}^{M,1} &:= \mathbb{E}\Bigg[\prod_{a=M_{2n-1}^-}^{M_{2n-1}^+} \frac{\begin{pmatrix} L(2n-1,a) \\ \frac{L(2n-1,a)}{2}\end{pmatrix}}{2^{L(2n-1,a)}} \mathbb{I}_{|M_{2n-1}^-|\leq m_n, M_{2n-1}^+ \leq m_n} \\
&\qquad \times \Big(\mathbb{I}_{M_{2n-1}^+ > \frac{m_n}{\ln(n)}} + \mathbb{I}_{|M_{2n-1}^-| > \frac{m_n}{\ln(n)}}\Big)\mathbb{I}_{\sharp A_\delta \geq c\frac{m_n}{\ln(n)}}\mathbb{I}_A\Bigg] \\
&\leq C\mathbb{E}[\prod_{a\in A_\delta} \frac{1}{L(2n-1,a)^{\frac12}} \mathbb{I}_{|M_{2n-1}^-|\leq m_n, M_{2n-1}^+ \leq m_n} \\
&\times \Big(\mathbb{I}_{M_{2n-1}^+ > \frac{m_n}{\ln(n)}} + \mathbb{I}_{|M_{2n-1}^-| > \frac{m_n}{\ln(n)}}\Big)\mathbb{I}_{\sharp A_\delta \geq c\frac{m_n}{\ln(n)}}\mathbb{I}_A] \\
&\leq C(\frac{1}{n^{\frac{\delta}{2}}})^{c\frac{m_n}{\ln(n)}} = C\exp(-\frac{\delta}{2}\ln(n) \times c\frac{m_n}{\ln(n)}) = C\exp(-\frac{\delta}{2}cm_n),
\end{aligned}$$

where on the event A_δ, we used the Stirling formula for the first inequality. It remains to handle:

$$\begin{aligned}
T_{2n}^{M,2} &:= \mathbb{E}\Bigg[\prod_{a\in [\![M_{2n-1}^-, M_{2n-1}^+]\!]} \frac{\begin{pmatrix} L(2n-1,a) \\ \frac{L(2n-1,a)}{2}\end{pmatrix}}{2^{L(2n-1,a)}} \mathbb{I}_{|M_{2n-1}^-|\leq m_n, M_{2n-1}^+ \leq m_n} \\
&\qquad \Big(\mathbb{I}_{M_{2n-1}^+ > \frac{m_n}{\ln(n)}} + \mathbb{I}_{|M_{2n-1}^-| > \frac{m_n}{\ln(n)}}\Big)\mathbb{I}_{\sharp A_\delta < c\frac{m_n}{\ln(n)}}\mathbb{I}_A\Bigg].
\end{aligned}$$

The first point to note is that, on the event $\{\sharp A_\delta < c\frac{m_n}{\ln(n)}\} \cap \{|M_{2n-1}^-| \leq m_n, |M_{2n-1}^+| \leq m_n\}$, necessarily the occupation measure of A_δ is *large*. Precisely, we have that defining:

$$A_\delta^C := \{a \in [\![-m_n, m_n]\!], a \notin A_\delta\},\ \sharp A_\delta^C \geq 2m_n - c\frac{m_n}{\ln(n)}.$$

On the other hand, the total *local time* generated by the points in A_δ^C is less than $2m_n n^\delta = 2n^{\frac13+\delta}\ln(n)^{\frac23} < n$, for $\delta \in (0, \frac23)$ and n large enough. Hence, the occupation time of A_δ is s.t.:

$$|\{i \in [\![1, 2n]\!] : S_i \in A_\delta\}| > n.$$

Since we also know that on the considered event $\{\sharp A_\delta < c\frac{m_n}{\ln(n)}\}$, we derive that there necessarily exists a level $a \in A_\delta$ s.t.

$$L(2n-1, a) > \frac{n}{c\frac{m_n}{\ln(n)}}.$$

We obtain:

$$
\begin{aligned}
T_{2n}^{M,2} &\le \mathbb{P}[|\{i \in [\![1,2n]\!] : S_i \in A_\delta\}| > n, \sharp A_\delta < c\frac{m_n}{\ln(n)}, |M_{2n-1}^-| \le m_n, M_{2n-1}^+ \le m_n] \\
&\le \mathbb{P}[\exists a \in A_\delta, L(2n-1,a) > c^{-1} n^{\frac{2}{3}} \ln(n)^{\frac{1}{3}}, \\
&\qquad \sharp A_\delta < c\frac{m_n}{\ln(n)}, |M_{2n-1}^-| \le m_n, M_{2n-1}^+ \le m_n] \\
&\le \sum_{a=-m_n}^{m_n} \mathbb{P}[L(2n-1,a) > c^{-1} n^{\frac{2}{3}} \ln(n)^{\frac{1}{3}}] \le C \exp(-c n^{\frac{1}{3}} \ln(n)^{\frac{2}{3}}),
\end{aligned}
$$

using again Lemma 3.3 for the last inequality. □

PROOF OF LEMMA 3.3 (TUBES FOR THE RANDOM WALK). Let us begin the proof observing that since,

$$
\begin{aligned}
\mathbb{P}[\forall k \in [1,2n]\!], |S_k| &\le \frac{m_n}{\ln(n)}] \\
&\le \mathbb{P}[\exists a \in [\![-\frac{m_n}{\ln(n)}, \frac{m_n}{\ln(n)}]\!], L(2n,a) \ge \frac{n}{\frac{m_n}{\ln(n)}}] \\
&\le \sum_{a=-\frac{m_n}{\ln(n)}}^{\frac{m_n}{\ln(n)}} \mathbb{P}[L(2n,a) \ge n^{\frac{2}{3}} \ln^{\frac{1}{3}}(n)],
\end{aligned}
$$

it suffices to prove the second statement of the Lemma. To this end, observe first that from Theorem 9.4 in Revesz [**Rev05**], we get for all $a > 0, k \in \mathbb{N}$:

$$
\mathbb{P}[L(2n,a) = k] = \begin{cases} \frac{1}{2^{2n-k+1}} \begin{pmatrix} 2n-k+1 \\ (2n+a)/2 \end{pmatrix}, \text{if } a \text{ is even,} \\ \frac{1}{2^{2n-k}} \begin{pmatrix} 2n-k \\ (2n+a-1)/2 \end{pmatrix}, \text{if } a \text{ is odd.} \end{cases} \tag{3.7}
$$

By symmetry we also derive that for $a < 0$, the above expression holds replacing a by $|a|$ (recall indeed that $L(2n,a) \overset{(\text{law})}{=} L(2n,-a)$). Eventually, for $a = 0$, Theorem 9.3 in [**Rev05**] yields:

$$
\mathbb{P}[L(2n,0) = k] = 2^{-2n+k} \begin{pmatrix} 2n-k \\ n \end{pmatrix}. \tag{3.8}
$$

Hence,

$$
\begin{aligned}
\mathcal{P}_{m_n} &:= \sum_{a=-m_n}^{m_n} \mathbb{P}[L(2n,a) > c^{-1} n^{\frac{2}{3}} \ln(n)^{\frac{1}{3}}] \\
&= \mathbb{P}[L(2n,0) > c^{-1} n^{\frac{2}{3}} \ln(n)^{\frac{1}{3}}] \\
&+ 2 \sum_{a=1}^{m_n} \mathbb{P}[L(2n,a) > c^{-1} n^{\frac{2}{3}} \ln(n)^{\frac{1}{3}}].
\end{aligned}
$$

Note as well from (3.7) that, in agreement with the intuition, $\mathbb{P}[L(2n,0) = k] > \mathbb{P}[L(2n,a) = k], a > 0, k \in \mathbb{N}$. We therefore derive:

$$
\mathcal{P}_{m_n} \le (1 + 2m_n) \mathbb{P}[L(2n,0) > c^{-1} n^{\frac{2}{3}} \ln(n)^{\frac{1}{3}}].
$$

Write now from (3.8):

$$\mathcal{P}_{m_n} \le (1+2m_n) \sum_{k=\lfloor c^{-1} n^{\frac{2}{3}} \ln(n)^{\frac{1}{3}} \rfloor}^{n} 2^{-2n+k} \binom{2n-k}{n}. \tag{3.9}$$

By the Stirling formula, we obtain that for $k \in [\![\lfloor c^{-1} n^{\frac{2}{3}} \ln(n)^{\frac{1}{3}} \rfloor, n-1]\!]$,

$$\begin{aligned}\mathbb{P}[L(2n,0)=k] &= 2^{-2n+k}\binom{2n-k}{n}\\ &\le \frac{e}{\pi\sqrt{2n}} \frac{\sqrt{n-\frac{k}{2}}}{\sqrt{n-k}} \exp\left((2n-k)\ln(1-\frac{k}{2n}) - (n-k)\ln(1-\frac{k}{n})\right).\end{aligned}$$

The contribution for $k=n$ gives $\mathbb{P}[L(2n,0)=k] = 2^{-n}$ and therefore a negligible term in the r.h.s. of (3.9). We will now split the summation in (3.9) according to $k \in [\![\lfloor c^{-1} n^{\frac{2}{3}} \ln(n)^{\frac{1}{3}} \rfloor, n^{1-\eta}]\!]$ and $k \in [\![n^{1-\eta}, n]\!]$ for $\eta > 0$ small enough to be specified later on. Observing that $\mathbb{P}[L(2n,0)=k]$ is a decreasing function of k we obtain:
(3.10)

$$\mathcal{P}_{m_n} \le (1+2m_n)\Bigg(\sum_{k=\lfloor c^{-1} n^{\frac{2}{3}} \ln(n)^{\frac{1}{3}} \rfloor}^{n^{1-\eta}} \mathbb{P}[L(2n,0)=k] + (n-n^{1-\eta})\mathbb{P}[L(2n,0)=n^{1-\eta}]\Bigg).$$

From (??) it can be deduced from usual computations that there exists $C>0$ s.t. uniformly on $k \in [\![\lfloor c^{-1} n^{\frac{2}{3}} \ln(n)^{\frac{1}{3}} \rfloor, n^{1-\eta}]\!]$, for n large enough:

$$\mathbb{P}[L(2n,0)=k] \le \frac{C}{\sqrt{n}} \exp\left(-\frac{k^2}{5n}\right).$$

Plugging this estimate in (3.10) yields:

$$\begin{aligned}\mathcal{P}_{m_n} &\le C(1+2m_n)\Bigg(\sum_{c^{-1} n^{\frac{1}{6}} \ln(n)^{\frac{1}{3}} < \frac{k}{\sqrt{n}} \le n^{1/2-\eta}} \frac{1}{\sqrt{2\pi n}} \exp\left(-\frac{1}{5}\Big(\frac{k}{\sqrt{n}}\Big)^2\right)\\ &\qquad + (n^{\frac{1}{2}} - n^{\frac{1}{2}-\eta}) \exp\left(-\frac{n^{1-2\eta}}{5}\right)\Bigg)\\ &\le C(1+2m_n)\left(\frac{1}{\sqrt{2\pi}} \int_{c^{-1} n^{\frac{1}{6}} \ln(n)^{\frac{1}{3}}}^{+\infty} \exp(-\frac{x^2}{5})dx + \exp\left(-\frac{n^{1-2\eta}}{6}\right)\right)\\ &\le C(1+2m_n)\left(\exp(-c^{-1} n^{\frac{1}{3}} \ln(n)^{\frac{2}{3}}) + \exp\left(-\frac{n^{1-2\eta}}{6}\right)\right)\\ &\le C\exp(-c^{-1} n^{\frac{1}{3}} \ln(n)^{\frac{2}{3}}),\end{aligned}$$

taking $\eta \in (0, \frac{1}{3})$ and up to modifications of C, c for the last inequality. This completes the proof. □

4. Quasi-Local Theorems

We first mention that the integral theorem (which is an obvious corollary of the functional Donsker-Prokhorov Central Limit Theorem (CLT) for the random walks) of course applies. Namely, we have the following result.

PROPOSITION 4.1 (Donsker-Prokhorov approximation). *Fix $t > 0$. If $\varepsilon \to 0$, $n^\varepsilon(t) := \lfloor \frac{t}{\varepsilon^2} \rfloor \to +\infty$, then*

$$(a^\varepsilon_{\lfloor sn^\varepsilon(t)\rfloor}, b^\varepsilon_{\lfloor sn_\varepsilon(t)\rfloor})_{s\in[0,1]} \underset{\varepsilon\to 0}{\overset{(\text{law})}{\longrightarrow}} (a(st), b(st))_{s\in[0,1]},$$

where a and b are defined in (2.4).

On the other hand, we are going to prove that some *quasi*-local Theorems as well hold. By *quasi*-local Theorem, we mean here that we consider a suitable renormalization of a neighborhood of the origin. Our main result in that direction is the following Theorem.

THEOREM 4.2. *Let ϕ be a smooth test function s.t. its Fourier transform is compactly supported in $[-1, 1]$ and s.t. $\int_{\mathbb{R}} \phi(x)dx = 1$. Denote, for a given $\delta > 0$, by $\phi_\delta(x) := \frac{1}{\delta}\phi(\frac{x}{\delta})$ its rescaling. Fix $t > 0$, possibly large, and define for $n \in 2\mathbb{N}$, $\varepsilon_n = \left(\frac{t}{n}\right)^{\frac{1}{2}}$. Then, for $\delta_n := t^{\frac{1}{2}} n^{-\frac{1}{2}+\gamma}, \gamma \in (0, \frac{1}{2})$, we have:*

$$\mathbb{E}\left[\mathbb{I}_{S_n=0}\, \phi_{\delta_n}\Big(\varepsilon_n \sum_{j=1}^{n} Y_j \exp(\varepsilon_n S_{j-1})\Big)\right] \sim_n \frac{2\varepsilon_n}{t^{\frac{1}{2}}\sqrt{2\pi}} \cdot p_2(t, 0). \tag{4.1}$$

Here, we denote for $t > 0$ by $p_2(t, \cdot)$ the density of the random variable $\tilde{b}_t := \int_0^t e^{\tilde{B}^1_s} dB^2_s$ where $\left(\tilde{B}^1_s\right)_{s\in[0,t]}$ is a usual Brownian Bridge independent of the Brownian motion B^2. The subscript 2 in $p_2(t, \cdot)$, is here to recall the considered random variable is associated with the second component of the Brownian motion on the group.

Also,

$$p_2(t, 0) = \mathbb{E}\Big[\frac{1}{\sqrt{2\pi \int_0^t e^{2\tilde{B}^1_s} ds}}\Big] \sim_{t\to+\infty} \frac{\pi}{t},\quad \frac{1}{\sqrt{2\pi t}} p_2(t, 0) = p_{\mathrm{Aff}(\mathbb{R})}(t, e, e). \tag{4.2}$$

Hence, we find the expected asymptotics in large time. We have a normalization in ε_n and not in ε_n^3 in (4.1), because we have already normalized our approximation of the stochastic integral in our scheme (3.2). We also specify that the threshold δ_n has the above form, which equivalently rewrites $\frac{\varepsilon_n}{\delta_n} = n^{-\gamma}$, in order that some remainder terms in the analysis can be neglected w.r.t. the intrinsic scaling of the limit theorem in $t^{-3/2}$. The previous condition equivalently expresses that the ratio $\frac{\varepsilon_n}{\delta_n}$ between the *time step* ε_n and the *window size* δ_n for the approximating Dirac mass is negligible in n (window bigger than time step). We refer to the proof in Section 4.2 below for details (see in particular equation (4.20)).

To illustrate the phenomenon that appears on Aff($\mathbb{R}$), i.e. the tremendous different rates between the pointwise return probabilities, and the *quasi*-local Theorem, we consider a rather simple model which already enjoys such properties. Basically, this dichotomy emphasizes that, the discrete subgroups are somehow *very* dense, in the sense that they allow to have the expected convergence rates towards the densities of the limiting objects when integrated on a suitable neighborhood.

4.1. Quasi-local CLT: the toy model. We discuss in this section some points related to the local CLT on a dense subgroup G_ε of a Lie group G in the simplest possible case, taking $G = \mathbb{R}$, $G_1 = \{x : x = \sum_{i=1}^N n_i\alpha_i\}$ (or more generally $G_\varepsilon = \{x : x = \varepsilon \sum_{i=1}^N d_i\alpha_i\}$, $\varepsilon > 0$). Here, $N \in \mathbb{N}$ is a fixed given integer, $\alpha = (\alpha_1, \cdots, \alpha_N)$ is s.t. the $\{\alpha_i, i \in [\![1, N]\!]\}$ are rationally independent real numbers and $d=(d_1,\cdots,d_N)\in\mathbb{Z}^N$ encodes the coordinates/displacements associated with the entries of α.

The subgroup G_1 is not only dense in $\mathbb{R}$ but is also in some sense *locally uniformly distributed*. This can for instance be seen from Herman Weyl's classical result (see e.g. [**SS03**]). Consider for a fixed non negative integer L, the sequence

$$\tilde{x}_d=\Big(\sum_{i=1}^N\alpha_i d_i\Big)\ \mathrm{Mod}\ L=\langle\alpha,d\rangle\ \mathrm{Mod}\ L,$$

where here the notation Mod L stands for the remainder term of the division by L. Then, for an arbitrary continuous and L periodic function f we have:

$$\lim_{M\to+\infty}\frac{\sum_{d\in\mathbb{Z}^N:|d|\le M}f(\tilde{x}_d)}{\sharp\{d\in\mathbb{Z}^N:|d|\le M\}}=\frac{1}{L}\int_0^L f(x)dx, \tag{4.3}$$

where $|\cdot|$ stands here for the Euclidean norm of $\mathbb{R}^N$.

Consider now the symmetric random walk $(x_n)_{n\in\mathbb{N}}$ on $\mathbb{R}$, s.t. $x_0=0$, $x_n=\sum_{j=1}^n u_j$ where the $(u_j)_{j\in\mathbb{N}^*}$ are i.i.d. real-valued discrete random variables with law:

$$u_1\overset{(\mathrm{law})}{=}p_0\delta_0+\frac12\sum_{i=1}^N p_i(\delta_{\alpha_i}+\delta_{-\alpha_i}),\ \forall i\in[\![1,N]\!],0<p_i<1,\ \sum_{i=0}^N p_i=1. \tag{4.4}$$

We can as well consider the auxiliary random walk $(X_n)_{n\in\mathbb{N}}$ on $\mathbb{R}^N$ s.t. $X_0=0$, $X_n=\sum_{j=1}^n U_j$ where the $(U_j)_{j\in\mathbb{N}^*}$ are i.i.d. $\mathbb{R}^N$-valued discrete random variables with law:

$$U_1\overset{(\mathrm{law})}{=}p_0\delta_{0_{\mathbb{R}^N}}+\frac12\sum_{i=1}^N p_i(\delta_{\alpha_i e_i}+\delta_{-\alpha_i e_i}),\ \forall i\in[\![1,N]\!],0<p_i<1,\ \sum_{i=0}^N p_i=1.$$

In the above expression the $(e_i)_{i\in[\![1,N]\!]}$ denote the canonical basis vectors of $\mathbb{R}^N$.

Observe that the relation between the random variables $(u_j)_{j\in\mathbb{N}^*}$ and $(U_j)_{j\in\mathbb{N}^*}$, and therefore between x and X is summarized as follows:

$$\forall j\in\mathbb{N}^*,\ u_j=\langle U_j,\sum_{k=1}^N e_k\rangle=\langle U_j,\mathbf{1}\rangle,\ x_n=\langle X_n,\sum_{k=1}^N e_k\rangle=\langle X_n,\mathbf{1}\rangle, \tag{4.5}$$

where $\mathbf{1}:=\sum_{k=1}^N e_k=(1,\cdots,1)^*$.

Introduce now for notational convenience:

$$\mathbb{P}[x_n=0]=r_n,$$

i.e. r_n denotes the *return* probability to 0 at time n. We want to emphasize the following fact. Even though, from the standard CLT:

$$\frac{x_n}{\sqrt{n}}\overset{(\mathrm{law})}{\underset{n}{\longrightarrow}}\mathcal{N}(0,\sigma^2),\ \sigma^2=\mathbb{E}[u_1^2]=\sum_{i=1}^N p_i\alpha_i^2, \tag{4.6}$$

we **do not have** $r_n\sim_n\frac{c}{\sqrt{n}}$ but instead $r_n\sim_n\frac{c}{n^{N/2}}$. The result can be intuitively justified from the fact that from the rational independence of the $\{\alpha_i\}_{i\in[\![1,N]\!]}$,

$$r_n=\mathbb{P}[x_n=0]=\mathbb{P}[X_n=0_{\mathbb{R}^N}]. \tag{4.7}$$

For the latter event, this means that in each direction the number of positive and negative transitions are the same, and the asymptotics for this return probability corresponds to the product of the return probabilities in each direction. This fact can be formalized with the following proposition.

PROPOSITION 4.3 (Asymptotics for the return probability). *As $n \to +\infty$, the following result holds:*

- *If $p_0 > 0$, then:*

$$r_n = \mathbb{P}[x_n = 0] \sim_n \frac{C(p)}{n^{\frac{N}{2}}},\ C(p) := \prod_{i=1}^{N} \frac{1}{\sqrt{2\pi p_i}}.$$

- *If $p_0 = 0$, then: $r_n = 0$ if n is odd and for n even:*

$$r_n = \mathbb{P}[x_n = 0] \sim_n \frac{2C(p)}{n^{\frac{N}{2}}}.$$

We point out that, since the underlying random walk X_n is actually a random walk on $\mathbb{Z}^N$ with steps distributed on the standard generators, the results of Proposition 4.3 can be directly derived from the *standard* local limit theorem, see e.g. Lawler and Limic [**LL10**]. We anyhow provide below a proof which together with the deviation result of Proposition 4.4 gives a very simple *one dimensional* analytic proof of a more general quasi-local theorem and also emphasizes the point we want to stress: namely, for the considered scalar walk the dramatic dichotomy between the behavior of the return probabilities and the integration on a neighborhood of Theorem 4.5 below.

PROOF. Observe that X_n is lattice valued. For a given $n \in \mathbb{N}$, defining $\mathscr{L}_n := \{(\xi_1, \cdots, \xi_N),\ \forall i \in [\![1, N]\!], \xi_i \in \{-n\alpha_i, \cdots, n\alpha_i\}\}$, we have $\mathbb{P}[X_n \in \mathscr{L}_n] = 1$. Actually $\mathrm{supp}(X_n) \subset \mathscr{L}_n$, where the inclusion is strict. Write then for all $t \in \mathbb{R}^N$:

$$\mathbb{E}[\exp(i\langle t, X_n\rangle)] = \sum_{\xi \in \mathscr{L}_n} \mathbb{P}[X_n = \xi] \exp(i\langle t, \xi\rangle). \tag{4.8}$$

Introducing the *rescaled torus* $\mathbb{T}_N^\alpha := [-\frac{\pi}{\alpha_i}, \frac{\pi}{\alpha_i}]$, we get that for all

$$(\xi, \zeta) \in \mathscr{L}_n, \frac{1}{|\mathbb{T}_N^\alpha|} \int_{\mathbb{T}_N^\alpha} \exp(-i\langle t, \xi\rangle) \exp(+i\langle t, \zeta\rangle) dt = \delta_{\xi,\zeta}.$$

Hence, for any $\xi_0 \in \mathrm{supp}(X_n)$:

$$\begin{aligned} \mathbb{P}[X_n = \xi_0] &= \frac{1}{|\mathbb{T}_N^\alpha|} \int_{\mathbb{T}_N^\alpha} \exp(-i\langle t, \xi_0\rangle) \mathbb{E}[\exp(i\langle t, X_n\rangle)] dt \\ &= \frac{\prod_{j=1}^N \alpha_j}{(2\pi)^N} \int_{\mathbb{T}_N^\alpha} \exp(-i\langle t, \xi_0\rangle) \varphi^n(t) dt, \end{aligned} \tag{4.9}$$

where $\varphi(t) := \mathbb{E}[\exp(i\langle t, U_1\rangle)] = p_0 + \sum_{j=1}^N p_j \cos(t_j\alpha_j) = 1 + \sum_{j=1}^N p_j\big(\cos(t_j\alpha_j) - 1\big) = 1 - 2\sum_{j=1}^N p_j \sin^2\left(\frac{t_j\alpha_j}{2}\right)$.

Recalling (4.7), we thus readily get from the inversion formula (4.9) taking $\xi_0 = 0$, and changing variable to $s_j = \alpha_j t_j,\ j \in \{1, \cdots, N\}$

$$\begin{aligned} r_n &= \mathbb{P}[X_n = 0_{\mathbb{R}^N}] \\ &= \frac{1}{(2\pi)^N} \int_{\mathbb{T}_N} \varphi^n\Big(\frac{s_1}{\alpha_1}, \cdots, \frac{s_N}{\alpha_N}\Big) ds \\ &= \frac{1}{(2\pi)^N} \int_{\mathbb{T}_N} \Big(1 - 2\sum_{j=1}^N p_j \sin^2\Big(\frac{s_j}{2}\Big)\Big)^n ds, \end{aligned}$$

where $\mathbb{T}_N := [-\pi, \pi]^N$. For *small values* of $|s|$, and recalling that X_n has zero third moments, we then get that:

$$\varphi\Big(\frac{s_1}{\alpha_1}, \cdots, \frac{s_N}{\alpha_N}\Big) = 1 + \sum_{j=1}^N p_j\Big(-\frac{s_j^2}{2} + O(s_j^4)\Big) = \exp\Big(-\frac{1}{2}\sum_{j=1}^N p_j s_j^2 + O(|s|^4)\Big). \tag{4.10}$$

Set $\delta_n := c\left(\frac{\ln(n)}{n}\right)^{1/2},\ c > \left(\frac{N}{\min_{j\in[\![1,N]\!]} p_j}\right)^{1/2}$. We now introduce $B_N(\delta_n) := \{s \in \mathbb{T}_N : |s|_\infty \le \delta_n\}$ (ball of radius δ_n around the origin) and $C_N(\delta_n) := \{s \in \mathbb{T}_N : \forall j \in [\![1, N]\!],\ s_j \in [-\pi, -\pi + \delta_n] \cup [\pi - \delta_n, \pi]\}$ (corners of radius δ_n of the torus $\mathbb{T}_N$). Set $M_N(\delta_n) := B_N(\delta_n) \cup C_N(\delta_n)$. Observe that for $s \in \mathbb{T}_N \backslash M_N(\delta_n)$, we have either:

(a) $\exists j_0 \in [\![1, N]\!],\ \cos(s_{j_0}) - 1 = -2\sin^2(\frac{s_{j_0}}{2}) \in [-2 + \frac{\delta_n^2}{2} + o(\delta_n^2), -\frac{\delta_n^2}{2} + o(\delta_n^2)]$.
(b) $K_S := \{j \in [\![1, N]\!] : |s_j| \le \delta_n\}$ and $K_L := \{j \in [\![1, N]\!] : (\pi - |s_j|) \le \delta_n\}$ are non empty.

In case (a), we readily get $|1 - 2\sum_{j=1}^N p_j \sin^2\left(\frac{s_j}{2}\right)| \le \Big(1 - p_{j_0}\frac{\delta_n^2}{2} + o(\delta_n^2)\Big)$. In case (b), we derive:

$$|1 - 2\sum_{j=1}^N p_j \sin^2\Big(\frac{s_j}{2}\Big)| \le |1 - 2\sum_{k\in K_L} p_k| + \frac{\delta_n^2}{2} + o(\delta_n^2) := c_{L,S}(n) \le 1 - \frac{1}{2}\min_{j\in[\![1,N]\!]} p_j,$$

for n large enough. We can therefore rewrite:

$$\begin{aligned} r_n &= \frac{\mathbf{1}}{(2\pi)^N} \int_{M_N(\delta_n)} \Big(1 - 2\sum_{j=1}^N p_j \sin^2\Big(\frac{s_j}{2}\Big)\Big)^n ds + R_N^n, \\ |R_N^n| &\le C \int_{\mathbb{T}_N\backslash M_N(\delta_n)} \Big\{\Big(1 - p_{j_0}\frac{\delta_n^2}{2} + o(\delta_n^2)\Big)^n + c_{L,S}(n)^n\Big\} ds \\ &\le Cn^{-\frac{p_{j_0}c^2}{2}} = o(n^{-N/2}). \end{aligned} \tag{4.11}$$

Let us discuss now the contribution associated with $C_N(\delta_n)$. For $s \in C_N(\delta_n)$, one has for all $j \in [\![1, N]\!]$:

$$-2\sin^2\Big(\frac{s_j}{2}\Big) = -2\Big(1 - \Big(\frac{\pi - |s_j|}{2}\Big)^2\Big) + O\big((\pi - |s_j|)^4\big),$$

so that $1 - 2\sum_{i=1}^N p_j \sin^2\left(\frac{s_j}{2}\right) = -1 + 2p_0 + \sum_{j=1}^N p_j \frac{(\pi - |s_j|)^2}{2} + O\big((\pi - |s_j|)^4\big)$. Hence,

- if $p_0 \neq 0$, we thus readily get $\frac{\mathbf{1}}{(2\pi)^N}\int_{C_N(\delta_n)} \Big(1 - 2\sum_{j=1}^N p_j \sin^2\left(\frac{s_j}{2}\right)\Big)^n ds = o(n^{-N/2})$.

- if $p_0 = 0$, by symmetry, we get $r_n = 0$ if n is odd and

$$\frac{1}{(2\pi)^N}\int_{M_N(\delta_n)}\Big(1-2\sum_{j=1}^N p_j\sin^2\Big(\frac{s_j}{2}\Big)\Big)^n ds$$
$$= \frac{2}{(2\pi)^N}\int_{B_N(\delta_n)}\Big(1-2\sum_{j=1}^N p_j\sin^2\Big(\frac{s_j}{2}\Big)\Big)^n ds,$$

if n is even. Recall now from (4.10) that:

$$\frac{1}{(2\pi)^N}\int_{B_N(\delta_n)}\Big(1-2\sum_{j=1}^N p_j\sin^2\Big(\frac{s_j}{2}\Big)\Big)^n ds$$
$$= \frac{1}{(2\pi)^N}\int_{B_N(\delta_n)}\exp\Big(-n\Big\{\sum_{j=1}^N p_j\frac{s_j^2}{2}+O(|s|^4)\Big\}\Big)ds$$
$$= \frac{1}{(2\pi n)^{\frac{N}{2}}\prod_{j=1}^N\sqrt{p_j}}\int_{\prod_{j=1}^N\left\{|\tilde{s}_j|\le \ln(n)^{1/2}p_j^{1/2}\right\}}\exp\Big(-\frac{1}{2}\sum_{j=1}^N\tilde{s}_j^2+O\big(\frac{\ln(n)^2}{n}\big)\Big)\frac{d\tilde{s}}{(2\pi)^{\frac{N}{2}}}$$
$$\sim_n \frac{1}{n^{\frac{N}{2}}}\prod_{i=1}^N\frac{1}{\sqrt{2\pi p_i}} = \frac{C(p)}{n^{\frac{N}{2}}}.$$

This gives the stated result.

We can as well refer more generally to the proof of the classical local CLT (see e.g. [**Pet05**], Chapter 5 in [**BR76**] for the multidimensional case or again Lawler and Limic [**LL10**]).

Observe that the asymptotic of the return probability r_n does not depend on the rationally independent numbers $(\alpha_j)_{j\in[\![1,N]\!]}$ chosen. We simply used the fact that, to return to 0, we must have over the considered time interval, for all $j \in [\![1,N]\!]$, the same numbers of random variables taking the values $-\alpha_j e_j$ and $\alpha_j e_j$. □

Hence, the bigger N, the smaller the *exact* return probability. Similarly, from (4.9) we can extend the previous proposition with the following result.

PROPOSITION 4.4 (Deviation bounds for the LLT). *Let $n \to +\infty$ and $y \in \mathbb{R}^N \cap \operatorname{supp}(X_n)$ be s.t. its Euclidean norm $|y| \le n^{\frac{3}{4}-\gamma}$, $\gamma > 0$ (which is meant to be small). Then, for $p_0 > 0$, recalling as well that $X_0 = 0$, we obtain:*

$$\mathbb{P}[X_n = y] \sim_n \prod_{j=1}^N\left(\frac{\exp(-\frac{y_j^2}{2\alpha_j^2 p_j n})}{(2\pi p_j n)^{\frac{1}{2}}}\right).$$

PROOF. We indicate that starting from (4.9), proceeding as in the previous proof of Proposition 4.3 and considering a localization with respect to a ball of radius $\delta_n = n^{-(1/4+\gamma/2)}$, we derive:

(4.12)
$$\mathbb{P}[X_n = y] = \frac{1}{(2\pi)^N}\int_{B(\delta_n)} \exp\Big(-i\Big\langle\Big(\frac{y_1}{\alpha_1},\cdots,\frac{y_N}{\alpha_N}\Big),s\Big\rangle\Big) \times \exp\Big(-n\big(\frac{1}{2}\sum_{j=1}^N p_j s_j^2 + O(|s|^4)\big)\Big)ds + R_N^n$$

(4.13)
$$= \frac{1}{(2\pi n)^{\frac{N}{2}}\prod_{j=1}^N \sqrt{p_j}}\int_{\prod_{j=1}^N\{|\tilde{s}_j|\le n^{1/4-\gamma/2}p_j^{1/2}\}} \times \exp\Big(-i\Big\langle\Big(\frac{y_1}{\alpha_1(p_1 n)^{1/2}},\cdots,\frac{y_N}{\alpha_N(p_N n)^{1/2}}\Big),\tilde{s}\Big\rangle\Big) \times \exp\Big(-\frac{1}{2}\sum_{j=1}^N \tilde{s}_j^2 + O(\frac{|\tilde{s}|^4}{n})\Big)\frac{d\tilde{s}}{(2\pi)^{N/2}} + R_N^n,$$

where, as in (4.11),
(4.14)
$$|R_N^n| \le C\int_{\mathbb{T}_N\setminus M_N(\delta_n)}\Big\{\Big(1-p_{j_0}\frac{\delta_n^2}{2}+o(\delta_n^2)\Big)^n + c_{L,S}(n)^n\Big\}ds \le C\exp(-cn^{1/2-\gamma}),$$

using the current choice of δ_n for the last inequality. Hence,

(4.15)
$$\mathbb{P}[X_n = y] \sim_n \frac{1}{(2\pi n)^{\frac{N}{2}}\prod_{j=1}^N \sqrt{p_j}}\int_{\prod_{j=1}^N\{|\tilde{s}_j|\le n^{1/4-\gamma/2}p_j^{1/2}\}} \times \exp\Big(-i\Big\langle\Big(\frac{y_1}{\alpha_1(p_1 n)^{1/2}},\cdots,\frac{y_N}{\alpha_N(p_N n)^{1/2}}\Big),\tilde{s}\Big\rangle\Big) \times \exp\Big(-\frac{1}{2}\sum_{j=1}^N \tilde{s}_j^2\Big)\frac{d\tilde{s}}{(2\pi)^{N/2}} + R_N^n =: \mathcal{P}_n(y) + R_N^n.$$

Write then,

$$\mathcal{P}_n(y) = \frac{1}{(2\pi n)^{\frac{N}{2}}\prod_{j=1}^N \sqrt{p_j}}\prod_{j=1}^N\int_{\mathbb{R}}\exp\Big(-i\frac{y_j}{\alpha_j(p_j n)^{1/2}}s\Big)\exp(-\frac{s^2}{2})ds + \mathcal{R}_N^n$$

(4.16)
$$= \frac{1}{(2\pi n)^{\frac{N}{2}}\prod_{j=1}^N \sqrt{p_j}}\exp\left(-\sum_{j=1}^N\frac{y_j^2}{2\alpha_j^2 p_j n}\right) + \mathcal{R}_N^n,$$

$$|\mathcal{R}_N^n| \le \frac{C}{n^{\frac{N}{2}}}\int_{(\prod_{j=1}^N\{|\tilde{s}_j|\le n^{1/4-\gamma/2}p_j^{1/2}\})^C}\exp(-\frac{|\tilde{s}|^2}{2})d\tilde{s} \le \frac{C}{n^{\frac{N}{2}}}\exp(-cn^{1/2-\gamma}).$$

On the considered range set for y, i.e. $|y| \le n^{3/4-\gamma}$, since $\exp(-\sum_{j=1}^N\frac{y_j^2}{2\alpha_j^2 p_j n}) \ge \exp(-c_0 n^{1/2-2\gamma})$ and $|\mathcal{R}_N^n| \le C\exp(-cn^{1/2-\gamma})$, the term $\mathcal{R}_N^n$ can indeed be seen as a global remainder uniformly in y. Equations (4.15) and (4.17), (4.14) then yield the result. Observe as well that for $y \in \mathrm{supp}(X_n)$, $\Big(\frac{y_1}{\alpha_1},\cdots,\frac{y_N}{\alpha_N}\Big) \in \mathbb{Z}^N$. □

Again those deviation results can be deduced from the proof of the more complex Theorem 2.3.11 in [**LL10**]. We keep the proof for the sake of completeness and its simplicity.

Observe now that from the previous definition of x_n, for any $\Gamma \subset \mathbb{R}$,

$$\mathbb{P}[x_n \in \Gamma] = \mathbb{P}[\langle X_n, \mathbf{1}\rangle \in \Gamma] = \sum_{y\in\mathbb{Z}^N, \langle y,\alpha\rangle\in\Gamma} \mathbb{P}[X_n = \sum_{i=1}^N \alpha_i y_i e_i]. \tag{4.17}$$

From equation (4.17) in Proposition 4.4, we derive the following theorem.

THEOREM 4.5. *For a given $\gamma \in (0, \frac{1}{2})$, and a positive sequence $\delta_n \to_n 0$ and s.t. $\delta_n \geq n^{-(\frac{1}{2}-\gamma)}$, we have for $p_0 > 0$:*

$$\mathbb{P}[x_{2n} \in (-\delta_n^{-1}, \delta_n^{-1})] \sim_n 2\delta_n^{-1} \frac{1}{\sqrt{2\pi(2n)}\sigma},$$

where as in the usual CLT stated in (4.6), $\sigma^2 = \sum_{i=1}^N p_i \alpha_i^2$.

From Proposition 4.3 and Theorem 4.5, we precisely see that, the integrated probability gives the expected usual rate in $n^{-1/2}$. Actually, this is precisely due to the last part of Proposition 4.4, we integrate in a neighborhood of a hyperplane of $\mathbb{R}^N$, whereas the pointwise return probabilities might have arbitrarily polynomial decay in function of the chosen N. We will show in the next subsection a similar behavior for our random walk on Aff($\mathbb{R}$).

REMARK 4.1 (Alternative formulation of Theorem 4.5). Note that, we can as well provide an upper bound of Theorem 4.5 not only around 0 but also for points a belonging to intervals whose size can as well go to infinity with n. Namely, for $\delta_n \to_n 0$ s.t. $\delta_n n^{\frac{1}{2}} \to_n +\infty$ and $\frac{|a|}{\ln(\delta_n^{\frac{1}{2}} n^{\frac{1}{2}})} \to_n 0$, one has:

$$\mathbb{P}[\frac{x_{2n}}{\sigma\sqrt{2n}} \in (a, a+\delta_n)] \sim_n \delta_n \frac{1}{\sqrt{2\pi}} \exp\left(-\frac{a^2}{2}\right). \tag{4.18}$$

Defining for $x \in \mathbb{R}$, $F_n(x) := \mathbb{P}[\frac{x_{2n}}{\sigma\sqrt{2n}} \leq x]$ and $F(x) := \frac{1}{\sqrt{2\pi}} \int_{-\infty}^x \exp(-\frac{y^2}{2})dy$, write from the Berry-Essen theorem:

$$\begin{aligned}
\mathbb{P}[\frac{x_{2n}}{\sigma\sqrt{2n}} \in (a, a+\delta_n)] &= F_n(a+\delta_n) - F_n(a) = \big(F_n(a+\delta_n) - F(a+\delta_n)\big) \\
&\quad +\big(F(a+\delta_n) - F(a)\big) \\
&\quad +\big(F(a) - F_n(a)\big) =: F(a+\delta_n) - F(a) + R_n, \\
|R_n| &\leq \frac{2C\mu_3}{\sigma^3\sqrt{2n}}, \mu_3 = \mathbb{E}[|u_1|^3],
\end{aligned}$$

with u_1 defined in (4.4). On the considered ranges for δ_n and a, R_n is indeed a remainder and (4.18) readily follows from the above equation and a first order Taylor expansion.

4.2. Proof of Theorem 4.2. We first need the following auxiliary lemma concerning the maximum of the conditioned random walk.

LEMMA 4.6 (Maximum and Minimum of the conditioned random walk). *Let $n \geq 0$ be given and consider the conditioned random walk $(\tilde{S}_j)_{j\in[\![0,n]\!]}$, $\tilde{S}_j = \sum_{i=1}^j X_i$ s.t. $\tilde{S}_0 = \tilde{S}_n = 0$. We recall here that $(X_i)_{i\in\mathbb{N}^*}$ is a sequence of i.i.d. Bernoulli* *random variables. Denoting by* $\tilde M_n^+ := \max_{i\in[\![0,n]\!]} \tilde S_i$, $\tilde M_n^- := \min_{i\in[\![0,n]\!]} \tilde S_i$ *we have that for all* $\theta > 0$ *there exists* $c := c(\theta) \ge 1$ *s.t.*

$$\mathbb{E}\Big[\exp\Big(\theta\frac{\tilde M_n^+}{\sqrt{n}}\Big)\Big] + \mathbb{E}\Big[\exp\Big(\theta\frac{\tilde M_n^-}{\sqrt{n}}\Big)\Big] \le c\exp(c\theta^2). \tag{4.19}$$

PROOF. It is well known from the Donsker invariance principle that $\tilde M_n^+, \tilde M_n^-$ respectively converge in law towards the maximum and the minimum of a standard Brownian bridge on $[0,1]$ (see e.g. Liggett [**Lig68**] or Vervaat [**Ver79**]). For the rest of the proof we focus on $\tilde M_n^+$, the results for $\tilde M_n^-$ can be derived similarly by symmetry.

For any $A > 0$, denoting by $\tilde M^+ := \sup_{s\in[0,1]} \tilde B_s$ where $\big(\tilde B\big)_{s\in[0,1]}$ is a standard Brownian bridge, we get that for all $\theta \ge 0$:

$$\mathbb{E}\Bigg[\exp\Big(\theta\frac{\tilde M_n^+}{\sqrt{n}}\Big)\mathbb{I}_{\big|\frac{\tilde M_n^+}{\sqrt{n}}\big|\le A}\Bigg] \underset{n}{\longrightarrow} \mathbb{E}\Big[\exp(\theta\tilde M^+)\mathbb{I}_{|\tilde M^+|\le A}\Big] \le \mathbb{E}\Big[\exp(\theta\tilde M^+)\Big].$$

Letting $A \to \infty$, we then obtain by usual uniform integrability arguments that:

$$\mathbb{E}\Bigg[\exp\Big(\theta\frac{\tilde M_n^+}{\sqrt{n}}\Big)\Bigg] \underset{n}{\longrightarrow} \mathbb{E}\Big[\exp(\theta\tilde M^+)\Big].$$

Therefore, there exists $C := C(\theta) \ge 1$ s.t. for all $n \ge 0$,

$$\mathbb{E}\Bigg[\exp\Big(\theta\frac{\tilde M_n^+}{\sqrt{n}}\Big)\Bigg] \le C\mathbb{E}\Big[\exp(\theta\tilde M^+)\Big] \le C\exp(c\theta^2),$$

where the last inequality simply follows from the exact expression of the joint law of the Brownian motion and its running maximum, see e.g. [**RY99**]. □

PROOF OF THEOREM 4.2. We have first, for even n:

$$\begin{aligned}
&\mathbb{E}\Bigg[\mathbb{I}_{S_n=0}\ \phi_{\delta_n}\Big(\varepsilon_n\sum_{j=1}^n Y_j\exp(\varepsilon_n S_{j-1})\Big)\Bigg]\\
&= \mathbb{P}[S_n=0]\mathbb{E}\Bigg[\phi_{\delta_n}\Big(\varepsilon_n\sum_{j=1}^n Y_j\exp(\varepsilon_n S_{j-1})\Big)|S_n=0\Bigg]\\
&= \mathbb{P}[S_n=0]\mathbb{E}\Bigg[\phi_{\delta_n}\Big(\varepsilon_n\sum_{j=1}^n Y_j\exp(\varepsilon_n \tilde S_{j-1})\Big)\Bigg],
\end{aligned}$$

where $(\tilde S_j)_{j\in[\![1,n]\!]}$ stands for the random walk conditioned to be at 0 at time n. Then:

$$\begin{aligned}
&\mathbb{E}\Bigg[\mathbb{I}_{S_n=0}\ \phi_{\delta_n}\Big(\varepsilon_n\sum_{j=1}^n Y_j\exp(\varepsilon_n S_{j-1})\Big)\Bigg]\\
&\qquad\sim_n \frac{2}{\sqrt{2\pi n}}\mathbb{E}\Bigg[\frac{1}{2\pi}\int_{\mathbb{R}}\hat\phi(\delta_n x)\exp\Big(-i\varepsilon_n x\sum_{j=1}^n Y_j\exp(\varepsilon_n\tilde S_{j-1})\Big)dx\Bigg].
\end{aligned}$$

Taking the conditional expectation w.r.t. to $(\tilde{S}_j)_{j\in[\![1,n]\!]}$ and using the symmetry of the i.i.d random variables $(Y_j)_{j\in[\![1,n]\!]}$, we derive:

$$\mathbb{E}\left[\mathbb{I}_{S_n=0}\ \phi_{\delta_n}\left(\varepsilon_n\sum_{j=1}^n Y_j\exp(\varepsilon_n S_{j-1})\right)\right]$$
$$\sim_n \frac{2}{\sqrt{2\pi n}}\frac{1}{2\pi}\int_{\mathbb{R}}\hat{\phi}(\delta_n x)\mathbb{E}\left[\prod_{j=1}^n\cos\left(\varepsilon_n x\exp(\varepsilon_n\tilde{S}_{j-1})\right)\right]dx.$$

Let now $\widetilde{M_n^-},\widetilde{M_n^+}$ denote the respective minimum and maximum values of the conditioned random walk (bridge) $(\tilde{S}_j)_{j\in[\![1,n]\!]}$. We can assume w.l.o.g. that $|\widetilde{M_n^{\pm}}|\le cn^{\frac12}\ln(n)^{\frac12}$ for a sufficiently large constant c. Indeed,

$$\mathbb{P}[|\widetilde{M_n^{\pm}}|\ge cn^{\frac12}\ln(n)^{\frac12}]=\frac{\mathbb{E}[\mathbb{I}_{|M_n^{\pm}|\ge cn^{\frac12}\ln(n)^{\frac12}}\mathbb{I}_{S_n=0}]}{\mathbb{P}[S_n=0]}$$
$$\le Cn^{\frac12}\mathbb{P}[|M_n^{\pm}|\ge cn^{\frac12}\ln(n)^{\frac12}]^{\frac1p}\mathbb{P}[S_n=0]^{\frac1q},\ p,q>1,\frac1p+\frac1q=1,$$

using the lower bound of the control

$$\frac{C^{-1}}{\sqrt{n}}\le\mathbb{P}[S_n=0]=\binom{n}{n/2}\frac{1}{2^n}\le\frac{C}{\sqrt{n}},\ C\ge 1,$$

which follows from the Stirling formula, for the last inequality. The upper bound and the Bernstein inequality[‡] for the standard random walk on $\mathbb{Z}$ then yield:

$$\mathbb{P}[|\widetilde{M_n^{\pm}}|\ge cn^{\frac12}\ln(n)^{\frac12}]\ \le\ C\exp\left(-\frac{c^2}{2p}\ln(n)\right)n^{\frac12(1-\frac1q)}=Cn^{-\frac{c^2}{2p}+\frac12(1-\frac1q)},$$

which again gives a negligible contribution w.r.t. to the scale $n^{-\frac32}$ for c large enough.

Recalling as well that we have assumed $\hat{\phi}$ to be compactly supported in $[-1,1]$, we get that we only have to consider the integration variable x in the range $|x|\le\frac{1}{\delta_n}$. Recall from the statement of Theorem 4.2 that $\frac{\varepsilon_n}{\delta_n}=n^{-\gamma}$ for $0<\gamma<\frac12$. Then, for all $j\in[\![1,n]\!]$, on the event $\{\widetilde{M_n^+}\le cn^{\frac12}\ln(n)^{\frac12}\}$:

$$\begin{aligned}\varepsilon_n|x|\exp(\varepsilon_n\tilde{S}_{j-1})&\le\frac{\varepsilon_n}{\delta_n}\exp(\varepsilon_n\tilde{S}_{j-1})\\&\le n^{-\gamma}\exp(\varepsilon_n\widetilde{M_n^+})\\&\le n^{-\gamma}\exp\left(ct^{\frac12}\left(\ln(n)\right)^{\frac12}\right)\to_n 0.\end{aligned}\tag{4.20}$$

[‡]We can also refer here to formula (2.16) of Theorem 2.13 in [**Rev05**] for a more precise result which is not needed for our current purpose.

On the associated sets, we will therefore obtain that the arguments in the cosines are uniformly small. Precisely:

$$\begin{aligned}
&\mathbb{E}\Bigg[\prod_{j=1}^{n}\cos\big(\varepsilon_n xY_j\exp(\varepsilon_n\tilde S_{j-1})\big)\mathbb{I}_{\widetilde{M_n^+}\le cn^{\frac12}\ln(n)^{\frac12}}\Bigg]\\
&=\mathbb{E}\Bigg[\prod_{j=1}^{n}\Bigg(1-\frac{(\varepsilon_n x)^2\exp(2\varepsilon_n\tilde S_{j-1})}{2}+O\Big((\varepsilon_n x)^4\exp(4\varepsilon_n\tilde S_{j-1})\Big)\Bigg)\\
&\qquad\qquad\times\mathbb{I}_{\widetilde{M_n^+}\le cn^{\frac12}\ln(n)^{\frac12}}\Bigg]\\
&=\mathbb{E}\Bigg[\exp\Bigg(-\sum_{j=1}^{n}\Big\{\frac{(\varepsilon_n x)^2\exp(2\varepsilon_n\tilde S_{j-1})}{2}+O\Big((\varepsilon_n x)^4\exp(4\varepsilon_n\tilde S_{j-1})\Big)\Big\}\Bigg)\\
&\qquad\qquad\times\mathbb{I}_{\widetilde{M_n^+}\le cn^{\frac12}\ln(n)^{\frac12}}\Bigg].
\end{aligned}$$

Hence,

$$\begin{aligned}
&\frac{1}{2\pi}\int\hat\phi(\delta_n x)\mathbb{E}\Bigg[\prod_{j=1}^{n}\cos\big(\varepsilon_n xY_j\exp(\varepsilon_n\tilde S_{j-1})\big)\Bigg]dx\\
&\sim_n\frac{1}{2\pi}\int\hat\phi(\delta_n x)\mathbb{E}\Bigg[\exp\Bigg(-\frac{x^2}{2}(\tilde A_n(t)+\tilde R_n(t))\Bigg)\mathbb{I}_{\widetilde{M_n^+}\le cn^{\frac12}\ln(n)^{\frac12}}\Bigg]dx=:I_n,
\end{aligned}\tag{4.21}$$

where,

$$\tilde A_n(t):=\varepsilon_n^2\sum_{j=1}^{n}\exp(2\varepsilon_n\tilde S_{j-1}),\ |\tilde R_n(t)|\le C\Big(\varepsilon_n^2\sum_{j=1}^{n}\exp(4\varepsilon_n\tilde S_{j-1})x^2\varepsilon_n^2\Big),\tag{4.22}$$

where the constant C in *absolute* constant, which in particular does not depend on x, t or n. Now, we derive from (4.20) that

$$x^2\varepsilon_n^2\exp\big(2\varepsilon_n\tilde S_{j-1}\big)\le Cn^{-2\gamma}\exp\big(2ct^{\frac12}(\ln(n))^{\frac12}\big)\underset{n}{\to}0.$$

Thus,

$$\begin{aligned}
|\tilde R_n(t)|&\le C\left(\varepsilon_n^2\sum_{j=1}^{n}\exp(4\varepsilon_n\tilde S_{j-1})x^2\varepsilon_n^2\right)\\
&\le C\tilde A_n(t)n^{-2\gamma}\exp\big(2ct^{\frac12}(\ln(n))^{\frac12}\big)\\
&:=C\tilde A_n(t)\beta_n,\ \beta_n\to_n 0.
\end{aligned}$$

We get that:

$$I_n\sim_n\frac{1}{2\pi}\int\hat\phi(\delta_n x)\mathbb{E}\Bigg[\exp\Bigg(-\frac{x^2}{2}\tilde A_n(t)\Bigg)\mathbb{I}_{\widetilde{M_n^+}\le cn^{\frac12}\ln(n)^{\frac12}}\Bigg]dx.\tag{4.23}$$

Indeed, for all $\lambda\in[0,1]$,

$$\widetilde{A}_n(t)+\lambda\tilde R_n(t)\ge\varepsilon_n^2\sum_{j=1}^{n}\exp(2\varepsilon_n\tilde S_{j-1})-|\tilde R_n(t)|\ge\frac12\varepsilon_n^2\sum_{j=1}^{n}\exp(2\varepsilon_n\tilde S_{j-1})=\frac12\tilde A_n(t).\tag{4.24}$$

so that:

$$
\begin{aligned}
|\Delta_n(t,x)| &:= \left| \exp\left(-\frac{x^2}{2}(\tilde{A}_n(t)+\tilde{R}_n(t))\right) - \exp\left(-\frac{x^2}{2}\tilde{A}_n(t)\right)\right| \\
&\leq \int_0^1 \exp\left(-\frac{x^2}{2}(\tilde{A}_n(t)+\lambda\tilde{R}_n(t))\right)\frac{x^2}{2}|\tilde{R}_n(t)|d\lambda, \\
&\underset{(4.24)}{\leq} C\exp\left(-\frac{x^2}{4}\tilde{A}_n(t)\right)\frac{x^2}{2}\tilde{A}_n(t)\beta_n \leq C\exp\left(-\frac{x^2}{8}\tilde{A}_n(t)\right)\beta_n.
\end{aligned}
$$

Thus, exploiting that $\hat{\phi}$ is bounded we get:

$$
(4.25) \quad \left| \int \hat{\phi}(\delta_n x)\mathbb{E}\Big[\Delta_n(t,x)\mathbb{I}_{\widetilde{M_n^+}\leq cn^{\frac{1}{2}}\ln(n)^{\frac{1}{2}}}\Big]dx\right| \leq C\beta_n\mathbb{E}[|\tilde{A}_n(t)|^{-1/2}].
$$

We now state a useful Proposition, whose proof is postponed to the end of the section for the sake of clarity.

Proposition 4.7. *For $\theta\in\{\frac{1}{2},1\}$ and $n\geq 1$, there exists $C\geq 1$ s.t.:*

$$
(4.26) \qquad \mathbb{E}[|\widetilde{A}_n(t)|^{-\theta}] \leq Ct^{-1}.
$$

Let us now prove that Proposition 4.7 and (4.25) yield (4.23).

We first split the term I_n introduced in (4.21) and equivalent to the r.h.s. of (4.23) into two parts.

$$
\begin{aligned}
I_n^1 &:= \frac{1}{2\pi}\int_{|x|\leq\frac{1}{\sqrt{\delta_n}}}\hat{\phi}(\delta_n x)\mathbb{E}[\exp(-\frac{1}{2}x^2\tilde{A}_n(t))\mathbb{I}_{\widetilde{M_n^+}\leq c(n\ln(n))^{\frac{1}{2}}}]dx \\
&\sim_n \hat{\phi}(0)\frac{1}{2\pi}\int_{|x|\leq\frac{1}{\sqrt{\delta_n}}}\mathbb{E}[\exp(-\frac{1}{2}x^2\tilde{A}_n(t))\mathbb{I}_{\widetilde{M_n^+}\leq c(n\ln(n))^{\frac{1}{2}}}]dx =: \bar{I}_n.
\end{aligned}
$$

Now, from the Fubini theorem, we get:

$$
\begin{aligned}
\bar{I}_n &:= \hat{\phi}(0)\frac{1}{2\pi}\Bigg(\mathbb{E}\Bigg[\Big\{\int_{\mathbb{R}}\exp(-\frac{1}{2}x^2\tilde{A}_n(t))dx\Big\}\,\mathbb{I}_{\widetilde{M_n^+}\leq c(n\ln(n))^{\frac{1}{2}}}\Bigg] \\
&\qquad - \int_{|x|>\frac{1}{\sqrt{\delta_n}}}\mathbb{E}[\exp(-\frac{1}{2}x^2\tilde{A}_n(t))\mathbb{I}_{\widetilde{M_n^+}\leq c(n\ln(n))^{\frac{1}{2}}}]dx\Bigg) \\
&= \left(\mathbb{E}[\frac{1}{\sqrt{2\pi\tilde{A}_n(t)}}\mathbb{I}_{\widetilde{M_n^+}\leq c(n\ln(n))^{\frac{1}{2}}}]\right) + O\left(\mathbb{E}[\exp(-\frac{1}{4}\frac{\tilde{A}_n(t)}{\delta_n})\frac{1}{\tilde{A}_n(t)^{1/2}}]\right) \\
&= \left(\mathbb{E}[\frac{1}{\sqrt{2\pi\tilde{A}_n(t)}}\mathbb{I}_{\widetilde{M_n^+}\leq c(n\ln(n))^{\frac{1}{2}}}]\right) + O\left(\delta_n^{1/2}\mathbb{E}[(\tilde{A}_n(t))^{-1}]\right).
\end{aligned}
$$

From Propositions 4.1 and 4.7 and Fatou's lemma, we obtain:

$$
(4.27) \qquad \bar{I}_n \quad \sim_n \quad \mathbb{E}\left[\frac{1}{\sqrt{2\pi\tilde{A}(t)}}\right] = p_2(t,0) \sim_{t\to+\infty} \frac{\pi}{t},
$$

where $\tilde{A}(t) = \int_0^t \exp(2\tilde{B}_s^1)ds$ and $p_2(t,.)$ stands for the density of $\int_0^t \exp(\tilde{B}_s^1)dB_s^2$ at time t and point 0 (see (4.2)). Indeed, conditionally to $\{(\tilde{B}_s^1)_{s\in[0,t]}\}$ the law of

$\int_0^t \exp(\tilde{B}_s^1)dB_s^2$ is a centered Gaussian with variance $\tilde{A}(t)$ (Wiener integral). The last equivalence in (4.27) can be derived directly from Proposition 6.6 in [**MY05**]. Another derivation, exploiting the explicit large time behavior of the return probability on Aff($\mathbb{R}$) given in Theorem 2.3, is proposed in equation (4.31) below. The term $I_n^1 \sim_n \bar{I}_n$ is the main contribution of I_n. The other contribution is small and can be treated as the above remainder. Let us write:

$$\begin{aligned} |I_n^2| &:= \frac{1}{2\pi}\int_{|x|>\frac{1}{\sqrt{\delta_n}}} |\hat{\phi}(\delta_n x)|\mathbb{E}[\exp(-\frac{1}{2}x^2\tilde{A}_n(t))\mathbb{I}_{M_n^+\le c(n\ln(n))^{\frac12}}]dx \\ &\le C\mathbb{E}[\exp(-\frac{1}{4}\frac{\tilde{A}_n(t)}{\delta_n})\frac{1}{\tilde{A}_n(t)^{1/2}}] \le \delta_n^{1/2}\mathbb{E}[(\tilde{A}_n(t))^{-1}]. \end{aligned}$$

This completes the proof of Theorem 4.2. □

PROOF OF PROPOSITION 4.7. Recall from Donati-Martin *et al.* [**DMMY00**] (see also Chaumont *et al.* [**CHY01**]) that for a standard Brownian bridge $(b_u)_{u\in[0,1]}$ on $[0,1]$, it holds that for $\alpha \in \mathbb{R}^+$,

$$\mathbb{E}\left[\left(\int_0^1 \exp(\alpha b_u)du\right)^{-1}\right] = 1. \tag{4.28}$$

We now detail how the indicated convergence rate in time can be deduced for $\theta = 1$ and the limit Brownian bridge from (4.28). Recall that if $(\tilde{B}_u)_{u\in[0,t]}$ is a standard Brownian bridge on $[0,t]$, then

$$(\tilde{B}_u)_{u\in[0,t]} \overset{(\text{law})}{=} \left((t-u)\int_0^u \frac{dB_v}{t-v}\right)_{u\in[0,t]},$$

where $(B_u)_{u\ge 0}$ is a standard Brownian motion. Hence:

$$\begin{aligned} \mathbb{E}\left[\left(\int_0^t \exp(2\tilde{B}_u)du\right)^{-1}\right] &= \mathbb{E}\left[\left(\int_0^t \exp\left(2(t-u)\int_0^u \frac{dB_v}{t-v}\right)du\right)^{-1}\right] \\ &= t^{-1}\mathbb{E}\left[\left(\int_0^1 \exp\left(2t(1-u)\int_0^{ut} \frac{dB_v}{t-v}\right)du\right)^{-1}\right]. \end{aligned} \tag{4.29}$$

A usual covariance computation then shows that $\left((1-u)\int_0^{ut}\frac{dB_v}{t-v}\right)_{u\in[0,1]} \overset{(\text{law})}{=} \frac{1}{t^{1/2}}\left((1-u)\int_0^u \frac{dB_v}{1-v}\right)_{u\in[0,1]} \overset{(\text{law})}{=} \frac{1}{t^{1/2}}(b_u)_{u\in[0,1]}$. Thus, from (4.29) and (4.28):

$$\mathbb{E}\left[\left(\int_0^t \exp(2\tilde{B}_u)du\right)^{-1}\right] = t^{-1}\mathbb{E}\left[\left(\int_0^1 \exp\left(2t^{1/2}b_u\right)du\right)^{-1}\right] = t^{-1}. \tag{4.30}$$

On the other hand, recall that:

$$\begin{aligned} p_{\mathrm{Aff}(\mathbb{R})}(t,e,e) &= p_{\left(B_t^1,\int_0^t \exp(B_s^1)dB_s^2\right)}(0,0) = p_{B_t^1}(0)p_{\int_0^t \exp(B_s^1)dB_s^2}(0|B_t^1=0) \\ &= \frac{1}{\sqrt{2\pi t}}\mathbb{E}\left[\left(2\pi\int_0^t \exp(2B_s^1)ds\right)^{-1/2}\Big|B_t^1=0\right]. \end{aligned}$$

Hence, the asymptotic behavior of the return density for the Brownian motion on the group given in Theorem 2.3 (see also (4.2)) yields:

$$\mathbb{E}\left[\left(2\pi\int_0^t \exp(2\tilde{B}_u^1)du\right)^{-1/2}\right] \sim_{t\to+\infty} \frac{\pi}{t}. \tag{4.31}$$

Let us now detail how the statement (4.26) of Proposition 4.7 can be derived from the previous controls (4.31), (4.30) on the continuous objects through convergence in law arguments. Starting from our simple random walk $S_0 = 0, S_k = \sum_{j=1}^k X_j,\ k \geq 1$ we first introduce for any fixed $n \in \mathbb{N}$ the random polygonal function

$$x_n(u) := S_{\lfloor nu \rfloor} + (nu - \lfloor nu \rfloor)X_{\lfloor nu \rfloor},\ u \in [0,1],$$

where we recall that $\lfloor \cdot \rfloor$ stands for the integer part. Introducing the rescaled conditioned process $\big(\theta_n(u)\big)_{u\in[0,1]} := \frac{1}{\sqrt{n}}\big(x_n(u)|S_n = 0\big)_{u\in[0,1]}$, we derive from Theorem 2 in [**Ver79**] that $\big(\theta_n(u)\big)_{u\in[0,1]} \Rightarrow \big(b_u\big)_{u\in[0,1]}$, standard Brownian bridge on $[0,1]$ with canonical measure μ on $C([0,1])$. Considering now the stepwise constant approximation:

$$\tilde{x}_n(u) := S_{\lfloor nu \rfloor},\ u \in [0,1],$$

and its associated rescaled conditioned process $\big(\tilde{\theta}_n(u)\big)_{u\in[0,1]} := \frac{1}{\sqrt{n}}\big(\tilde{x}_n(u)|S_n = 0\big)_{u\in[0,1]}$, it is easily seen that the corresponding measures $\tilde{\mu}_n$ on $D([0,1])$ converge weakly in $D[0,1]$ to the distribution μ (canonical measure of the Brownian bridge on $C([0,1])$). From the definition of $\tilde{A}_n(t)$ in (4.22), recalling as well that $\varepsilon_n = \left(\frac{t}{n}\right)^{1/2}$, we thus rewrite:

$$\tilde{A}_n(t) := \varepsilon_n^2 \sum_{j=1}^n \exp(2\varepsilon_n \tilde{S}_{j-1}) = \frac{t}{n}\sum_{j=1}^n \exp\left(2t^{1/2}\frac{1}{\sqrt{n}}\tilde{x}_n\big(\tfrac{j}{n}\big)\right) = t\int_0^1 \exp\Big(2t^{1/2}\tilde{\theta}_n(u)\Big)du.$$

Hence, from the previous convergence in law $\tilde{A}_n(t) \overset{\text{(law)}}{\to} t\int_0^1 \exp(2t^{1/2}b_u)du \overset{\text{(law)}}{=} \int_0^t \exp(2\tilde{B}_u)du$ and for a given $A > 0$ and $\theta \in \{\frac{1}{2}, 1\}$:

$$\mathbb{E}[(\tilde{A}_n(t))^{-\theta}\mathbb{I}_{A^{-1}\leq \tilde{A}_n(t)\leq A}] \underset{n}{\longrightarrow} \mathbb{E}[(\tilde{A}_t)^{-\theta}\mathbb{I}_{A^{-1}\leq \tilde{A}(t)\leq A}].$$

The statement (4.26) now follows from the above equation and the previously established estimates (4.31), (4.30), noting as well that, since

$$\tilde{A}_n(t)^{-\theta} \leq \left(t\exp(2t^{1/2}\frac{M_n^-}{\sqrt{n}})\right)^{-\theta},$$

Lemma 4.6 gives that the sequence $\big(\tilde{A}_n(t)^{-\theta}\big)_{n\geq 0}$ is bounded in $L^2(\mathbb{P})$ and therefore uniformly integrable. The proof is complete. □

REMARK 4.2 (Balance of n and t for the approximation). Observe from the previous proof of Theorem 4.2 that one can actually consider at the same time n and t going to infinity provided inequality (4.20) holds. This control is needed in order to isolate the remainder terms, and basically imposes $t \leq c_0 \ln(n)$ for c_0 small enough which guarantees $n^{-\gamma}\exp\left(ct^{\frac{1}{2}}(\ln(n))^{1/2}\right) \to_n 0$.

4.3. The Mixed Case. We consider in this Section that the random variables $(Y_i)_{i\in\mathbb{N}}$ in the definition of the random walk approximation (3.2) are i.i.d. and have common standard Gaussian law, i.e. $Y_i \overset{(\text{law})}{=} \mathcal{N}(0,1)$. This modification is precisely enough to restore the "expected" local limit theorem.

THEOREM 4.8. *For the previously described random walk, taking* $\varepsilon_n = \left(\frac{t}{n}\right)^{\frac{1}{2}}$ *and for* $n \in 2\mathbb{N}$*:*

$$\begin{aligned}\mathbb{P}[a_{\varepsilon_n} = 1, b_{\varepsilon_n} \in [0, dx)] \\ = \mathbb{P}[S_n = 0, \varepsilon_n \sum_{j=1}^{n} Y_j \exp(\varepsilon_n S_{j-1}) \in [0, dx)] \sim_n 2\varepsilon_n \cdot p_{\mathrm{Aff}(\mathbb{R})}(t, e, e)dx.\end{aligned}$$

We indeed have a result similar to Theorem 4.2, except that no integration with respect to the previous mollifyer ϕ_{δ_n} is needed.

PROOF. Note that the random variable $b_\varepsilon(n)$ now has a conditional Gaussian density (for fixed trajectory $(S_k)_{k\in\mathbb{N}}$). We thus readily get:

$$\mathbb{P}[a_{\varepsilon_n} = 1, b_{\varepsilon_n} \in [0, dx)] \sim_n \frac{2}{\sqrt{2\pi n}} \mathbb{E}\left[\frac{1}{\sqrt{2\pi\varepsilon_n^2 \sum_{j=1}^{n} \exp(2\varepsilon_n S_{j-1})}} \middle| S_n = 0 \right] dx.$$

Proposition 4.7 remains valid taking $\tilde{R}_n = 0$ in the definition (4.24). With the notations used therein, this precisely gives $\tilde{A}_n(t) = \varepsilon_n^2 \sum_{j=1}^{n} \exp(2\varepsilon_n S_{j-1})$. We then derive the statement from Propositions 4.1, 4.7 and Fatou's lemma. □

Acknowledgments

We would like to thank the referee for his careful reading and helpful comments and suggestions.

References

[Ale02] Georgios K. Alexopoulos, *Random walks on discrete groups of polynomial volume growth*, Ann. Probab. **30** (2002), no. 2, 723–801, DOI 10.1214/aop/1023481007. MR1905856

[Bou83] Philippe Bougerol, *Exemples de théorèmes locaux sur les groupes résolubles* (French, with English summary), Ann. Inst. H. Poincaré Sect. B (N.S.) **19** (1983), no. 4, 369–391. MR730116

[Bou15] Philippe Bougerol, *Matsumoto-Yor process and infinite dimensional hyperbolic space*, In memoriam Marc Yor—Séminaire de Probabilités XLVII, Lecture Notes in Math., vol. 2137, Springer, Cham, 2015, pp. 521–559, DOI 10.1007/978-3-319-18585-9_23. MR3444313

[BR76] R. N. Bhattacharya and R. Ranga Rao, *Normal approximation and asymptotic expansions: Wiley Series in Probability and Mathematical Statistics*, John Wiley & Sons, New York-London-Sydney, 1976. MR0436272

[Bre05] E. Breuillard, *Local limit theorems and equidistribution of random walks on the Heisenberg group*, Geom. Funct. Anal. **15** (2005), no. 1, 35–82, DOI 10.1007/s00039-005-0501-3. MR2140628

[CHY01] L. Chaumont, D. G. Hobson, and M. Yor, *Some consequences of the cyclic exchangeability property for exponential functionals of Lévy processes*, Séminaire de Probabilités, XXXV, Lecture Notes in Math., vol. 1755, Springer, Berlin, 2001, pp. 334–347, DOI 10.1007/978-3-540-44671-2_23. MR1837296

[DMMY00] Catherine Donati-Martin, Hiroyuki Matsumoto, and Marc Yor, *On striking identities about the exponential functionals of the Brownian bridge and Brownian motion*, Period. Math. Hungar. **41** (2000), no. 1-2, 103–119, DOI 10.1023/A:1010308203346. Endre Csáki 65. MR1812799

[Gro81] Mikhael Gromov, *Groups of polynomial growth and expanding maps*, Inst. Hautes Études Sci. Publ. Math. **53** (1981), 53–73. MR623534

[Gru96] J.-C. Gruet, *Semi-groupe du mouvement brownien hyperbolique* (French, with French summary), Stochastics Stochastics Rep. **56** (1996), no. 1-2, 53–61. MR1396754

[Ibé76] Michel Ibéro, *Intégrales stochastiques multiplicatives et construction de diffusions sur un groupe de Lie* (French), Bull. Sci. Math. (2) **100** (1976), no. 2, 175–191. MR0517986

[IL71] I. A. Ibragimov and Yu. V. Linnik, *Independent and stationary sequences of random variables*, Wolters-Noordhoff Publishing, Groningen, 1971. With a supplementary chapter by I. A. Ibragimov and V. V. Petrov; Translation from the Russian edited by J. F. C. Kingman. MR0322926

[IW80] Nobuyuki Ikeda and Shinzo Watanabe, *Stochastic differential equations and diffusion processes*, North-Holland Mathematical Library, vol. 24, North-Holland Publishing Co., Amsterdam-New York; Kodansha, Ltd., Tokyo, 1981. MR637061

[Kes59] Harry Kesten, *Symmetric random walks on groups*, Trans. Amer. Math. Soc. **92** (1959), 336–354, DOI 10.2307/1993160. MR0109367

[KMM11] V. D. Konakov, S. Menozzi, and S. A. Molchanov, *Diffusion processes on solvable groups of upper triangular* (2×2)*-matrices and their approximations* (Russian), Dokl. Akad. Nauk **439** (2011), no. 5, 585–588, DOI 10.1134/S1064562411050036; English transl., Dokl. Math. **84** (2011), no. 1, 527–530. MR2883788

[KMM17] V. Konakov, S. Menozzi, and S. Molchanov. Approximation of diffusion processes on solvable Lie groups by random walks. Local and quasi-local theorems. *Analytical and computational methods in probability theory and its application (ACMPT-17)*, pages 202–205, 2017.

[KTS59] F. I. Karpelevič, V. N. Tutubalin, and M. G. Šur, *Limit theorems for compositions of distributions in the Lobačevskiĭ plane and space* (Russian, with English summary), Teor. Veroyatnost. i Primenen. **4** (1959), 432–436. MR0114235

[Lig68] Thomas M. Liggett, *An invariance principle for conditioned sums of independent random variables*, J. Math. Mech. **18** (1968), 559–570, DOI 10.1512/iumj.1969.18.18043. MR0238373

[LL10] Gregory F. Lawler and Vlada Limic, *Random walk: a modern introduction*, Cambridge Studies in Advanced Mathematics, vol. 123, Cambridge University Press, Cambridge, 2010. MR2677157

[McK69] H. P. McKean Jr., *Stochastic integrals*, Probability and Mathematical Statistics, No. 5, Academic Press, New York-London, 1969. MR0247684

[Mel02] Camillo Melzi, *Large time estimates for non-symmetric heat kernel on the affine group*, Ann. Math. Blaise Pascal **9** (2002), no. 1, 63–78. MR1914261

[Mil68] John Milnor, *Growth of finitely generated solvable groups*, J. Differential Geometry **2** (1968), 447–449. MR0244899

[Mol75] S. A. Molčanov, *Diffusion processes, and Riemannian geometry* (Russian), Uspehi Mat. Nauk **30** (1975), no. 1(181), 3–59. MR0413289

[MV08] S. Molchanov and B. Vainberg. Estimates for the counting function of the Laplace operator on domains with rough boundaries. *Around the Research of Vladimir Mazya III: Analysis and Applications*, 3, 2008.

[MY05] Hiroyuki Matsumoto and Marc Yor, *Exponential functionals of Brownian motion. I. Probability laws at fixed time*, Probab. Surv. **2** (2005), 312–347, DOI 10.1214/154957805100000159. MR2203675

[Pet05] V. V. Petrov, *Summy nezavisimykh sluchaĭnykh velichin* (Russian), Izdat. "Nauka", Moscow, 1972. MR0322927

[Pin95] R.G. Pinsky. *Positive Harmonic Functions and Diffusion*, volume 45. Cambridge Studies Advanced Mathematics, 1995.

[PSC02] C. Pittet and L. Saloff-Coste, *On random walks on wreath products*, Ann. Probab. **30** (2002), no. 2, 948–977, DOI 10.1214/aop/1023481013. MR1905862

[Rev05] Pál Révész, *Random walk in random and non-random environments*, 2nd ed., World Scientific Publishing Co. Pte. Ltd., Hackensack, NJ, 2005. MR2168855

[RW85] D. Rogers and D. Williams. *Diffusions and Markov processes, II.* C.U.P., 1985.

[RY99] Daniel Revuz and Marc Yor, *Continuous martingales and Brownian motion*, 3rd ed., Grundlehren der Mathematischen Wissenschaften [Fundamental Principles of Mathematical Sciences], vol. 293, Springer-Verlag, Berlin, 1999. MR1725357

[SS03] Elias M. Stein and Rami Shakarchi, *Fourier analysis*, Princeton Lectures in Analysis, vol. 1, Princeton University Press, Princeton, NJ, 2003. An introduction. MR1970295

[Tes13] Romain Tessera, *Isoperimetric profile and random walks on locally compact solvable groups*, Rev. Mat. Iberoam. **29** (2013), no. 2, 715–737, DOI 10.4171/RMI/736. MR3047434

[Ver79] Wim Vervaat, *A relation between Brownian bridge and Brownian excursion*, Ann. Probab. **7** (1979), no. 1, 143–149. MR515820

[Wol68] Joseph A. Wolf, *Growth of finitely generated solvable groups and curvature of Riemannian manifolds*, J. Differential Geometry **2** (1968), 421–446. MR0248688

[Yor92] Marc Yor, *On some exponential functionals of Brownian motion*, Adv. in Appl. Probab. **24** (1992), no. 3, 509–531, DOI 10.2307/1427477. MR1174378

Higher School of Economics, National Research University, Shabolovka 28, Moscow, Russian Federation.
Email address: vkonakov@hse.ru

Laboratoire de Modélisation Mathématique d'Evry (LaMME), UMR CNRS 8071, Université d'Evry Val-d'Essonne, 23 Boulevard de France, 91037 Evry and Higher School of Economics, National Research University, Shabolovka 28, Moscow, Russian Federation.
Email address: stephane.menozzi@univ-evry.fr

University of North Carolina at Charlotte, USA and Higher School of Economics, National Research University, Shabolovka 28, Moscow, Russian Federation
Email address: smolchan@uncc.edu

Contemporary Mathematics
Volume **739**, 2019
https://doi.org/10.1090/conm/739/14896

Quantum limits of Eisenstein series in $\mathbb{H}^3$

Niko Laaksonen

ABSTRACT. We study the quantum limits of Eisenstein series off the critical line for $\mathrm{PSL}_2(\mathcal{O}_K)\backslash\mathbb{H}^2$, where K is an imaginary quadratic field of class number one. This generalises the results of Petridis, Raulf and Risager on $\mathrm{PSL}_2(\mathbb{Z})\backslash\mathbb{H}^2$. We observe that the measures $|E(p,\sigma_t+it)|^2 d\mu(p)$ become equidistributed only if $\sigma_t \to 1$ as $t\to\infty$. We use these computations to study measures defined in terms of the scattering states, which are shown to converge to the absolutely continuous measure $E(p,3)d\mu(p)$ under the GRH.

1. Introduction

Suppose M is a compact negatively curved Riemannian manifold (without boundary) with the unit tangent bundle $X = SM$, then the geodesic flow on X is ergodic [**1**] (see also [**3**, Appendix]). The problem then is to study the quantised flow, in terms of the eigenfunctions ϕ_j of Δ on M, in the large eigenvalue limit. Shnirelman [**27**], Zelditch [**32**] and Colin de Verdière [**5**] proved that there is a full density subsequence of the measures $\mu_j = |\phi_j|^2\mu$, which converges weakly to the uniform measure μ on M. It is not known in general whether μ is the unique limit. When $M = \Gamma\backslash\mathbb{H}^2$, and Γ is arithmetic, more tools are available such as Hecke operators and explicit Fourier expansions of Eisenstein series. Rudnick [**25**] conjectured that for compact M with constant negative curvature the limit μ is unique. This is the Quantum Unique Ergodicity (QUE) conjecture. For Γ of arithmetic type the distribution of the eigenstates is well-understood. In 1995, Luo and Sarnak [**17**] proved the conjecture for Eisenstein series for non-compact arithmetic Γ and, in particular, for $\Gamma = \mathrm{PSL}_2(\mathbb{Z})$. The precise result is that given Jordan measurable subsets A and B of M, then

$$\lim_{t\to\infty}\frac{\int_A |E(z,\tfrac12+it)|^2\,d\mu(z)}{\int_B |E(z,\tfrac12+it)|^2\,d\mu(z)} = \frac{\mu(A)}{\mu(B)}, \tag{1.1}$$

where $\mu(B)\neq 0$. They compute the asymptotics explicitly as

$$\int_A |E(z,\tfrac12+it)|^2\,d\mu(z) \sim \frac{6}{\pi}\mu(A)\log t,$$

2010 *Mathematics Subject Classification.* Primary 11F72; Secondary 35P25.

Key words and phrases. Quantum limits; Eisenstein series; scattering poles; Bianchi groups.

Part of the work was completed during the author's stay at the University of Copenhagen. This visit was supported by the 150th Anniversary Postdoctoral Mobility Grant from the London Mathematical Society.

as $t \to \infty$ (the actual erroneous constant in [**17**] is $48/\pi$, but it is not significant for their purposes). Jakobson [**12**] extended (1.1) to the unit tangent bundle. The result of Luo and Sarnak was also generalised to $\mathrm{PSL}_2(\mathcal{O}_K)\backslash\mathbb{H}^3$ by Koyama [**13**], where $\mathrm{PSL}_2(\mathcal{O}_K)$ is the ring of integers of an imaginary quadratic field of class number one, and to $\mathrm{PSL}_2(\mathcal{O}_K)\backslash(\mathbb{H}^2)^n$ with K a totally real field of degree n and narrow class number one by Truelsen [**30**]. In particular, the quantum limit in [**30**] for $\mu_{m,t} = |E(z, \frac{1}{2} + it, m)|^2\mu$ is

$$\mu_{t,m} \to \frac{(2\pi)^n nR}{2d_K\zeta_K(2)} \log t,$$

where $E(z, s, m)$ are a family of Eisenstein series parametrised by $m \in \mathbb{Z}^{n-1}$, ζ_K is the Dedekind zeta function and R and d_K are the regulator and the discriminant of K, respectively. The QUE for ϕ_j a Hecke–Maaß eigenform was proven by Lindenstrauss [**15**] in the compact case and up to a possible escape of mass at the cusp in the non-compact setting. Soundararajan [**28**] ruled out the escape of mass, thus completing the proof of the full QUE conjecture for all arithmetic surfaces. Holowinsky and Soundararajan [**11**] also study QUE in the holomorphic case. They consider holomorphic, L^2-normalised Hecke cusp forms f_k of weight k for $\mathrm{SL}_2(\mathbb{Z})$. They prove that the measures $|y^{k/2}f_k(z)|^2\mu$ converge weakly to μ as $k \to \infty$. Another interesting direction for the QUE of Eisenstein series has recently been proved by Young [**31**], who proves equidistribution of Eisenstein series for $\Gamma = \mathrm{PSL}_2(\mathbb{Z})$ when they are restricted to "thin sets", e.g. geodesics connecting 0 and ∞ (as opposed to restricting to compact Jordan measurable subsets of $\Gamma\backslash\mathbb{H}^2$ as in [**17**]). For a general cofinite $\Gamma \subset \mathrm{PSL}_2(\mathbb{R})$ it is not clear whether there are infinitely many cusp forms so that the limit of $|\phi_j|^2\mu$ might not be relevant [**16, 23**]. Petridis, Raulf and Risager [**19**] (see also [**18**]) propose to study the scattering states of Δ instead of the cuspidal spectrum. It is known that under small deformations of Γ, the cusp forms dissolve into scattering states as characterised by Fermi's Golden Rule [**20, 22**]. The scattering states are described as residues of Eisenstein series on the left half-plane ($\mathrm{Re}\, s < 1/2$) at the non-physical poles of the scattering matrix. These poles are called resonances. Let ρ_n be a sequence of poles of the scattering matrix. For $\mathrm{PSL}_2(\mathbb{Z})\backslash\mathbb{H}^2$ this corresponds to half a non-trivial zero of ζ. Petridis, Raulf and Risager define the measures

$$u_{\rho_n}(z) = (\operatorname*{Res}_{s=\rho_n} \varphi(s))^{-1} \operatorname*{Res}_{s=\rho_n} E(z, s).$$

The normalisation is chosen so that u_{ρ_n} has simple asymptotics $y^{1-\rho_n}$ for its growth at infinity. The result is that for any compact Jordan measurable subset A of $\Gamma\backslash\mathbb{H}^2$,

$$\int_A |u_{\rho_n}(z)|^2 d\mu(z) \to \int_A E(z, 2 - \gamma_\infty) d\mu(z),$$

where γ_∞ is the limit of the real part of the Riemann zeros. Under the RH the limit is $E(z, 3/2)d\mu(z)$. This is obtained by studying the quantum limits of Eisenstein series off the critical line.

We generalise their result to three dimensions $\Gamma\backslash\mathbb{H}^3$ for Γ a Bianchi group of class number one. Let ρ_n be a sequence of poles of the scattering matrix $\varphi(s)$ of $E(p, s)$ on $\Gamma\backslash\mathbb{H}^3$ and define

$$v_{\rho_n}(p) = (\operatorname*{Res}_{s=\rho_n} \varphi(s))^{-1} \operatorname*{Res}_{s=\rho_n} E(p, s).$$

From the explicit form of φ (2.3) we know that ρ_n is equal to a non-trivial zero of ζ_K. Define $s(t) = \sigma_t + it$, where $\sigma_t > 1$ is a sequence converging to $\sigma_\infty \geq 1$. Also, let γ_n be a sequence of real parts of the non-trivial zeros of ζ_K. We assume γ_n satisfies $\lim \gamma_n = \gamma_\infty < 1$. We will prove the following theorems.

THEOREM 1. *Let A be a compact Jordan measurable subset of $\Gamma\backslash\mathbb{H}^3$. Then*

$$\int_A |v_{\rho_n}(p)|^2 d\mu(p) \to \int_A E(p, 4-2\gamma_\infty) d\mu(p)$$

as $n \to \infty$.

Notice that $4 - 2\gamma_\infty > 2$ so that we are in the region of absolute convergence. Under the GRH the limit is $E(p,3)d\mu(p)$.

THEOREM 2. *Assume $\sigma_\infty = 1$ and $(\sigma_t - 1)\log t \to 0$. Let A and B be compact Jordan measurable subsets of $\Gamma\backslash\mathbb{H}^3$. Then*

$$\frac{\mu_{s(t)}(A)}{\mu_{s(t)}(B)} \to \frac{\mu(A)}{\mu(B)},$$

as $t \to \infty$. In fact, we have

$$\mu_{s(t)}(A) \sim \mu(A) \frac{2(2\pi)^2}{|\mathcal{O}^\times||d_K|\zeta_K(2)} \log t. \tag{1.2}$$

Let F be the fundamental domain of $\mathcal{O}_K$ as a lattice in $\mathbb{R}^2$. Since $|F| = \sqrt{|d_K|}/2$ and $\mathrm{vol}(\Gamma\backslash\mathbb{H}^3) = |d_K|^{3/2}\zeta_K(2)/(4\pi^2)$, [**26**, Proposition 2.1], it is also possible to express the constant in (1.2) in terms of the volumes.

REMARK 1. The constant for the QUE of Eisenstein series *on* the critical line in [**13**] is $2/\zeta_K(2)$. However, there is a small mistake in his computations on page 485, where the residue of the double pole of $\zeta_K^2(s/2)$ goes missing. After fixing this (and taking into account the number of units of $\mathcal{O}$ which is normalised away in [**13**]) his result agrees with our limit (1.2) for $\sigma_\infty = 1$.

THEOREM 3. *Assume $\sigma_\infty > 1$. Let A be a compact Jordan measurable subset of $\Gamma\backslash\mathbb{H}^3$. Then*

$$\mu_{s(t)}(A) \to \int_A E(p, 2\sigma_\infty) d\mu(p),$$

as $t \to \infty$.

Theorem 1 says that the measures v_ρ do *not* become equidistributed. We could of course renormalise the measures and use

$$d\nu_{s(t)}(p) = \left| \frac{E(p, s(t))}{\sqrt{E(p, 2\sigma_\infty)}} \right|^2 d\mu(p).$$

Then we have the following corollary.

COROLLARY 1. *Assume $\sigma_\infty > 1$. Let A be a compact Jordan measurable subset of $\Gamma\backslash\mathbb{H}^3$. Then*

$$\nu_{s(t)}(A) \to \mu(A),$$

as $t \to \infty$.

The measures ν_{ρ_n} are not eigenfunctions of Δ so their equidistribution is not directly related to the QUE conjecture.

REMARK 2. Dyatlov [**6**] investigated quantum limits of Eisenstein series and scattering states for more general Riemannian manifolds with cuspidal ends. He proves results analogous to Theorems 1 and 3. However, only the case of surfaces is explicitly written down and the limits are not identified as concretely for the arithmetic special cases such as in [**19**] or Theorem 1 and 3. Dyatlov uses a very different method of decomposing the Eisenstein series into plane waves and studying their microlocal limits, which does not use global properties of the surface, such as hyperbolicity.

2. Spectral Theory in $\mathrm{PSL}_2(\mathcal{O}_K)\backslash\mathbb{H}^3$

For the general spectral theory in hyperbolic three-space and the relevant facts about Bianchi groups and ζ_K we refer to [**7**]. Fix a square-free integer $D < 0$ and let $K = \mathbb{Q}(\sqrt{D})$ be the corresponding imaginary quadratic field of discriminant d_K. Let $\mathcal{O}$ be the ring of integers of K and let $\langle 1, \omega\rangle$, where

$$\omega = \frac{d_K + \sqrt{d_K}}{2},$$

be a $\mathbb{Z}$-basis for $\mathcal{O}$. Let $\Gamma = \mathrm{PSL}_2(\mathcal{O})$. For simplicity, restrict D so that K has class number one. This means that Γ has exactly one cusp (up to Γ-equivalence) which we may suppose is $\infty \in \mathbb{P}^1\mathbb{C}$. The Dedekind zeta function of K is defined for $\mathrm{Re}\, s > 1$ by

$$\zeta_K(s) = \sum_{\mathfrak{a}\subset\mathcal{O}}{}' \frac{1}{N\mathfrak{a}^s} = \prod_{\mathfrak{p}} \frac{1}{1 - N\mathfrak{p}^{-s}},$$

where the prime in the summation denotes that it is taken over nonzero ideals $\mathfrak{a}$, and the Euler product is taken over prime ideals $\mathfrak{p} \subset \mathcal{O}$. We define the completed zeta function by

$$\xi_K(s) = \left(\frac{\sqrt{|d_K|}}{2\pi}\right)^s \Gamma(s)\zeta_K(s).$$

Notice that this differs from the standard way of completing ζ_K due to the inclusion of the discriminant. We know that ξ_K satisfies a functional equation

$$\xi_K(s) = \xi_K(1-s), \tag{2.1}$$

and has an analytic continuation to all of $\mathbb{C}$ with a simple pole at $s = 1$ with residue

$$\operatorname*{Res}_{s=1} \zeta_K(s) = \frac{2\pi}{|\mathcal{O}^\times|}, \tag{2.2}$$

where $|\mathcal{O}^\times|$ is the number of units of $\mathcal{O}$.

Let $\mathbb{H}^3 = \{p = z + jy : z \in \mathbb{C}, y > 0\}$, where $z = x_1 + ix_2$, be the three dimensional hyperbolic space. The standard volume element on $\mathbb{H}^3$ is given by

$$d\mu(p) = \frac{dx_1\, dx_2\, dy}{y^3},$$

and the hyperbolic Laplacian is

$$\Delta = y^2\left(\frac{\partial^2}{\partial x_1^2} + \frac{\partial^2}{\partial x_2^2} + \frac{\partial^2}{\partial y^2}\right) - y\frac{\partial}{\partial y},$$

with the corresponding eigenvalue equation $\Delta f + \lambda f = 0$. We write the eigenvalues of Δ as $\lambda_j = s_j(2 - s_j) = 1 + t_j^2$. We know that for cofinite subgroups Γ of $\mathrm{PSL}_2(\mathbb{C})$ the Laplacian has both discrete and continuous spectrum. The continuous spectrum spans (in the λ aspect) the interval $[1, \infty)$ with the eigenpacket given by Eisenstein series on the critical line, $E(p, 1+it)$. The discrete spectrum consists of Maaß cusp forms and the eigenvalue $\lambda = 0$.

The Eisenstein series for Γ at the cusp at ∞ is given for $\operatorname{Re} s > 1$ by

$$E(p,s) = \sum_{\gamma \in \Gamma_\infty \backslash \Gamma} y(\gamma p)^s,$$

where Γ_∞ is the stabiliser of Γ at ∞. The Eisenstein series is an eigenfunction of Δ with the eigenvalue $\lambda = s(2-s)$, but it is not square integrable. Let

$$\varphi(s) = \frac{\xi_K(s-1)}{\xi_K(s)} = \frac{2\pi}{\sqrt{|d_K|}} \frac{1}{s-1} \frac{\zeta_K(s-1)}{\zeta_K(s)} \tag{2.3}$$

be the scattering matrix of $E(p,s)$. The Fourier expansion of $E(p,s)$ at the cusp is then given by

$$E(p,s) = y^s + \varphi(s) y^{2-s} + \frac{2y}{\xi_K(s)} \sum_{0 \neq n \in \mathcal{O}} |n|^{s-1} \sigma_{1-s}(n) K_{s-1}\left(\frac{4\pi |n| y}{\sqrt{|d_K|}}\right) e^{2\pi i \langle \frac{2\bar{n}}{\sqrt{d_K}}, z\rangle}, \tag{2.4}$$

where

$$\sigma_s(n) = \sum_{\substack{(d) \subset \mathcal{O} \\ d | n}} |d|^{2s}.$$

This form of the Fourier expansion can be found in [**13, 24**]. It also appears in a more general form in [**2**, (13)][**7**, §6 Theorem 2.11.]. The Eisenstein series $E(p,s)$ can be analytically continued to all of $\mathbb{C}$ as a meromorphic function of s. We can see from (2.3) that to the right of the critical line $s = 1$, $E(p,s)$ has only a simple pole at $s = 2$ with residue

$$\operatorname*{Res}_{s=2} E(v,s) = \frac{|\mathcal{F}_\infty|}{\operatorname{vol}(M)},$$

where $\mathcal{F}_\infty$ is the fundamental domain of Γ_∞ acting on the boundary $\mathbb{C}$, [**7**, §6 Theorem 1.11]. Moreover, since $\varphi(s)\varphi(2-s) = 1$, the Eisenstein series has a functional equation [**7**, §6 Theorem 1.2]

$$E(p,s) = \varphi(s) E(p, 2-s). \tag{2.5}$$

We will also use the incomplete Eisenstein series which are defined for a smooth $\psi(x)$ with compact support on $\mathbb{R}^+$ by

$$E(p|\psi) = \sum_{\gamma \in \Gamma_\infty \backslash \Gamma} \psi(y(\gamma p)).$$

As in two dimensions, it is possible to decompose $L^2(M)$ into the orthogonal spaces spanned by the closures of the spaces of incomplete Eisenstein series on one hand, and the Maaß cusp forms on the other hand.

Finally, for a Hecke–Maaß cusp form u_j we have the following Fourier expansion.

$$u_j(p) = y \sum_{0 \neq n \in \mathcal{O}_K^*} \rho_j(n) K_{it_j}(2\pi |n| y) e^{2\pi i \langle n, z\rangle}, \tag{2.6}$$

where $\mathcal{O}_K^*$ is the dual lattice,

$$\mathcal{O}_K^* = \{m : \langle m, n\rangle \in \mathbb{Z} \text{ for all } n \in \mathcal{O}_K\}.$$

Also, the Hecke eigenvalues satisfy $\rho_j(n) = \rho_j(1)\lambda_j(n)$, and in particular [**10**, Satz 16.8, pg. 119]

$$L(u_j, s) = \rho_j(1) \sum_{n \in \mathcal{O}_K} \frac{\lambda_j(n)}{N(n)^s} = \rho_j(1) \prod_{\mathfrak{p}} (1 - \lambda_j(\mathfrak{p}) N\mathfrak{p}^{-s} + N\mathfrak{p}^{1-2s})^{-1}. \tag{2.7}$$

We can split the space of Hecke–Maaß cusp forms further into even and odd cusp forms depending on the sign in $\rho(-n) = \pm\rho(n)$.

3. Proofs

Let $M = \Gamma\backslash\mathbb{H}^3$. Since any function in $L^2(M)$ can be decomposed in terms of the Hecke–Maaß cusp forms $\{u_j\}$ and the incomplete Eisenstein series $E(p\,\psi)$, it is sufficient to consider them separately.

3.1. Discrete Part.

We will first prove that the contribution of the discrete spectrum vanishes in the limit.

LEMMA 1. *Let u_j be a Hecke–Maaß cusp form. Then*

$$\int_M u_j(p) |E(p, s(t))|^2 d\mu(p) \to 0,$$

as $t \to \infty$.

PROOF. Denote the integral by

$$J_j(s(t)) = \int_M u_j(p) |E(p, s(t))|^2 d\mu(p).$$

We define

$$I_j(s) = \int_M u_j(p) E(p, s(t)) E(p, s) d\mu(p).$$

Unfolding the integral gives

$$I_j(s) = \int_0^\infty \int_F u_j(p) E(p, s(t)) y^s \frac{dx_1 dx_2 dy}{y^3}.$$

After a change of variables, we may suppose that the u_j in $I_j(s)$ is even as the integral over the odd cusp forms vanishes. Substituting Fourier expansions of the Eisenstein series (2.4) and the cusp forms (2.6) into the above integral gives

$$\begin{aligned} I_j(s) = \int_0^\infty \int_F \Bigg(2y \sum_{0 \neq n \in \mathcal{O}^*} \rho_j(n) K_{ir_j}(2\pi|n|y) \cos(2\pi\langle n, z\rangle) \Bigg) \\ \times \Bigg(y^{s(t)} + \varphi(s(t)) y^{2-s(t)} \\ + \frac{2y}{\xi_K(s(t))} \sum_{0 \neq m \in \mathcal{O}} |m|^{s(t)-1} \sigma_{1-s(t)}(m) K_{s(t)-1}\left(\frac{4\pi|m|y}{\sqrt{|d_K|}}\right) e^{2\pi i \langle \frac{2m}{\sqrt{d_K}}, z\rangle} \Bigg) y^s \frac{dx_1 dx_2 dy}{y^3}. \end{aligned}$$

By the definition of F and the formula $\cos(a+b) = \cos a \cos b - \sin a \sin b$ it is simple to see that

$$\int_F \cos(2\pi\langle n, z\rangle)\, dz = \begin{cases} 0, & \text{if } 0 \neq n \in \mathcal{O}^*, \\ 1, & \text{if } n = 0. \end{cases}$$

Evaluating the integral over F tells us that only the terms with $n = \pm 2m/\sqrt{|d_K|}$ remain and that the integral over the imaginary part goes to zero. Hence, with the identification $\mathcal{O} \to \mathcal{O}^*$ by $\alpha \mapsto (2/\sqrt{d_K})\overline{\alpha}$, we get

$$I_j(s) = \frac{4}{\xi_K(s(t))} \int_0^\infty \sum_{0 \neq n \in \mathcal{O}^*} |n|^{s(t)-1} \sigma_{1-s(t)}(n) \rho_j(n) K_{s(t)-1}(2\pi|n|y) K_{it_j}(2\pi|n|y) y^s \frac{dy}{y}.$$

The change of variables $y \mapsto y/|n|$ yields

$$I_j(s) = \frac{4}{\xi_K(s(t))} \sum_{0 \neq n \in \mathcal{O}^*} \frac{|n|^{s(t)-1} \sigma_{1-s(t)}(n) \rho_j(n)}{|n|^s} \int_0^\infty K_{s(t)-1}(2\pi y) K_{it_j}(2\pi y) y^s \frac{dy}{y}.$$

We can evaluate the integral by [**8**, 6.576 (4) and 9.100] to get

$$I_j(s) = \frac{4}{\xi_K(s(t))} \frac{2^{-3}\pi^{-s}}{\Gamma(s)} \prod \Gamma\left(\frac{s \pm (s(t)-1) \pm it_j}{2}\right) R(s),$$

where the product is taken over all combinations of $\pm$ and

$$R(s) = \sum_{0 \neq n \in \mathcal{O}^*} \frac{|n|^{s(t)-1} \sigma_{1-s(t)}(n) \rho_j(n)}{|n|^s}.$$

Since u_j is a Hecke eigenform, we can factorise $R(s)$ with (2.7) as

$$\begin{aligned} R(s) &= \rho_j(1) \prod_{(p):\text{prime ideal}} \sum_{k=0}^\infty \frac{\lambda_j(p^k)|p|^{k(s(t)-1)} \sigma_{1-s(t)}(p^k)}{|p|^{ks}} \\ &= \rho_j(1) \prod_{(p)} \sum_{k=0}^\infty \frac{\lambda_j(p^k)|p|^{k(s(t)-1)}}{|p|^{ks}} \frac{1-|p|^{2(1-s(t))(k+1)}}{1-|p|^{2(1-s(t))}}, \end{aligned}$$

and thus

$$R(s) = \rho_j(1) \prod_{(p)} \frac{1-|p|^{-2s}}{1-\lambda_j(p)|p|^{-(s-s(t)+1)} + |p|^{-2(s-s(t)+1)}} \times \frac{1}{1-\lambda_j(p)|p|^{-(s+s(t)-1)} + |p|^{-2(s+s(t)-1)}}. \tag{3.1}$$

We can identify the L-functions to get

$$R(s) = \rho_j(1) \frac{L(u_j, \frac{s-s(t)+1}{2}) L(u_j, \frac{s+s(t)-1}{2})}{\zeta_K(s)}.$$

Now,

$$J_j(t) = I_j(\overline{s(t)}),$$

so that

$$\begin{aligned} J_j(t) &= \frac{2^{-1}}{\xi_K(s(t))} \frac{\pi^{-\overline{s(t)}}}{\Gamma(\overline{s(t)})} \prod \Gamma\left(\tfrac{\overline{s(t)} \pm (s(t)-1) \pm it_j}{2}\right) \rho_j(1) \frac{L(u_j, \frac{1}{2}-it) L(u_j, \sigma_t - \frac{1}{2})}{\zeta_K(\overline{s(t)})} \\ &= \frac{2^{s(t)-1}\pi^{2it}\rho_j(1)}{|d_K|^{s(t)/2} |\zeta_K(s(t))|^2} L(u_j, \tfrac{1}{2}-it) L(u_j, \sigma_t - \tfrac{1}{2}) \frac{\prod \Gamma\left(\frac{\overline{s(t)} \pm (s(t)-1) \pm it_j}{2}\right)}{|\Gamma(s(t))|^2}. \end{aligned}$$

With Stirling asymptotics we see that the quotient of Gamma factors is $O(|t|^{1-2\sigma_t})$. We use the estimate

$$\log^{-2}|t| \ll \zeta_K(s(t)) \ll \log^2|t|, \tag{3.2}$$

which follows by adapting [**29**, (3.5.1) and Theorem 3.11] for $L(s,\chi)$ and the zero-free region [**14**]. For the L-functions we need a subconvex bound to guarantee vanishing. Petridis and Sarnak [**21**] show that there is a $\delta > 0$ such that

$$L(u_j, \tfrac{1}{2}+it) \ll_j |1+t|^{1-\delta}.$$

In fact, they have $\delta = 7/166$, although this is not crucial for us. Hence, $J_j(t) \to 0$, as $t \to \infty$. □

3.2. Continuous Part. Let $h(y) \in C^\infty(\mathbb{R}^+)$ be a rapidly decreasing function at 0 and ∞ so that $h(y) = O_N(y^N)$ for $0 < y < 1$ and $h(y) = O_N(y^{-N})$ for $y \gg 1$ for all $N \in \mathbb{N}$. Denote the Mellin transform of h by $H = \mathcal{M}h$, i.e.

$$H(s) = \int_0^\infty h(y) y^{-s} \frac{dy}{y}$$

and the Mellin inversion formula gives

$$h(y) = \frac{1}{2\pi i} \int_{(\sigma)} H(s) y^s \, ds,$$

for any $\sigma \in \mathbb{R}$. We consider the incomplete Eisenstein series denoted by

$$F_h(p) = E(p|h) = \sum_{\gamma \in \Gamma_\infty \backslash \Gamma} h(y(\gamma p)) = \frac{1}{2\pi i} \int_{(3)} H(s) E(p,s) \, ds,$$

where h is a smooth function on $\mathbb{R}^+$ with compact support. We prove the following lemma.

LEMMA 2. *Let h be a function satisfying the conditions stated above. Then*

$$\int_M F_h(p) |E(v,s(t))|^2 d\mu(p) \sim \begin{cases} \int_M F_h(p) E(p, 2\sigma_\infty) d\mu(p), & \text{if } \sigma_\infty > 1, \\ \frac{2(2\pi)^2}{|\mathcal{O}^\times||d_K|\zeta_K(2)} \log t \int_M F_h(p) d\mu(p), & \text{if } (\sigma_t - 1)\log t \to 0, \end{cases}$$

as $t \to \infty$.

Now, unfolding gives

$$\begin{aligned} \int_M F_h(p) |E(p,s(t))|^2 d\mu(p) &= \int_M \frac{1}{2\pi i} \int_{(3)} H(s) E(p,s) \, ds \, |E(p,s(t))|^2 d\mu(p) \\ &= \int_0^\infty \frac{1}{2\pi i} \int_{(3)} H(s) y^s \, ds \int_F |E(p,s(t))|^2 d\mu(p) \\ &= \int_0^\infty h(y) \operatorname{vol}(F) \left(\sum_{n \in \mathcal{O}} |a_n(y,s(t))|^2 \right) \frac{dy}{y^3}. \end{aligned}$$

We will deal separately with the contribution of the $n = 0$ term and the rest. We factor out the constant $\operatorname{vol}(F)$ in the analysis below.

3.2.1. *Contribution of the constant term.* We know that

$$|a_0(y, s(t))|^2 = y^{2\sigma_t} + 2\,\mathrm{Re}(\varphi(s(t))y^{2-2it}) + |\varphi(s(t))|^2 y^{4-2\sigma_t}.$$

The first term is

$$\int_0^\infty h(y) y^{2\sigma_t - 2} \frac{dy}{y} = H(2 - 2\sigma_t),$$

which converges to $H(2 - 2\sigma_\infty)$. For the second term we first have that

$$\varphi(s(t)) \int_0^\infty h(y) y^{-2it} \frac{dy}{y} = \varphi(s(t)) H(2it).$$

Since $H(s)$ is in Schwartz class in t, the function $H(2it)$ decays rapidly, whereas $\varphi(s(t))$ is bounded. By taking complex conjugates we see that the second term will also tend to zero. Finally, for the third expression in the constant term we get

$$|\varphi(s(t))|^2 \int_0^\infty h(y) y^{2-2\sigma_t} \frac{dy}{y} = |\varphi(s(t))|^2 H(2\sigma_t - 2).$$

If $\sigma_\infty \neq 1$ then

$$|\varphi(s(t))| = \left| \frac{2\pi}{s(t) - 1} \frac{\zeta_K(s(t) - 1)}{\zeta_K(s(t))} \right|.$$

To estimate this we need the convexity bound for ζ_K,

$$\zeta_K(s(t) - 1) = \zeta_K(\sigma_t - 1 + it) = O(|t|^{1 - \sigma_t/2 + \epsilon}),$$

and of course $1/(s(t) - 1) = O(|t|^{-1})$. Combining all of this with (3.2) we get

$$\varphi(s(t)) = O(|t|^{-\sigma_t/2 + \epsilon}),$$

and so

$$\varphi(s(t)) \to 0, \tag{3.3}$$

as $t \to \infty$, when $\sigma_\infty \neq 1$. So in summary, the contribution of the constant term converges to $H(2 - 2\sigma_\infty)$ if $\sigma_\infty \neq 1$ and is $O(1)$ otherwise.

3.2.2. *Contribution of the non-constant terms.* In this case the contribution equals

$$\begin{aligned} A(t) &= \int_0^\infty \frac{1}{2\pi i} \int_{(3)} H(s) y^s \, ds \frac{4y^2}{|\xi_K(s(t))|^2} \sum_{n \in \mathcal{O}}{}' |n|^{2\sigma_t - 2} |\sigma_{1-s(t)}(n)|^2 \left| K_{s(t)-1}\left(\tfrac{4\pi |n| y}{\sqrt{|d_K|}}\right) \right|^2 \frac{dy}{y^3} \\ &= \frac{4|\mathcal{O}^\times|}{|\xi_K(s(t))|^2} \frac{1}{2\pi i} \int_{(3)} H(s) \left(\tfrac{\sqrt{|d_K|}}{4\pi}\right)^s \sum_{n \in \mathcal{O}/\sim}{}' \frac{|\sigma_{1-s(t)}(n)|^2}{|n|^{s+2-2\sigma_t}} \int_0^\infty y^s |K_{s(t)-1}(y)|^2 \frac{dy}{y} \, ds, \end{aligned}$$

where $a \sim b$ if a and b generate the same ideal in $\mathcal{O}$ and prime in the summation denotes that it is taken over $n \neq 0$. We now need to evaluate the series. Keeping in mind that $N(p) = |p|^2$, we get by a standard calculation

$$\begin{aligned} \sum_{n \in \mathcal{O}/\sim}{}' \frac{\sigma_a(n)\sigma_b(n)}{|n|^s} &= \prod_{(p):\text{prime ideal}} \sum_{k=0}^\infty \frac{\sigma_a(p^k)\sigma_b(p^k)}{|p|^{ks}} \\ &= \prod_{(p)} \frac{1 - |p|^{2(a+b-s)}}{(1 - |p|^{-s})(1 - |p|^{2a-s})(1 - |p|^{2b-s})(1 - |p|^{2a+2b-s})}, \end{aligned}$$

and hence

$$\sum_{n \in \mathcal{O}/\sim}{}' \frac{\sigma_a(n)\sigma_b(n)}{|n|^s} = \frac{\zeta_K(\frac{s}{2})\zeta_K(\frac{s}{2} - a)\zeta_K(\frac{s}{2} - b)\zeta_K(\frac{s}{2} - a - b)}{\zeta_K(s - a - b)}.$$

For $a = \bar{b} = 1 - s(t)$ and $s = s - 2(\sigma_t - 1)$ this becomes

$$\sum_{n \in \mathcal{O}/\sim}' \frac{|\sigma_{1-s(t)}(n)|^2}{|n|^{s-2\sigma_t+2}} = \frac{\zeta_K(\frac{s}{2} - \sigma_t + 1)\zeta_K(\frac{s}{2} + it)\zeta_K(\frac{s}{2} - it)\zeta_K(\frac{s}{2} + \sigma_t - 1)}{\zeta_K(s)}.$$

Again, by [**8**, 6.576 (4)] we see that

$$\int_0^\infty y^s |K_{s(t)-1}(y)|^2 \frac{dy}{y} = \frac{2^{s-3}}{\Gamma(s)}\Gamma(\tfrac{s}{2} - \sigma_t + 1)\Gamma(\tfrac{s}{2} + it)\Gamma(\tfrac{s}{2} - it)\Gamma(\tfrac{s}{2} + \sigma_t - 1).$$

Hence, $A(t)$ becomes

$$\begin{aligned} A(t) &= \frac{|\mathcal{O}^\times|}{|\xi_K(s(t))|^2} \frac{1}{4\pi i} \int_{(3)} H(s) \frac{\xi_K(\frac{s}{2} - \sigma_t + 1)\xi_K(\frac{s}{2} - it)\xi_K(\frac{s}{2} + it)\xi_K(\frac{s}{2} + \sigma_t - 1)}{\xi_K(s)}\, ds \\ &= \frac{|\mathcal{O}^\times|}{|\xi_K(s(t))|^2} \frac{1}{4\pi i} \int_{(3)} B(s)\, ds, \end{aligned}$$

say. By the Dirichlet Class Number Formula (2.2) for ζ_K, the completed zeta function ξ_K has a simple pole at $s = 1$ with

$$\operatorname*{Res}_{s=1} \xi_K(s) = \frac{1}{|\mathcal{O}^\times|}.$$

There is also a simple pole at $s = 0$. It follows that the poles of $B(s)$ in the region $\operatorname{Re} s \geq 1$ are at $2 \pm 2it$, $2\sigma_t$, $2\sigma_t - 2$, and $4 - 2\sigma_t$. Moving the line of integration to $\operatorname{Re} s = 1$ gives

$$\begin{aligned} A(t) = \frac{|\mathcal{O}^\times|}{2|\xi_K(s(t))|^2} \Bigg(&\operatorname*{Res}_{s=2\pm 2it} B(s) + \operatorname*{Res}_{s=2\sigma_t} B(s) + \delta_t \operatorname*{Res}_{s=4-2\sigma_t} B(s) \\ &+ (1 - \delta_t) \operatorname*{Res}_{s=2\sigma_t - 2} B(s) + \frac{1}{2\pi i} \int_{(1)} B(s)\, ds \Bigg), \\ = A_1 + A_2 + \cdots + A_5, \end{aligned}$$

where $\delta_t = 1$ if $\sigma_t < 3/2$ and 0 otherwise. We deal with each of the residues A_i separately. For the first term we have

$$A_1 = \frac{H(2 \pm 2it)}{|\xi_K(\sigma_t + it)|^2} \frac{\xi_K(2 - \sigma_t \pm it)\xi_K(1 \pm 2it)\xi_K(\sigma_t \pm it)}{\xi_K(2 \pm 2it)}.$$

By Stirling asymptotics and convexity estimates for the Dedekind zeta functions, the quotient of the ξ_K functions is bounded by $|t|^{1-2\sigma_t} \log^{10}|t|$. By virtue of H being of rapid decay in t it follows that $A_1 \to 0$ as $t \to \infty$.

The second term is

$$A_2 = H(2\sigma_t) \frac{\xi_K(2\sigma_t - 1)}{\xi_K(2\sigma_t)}.$$

If $\sigma_\infty \neq 1$ then

$$A_2 \to H(2\sigma_\infty) \frac{\xi_K(2\sigma_\infty - 1)}{\xi_K(2\sigma_\infty)},$$

but if $\sigma_t \to 1$ then

$$A_2 \sim H(2) \frac{1}{2|\mathcal{O}^\times|\xi_K(2)(\sigma_t - 1)}.$$

Now, in the third term we use the form (2.3) of φ and the fact that ξ_K satisfies the functional equation (2.1). We can then write

$$A_3 = \delta_t H(4-2\sigma_t)|\varphi(s(t))|^2 \frac{\xi_K(3-2\sigma_t)}{\xi_K(4-2\sigma_t)}.$$

By (3.3) we have that $\varphi(s(t)) \to 0$ as $t \to \infty$ for $\sigma_\infty \neq 1$. Hence, if $\sigma_\infty \neq 1$, then $A_3 \to 0$. On the other hand, if $\sigma_\infty = 1$, then

$$A_3 \sim \frac{\delta_t}{2|\mathcal{O}^\times|} H(2)|\varphi(s(t))|^2 \frac{-1}{\xi_K(2)(\sigma_t - 1)},$$

which is bounded.

For the fourth term we have

$$A_4 = (1-\delta_t)\zeta_K(0)H(2\sigma_t - 2)|\varphi(s(t))|^2,$$

which clearly converges to 0 if $\sigma_\infty \neq 1$ and is bounded for $\sigma_\infty = 1$ as in the previous case.

Finally, the fifth term is

$$A_5 = \frac{|\mathcal{O}^\times|}{2|\xi_K(s(t))|^2} \frac{1}{2\pi i} \int_{(1)} B(s)\, ds = \frac{|\mathcal{O}^\times|}{2|\xi_K(\sigma_t + it)|^2} \mathcal{I},$$

where

$$\mathcal{I} = \frac{1}{2\pi} \int_{-\infty}^{\infty} H(1+i\tau) \frac{|\xi_K(\sigma_t - \frac{1}{2} + i\tau)|^2 \xi_K(\frac{1}{2} + i(\tau + t))\xi_K(\frac{1}{2} + i(\tau - t))}{\xi_K(1+2i\tau)}\, d\tau.$$

We now estimate the growth of A_5 in terms of t. The exponential contribution from the gamma functions in the integral is equal to

$$(e^{-\frac{\pi}{2}|t|})^2 e^{-\frac{\pi}{2}|\tau + t|} e^{-\frac{\pi}{2}|\tau - t|} e^{\frac{\pi}{2}|2\tau|} \ll e^{-\pi|t|}.$$

This cancels with the exponential growth of $|\xi_K(s(t))|^2$. Since $H(1+i\tau)$ decays rapidly, we can bound $\zeta_K(\sigma_t - \frac{1}{2} + i\tau)$ polynomially and absorb it into H. Hence

$$A_5 \ll \frac{\log^4|t|}{(|t|^{\sigma_t - 1/2})^2} \int_{-\infty}^{\infty} \widetilde{H}(\tau)|\zeta_K(\tfrac{1}{2} + i(\tau + t))\zeta_K(\tfrac{1}{2} + i(\tau - t))|\, d\tau.$$

where $\widetilde{H}$ is a function of rapid decay. The Dedekind zeta functions can be estimated with the subconvex bound

$$\zeta_K(\tfrac{1}{2} + it) \ll t^{1/3+\epsilon}$$

due to Heath-Brown [**9**] (he proves a more general bound for arbitrary number fields of degree n). We get

$$A_5 \ll |t|^{5/3 - 2\sigma_t + 2\epsilon} \log^4|t| \int_{-\infty}^{\infty} \widetilde{H}(\tau)(t^{-1} + |\tau t^{-1} + 1|)^{1/3+\epsilon}(t^{-1} + |\tau t^{-1} - 1|)^{1/3+\epsilon}\, d\tau,$$

which is $o(1)$ since $\sigma_t \geq 1$.

Hence we have proved that the integral

$$\int_M F_h(p)|E(p, s(t))|^2 d\mu(p)$$

converges to

$$\mathrm{vol}(F)\left(H(2-2\sigma_\infty) + H(2\sigma_\infty)\frac{\xi_K(2\sigma_\infty - 1)}{\xi_K(2\sigma_\infty)}\right),$$

if $\sigma_\infty \neq 1$. On the other hand, for $\sigma_\infty = 1$ the contribution is asymptotic to

$$\mathrm{vol}(F)H(2)\frac{1-|\varphi(s(t))|^2}{2|\mathcal{O}^\times|\xi_K(2)(\sigma_t-1)}+O(1). \tag{3.4}$$

To finish the proof, we apply Mellin inversion and unfold backwards to see that

$$\begin{aligned}\mathrm{vol}(F)\left(H(2-2\sigma_\infty)+H(2\sigma_\infty)\frac{\xi_K(2\sigma_\infty-1)}{\xi_K(2\sigma_\infty)}\right)&\\ &=\int_0^\infty h(y)\,\mathrm{vol}(F)(y^{2\sigma_\infty-2+2}+\varphi(2\sigma_\infty)y^{-2\sigma_\infty+2})\frac{dy}{y^3}\\ &=\int_0^\infty h(y)\left(\int_F E(z+jy,2\sigma_\infty)\,dz\right)\frac{dy}{y^3}\\ &=\int_M F_h(p)E(p,2\sigma_\infty)d\mu(p),\end{aligned}$$

and

$$\mathrm{vol}(F)H(2)=\int_M F_h(p)d\mu(p).$$

For the second case, we need to estimate the quotient with the scattering matrix. We will show that

$$\frac{1-|\varphi(s(t))|^2}{2|\mathcal{O}^\times|\xi_K(2)(\sigma_t-1)}\sim\frac{2(2\pi)^2}{|\mathcal{O}^\times||d_K|\zeta_K(2)}\log t.$$

Let $G(\sigma)=\varphi(\sigma+it)\varphi(\sigma-it)$ and notice that

$$G'(\sigma)=\frac{\varphi'}{\varphi}(\sigma\pm it)G(\sigma),$$

where the $\pm$ denotes the linear combination $\frac{\varphi'}{\varphi}(\sigma\pm it)=\frac{\varphi'}{\varphi}(\sigma+it)+\frac{\varphi'}{\varphi}(\sigma-it)$. We then apply the mean value theorem twice on the intervals $[1,\sigma]$ and $[1,\sigma']$, respectively. We get

$$\frac{G(1)-G(\sigma)}{1-\sigma}=\left(G(1)-(1-\sigma')G(\sigma'')\frac{\varphi'}{\varphi}(\sigma''\pm it)\right)\frac{\varphi'}{\varphi}(\sigma'\pm it),$$

where $1\le\sigma''\le\sigma'\le\sigma$. On noticing that $G(1)=1$, this gives

$$\frac{1-|\varphi(\sigma+it)|^2}{1-\sigma}=\left(1-(1-\sigma')|\varphi(\sigma''+it)|^2\frac{\varphi'}{\varphi}(\sigma''\pm it)\right)\frac{\varphi'}{\varphi}(\sigma'\pm it).$$

Using the asymptotics

$$\frac{\varphi'}{\varphi}(\sigma\pm it)\sim-4\log t, \tag{3.5}$$

and the fact that $|\varphi(\sigma+it)|$ is bounded for $\sigma\ge1$ proves the lemma. The estimate (3.5) follows immediately from the standard asymptotics for the digamma function,

$$\frac{\Gamma'}{\Gamma}(\sigma+it)=\log|t|+O(1),$$

and the Weyl bound

$$\frac{\zeta_K'}{\zeta_K}(\sigma+it)\ll\frac{\log t}{\log\log t},$$

for ζ_K'/ζ_K (see [**29**, Theorems 3.11 and 5.17] and [**4**]). □

PROOFS OF THEOREMS 2 AND 3. These follow now from Lemmas 1 and 2 by approximation arguments similar to [**17**] and [**13**]. □

Theorem 1 now follows easily.

PROOF OF THEOREM 1. By the functional equation (2.5) of $E(p,s)$, we get

$$\begin{aligned}|v_{o_n}|^2 d\mu(p) &= |(\operatorname*{Res}_{s=\rho_n} \varphi(s))^{-1} \operatorname*{Res}_{s=\rho_n} E(p,s)|^2 d\mu(p)\\ &= |(\operatorname*{Res}_{s=\rho_n} \varphi(s))^{-1} \operatorname*{Res}_{s=\rho_n} \varphi(s)E(p,2-s)|^2 d\mu(p)\\ &= |E(p,2-\rho_n)|^2 d\mu(p).\end{aligned}$$

We apply Theorem 3 with $\sigma_\infty = 2-\gamma_\infty$ (since $\gamma_\infty < 1$) to conclude the proof. □

References

[1] D. V. Anosov, *Geodesic flows on closed Riemann manifolds with negative curvature.*, Proceedings of the Steklov Institute of Mathematics, No. 90 (1967). Translated from the Russian by S. Feder, American Mathematical Society, Providence, R.I., 1969. MR0242194

[2] Tetsuya Asai, *On a certain function analogous to* $\log_\eta(z)$, Nagoya Math. J. **40** (1970), 193–211. MR0271038

[3] Werner Ballmann, *Lectures on spaces of nonpositive curvature*, DMV Seminar, vol. 25, Birkhäuser Verlag, Basel, 1995. With an appendix by Misha Brin. MR1377265

[4] M. D. Coleman, *A zero-free region for the Hecke L-functions*, Mathematika **37** (1990), no. 2, 287–304, DOI 10.1112/S0025579300013000. MR1099777

[5] Y. Colin de Verdière, *Ergodicité et fonctions propres du laplacien* (French, with English summary), Comm. Math. Phys. **102** (1985), no. 3, 497–502. MR818831

[6] Semyon Dyatlov, *Microlocal limits of Eisenstein functions away from the unitarity axis*, J. Spectr. Theory **2** (2012), no. 2, 181–202, DOI 10.4171/JST/26. MR2913877

[7] J. Elstrodt, F. Grunewald, and J. Mennicke, *Groups acting on hyperbolic space*, Springer Monographs in Mathematics, Springer-Verlag, Berlin, 1998. Harmonic analysis and number theory. MR1483315

[8] I. S. Gradshteyn and I. M. Ryzhik, *Table of integrals, series, and products*, 7th ed., Elsevier/Academic Press, Amsterdam, 2007. Translated from the Russian; Translation edited and with a preface by Alan Jeffrey and Daniel Zwillinger; With one CD-ROM (Windows, Macintosh and UNIX). MR2360010

[9] D. R. Heath-Brown, *The growth rate of the Dedekind zeta-function on the critical line*, Acta Arith. **49** (1988), no. 4, 323–339, DOI 10.4064/aa-49-4-323-339. MR937931

[10] Dieter Heitkamp, *Hecke-Theorie zur* $\mathrm{SL}(2;\mathfrak{o})$ (German), Schriftenreihe des Mathematischen Instituts der Universität Münster, 3. Serie [Series of the Mathematical Institute of the University of Münster, 3rd Series], vol. 5, Universität Münster, Mathematisches Institut, Münster, 1992. MR1158047

[11] Roman Holowinsky and Kannan Soundararajan, *Mass equidistribution for Hecke eigenforms*, Ann. of Math. (2) **172** (2010), no. 2, 1517–1528. MR2680499

[12] Dmitry Jakobson, *Quantum unique ergodicity for Eisenstein series on* $\mathrm{PSL}_2(\mathbf{Z})\backslash\mathrm{PSL}_2(\mathbf{R})$ (English, with English and French summaries), Ann. Inst. Fourier (Grenoble) **44** (1994), no. 5, 1477–1504. MR1313792

[13] Shin-ya Koyama, *Quantum ergodicity of Eisenstein series for arithmetic 3-manifolds*, Comm. Math. Phys. **215** (2000), no. 2, 477–486, DOI 10.1007/s002200000317. MR1799856

[14] Edmund Landau, *Über die Wurzeln der Zetafunktion* (German), Math. Z. **20** (1924), no. 1, 98–104, DOI 10.1007/BF01188073. MR1544664

[15] Elon Lindenstrauss, *Invariant measures and arithmetic quantum unique ergodicity*, Ann. of Math. (2) **163** (2006), no. 1, 165–219, DOI 10.4007/annals.2006.163.165. MR2195133

[16] Wenzhi Luo, *Nonvanishing of L-values and the Weyl law*, Ann. of Math. (2) **154** (2001), no. 2, 477–502, DOI 10.2307/3062104. MR1865978

[17] Wen Zhi Luo and Peter Sarnak, *Quantum ergodicity of eigenfunctions on* $\mathrm{PSL}_2(\mathbf{Z})\backslash\mathbf{H}^2$, Inst. Hautes Études Sci. Publ. Math. **81** (1995), 207–237. MR1361757

[18] Yiannis N. Petridis, Nicole Raulf, and Morten S. Risager, *Erratum to "Quantum limits of Eisenstein series and scattering states" [MR3121690]*, Canad. Math. Bull. **56** (2013), no. 4, 827–828, DOI 10.4153/CMB-2013-008-6. MR3121691

[19] Yiannis N. Petridis, Nicole Raulf, and Morten S. Risager, *Quantum limits of Eisenstein series and scattering states*, Canad. Math. Bull. **56** (2013), no. 4, 814–826, DOI 10.4153/CMB-2011-200-2. MR3121690

[20] Yiannis N. Petridis and Morten S. Risager, *Dissolving of cusp forms: higher-order Fermi's golden rules*, Mathematika **59** (2013), no. 2, 269–301, DOI 10.1112/S0025579312001118. MR3081772

[21] Yiannis N. Petridis and Peter Sarnak, *Quantum unique ergodicity for* $\mathrm{SL}_2(\mathcal{O})\backslash\mathbf{H}^3$ *and estimates for L-functions*, J. Evol. Equ. **1** (2001), no. 3, 277–290, DOI 10.1007/PL00001371. Dedicated to Ralph S. Phillips. MR1861223

[22] R. Phillips and P. Sarnak, *Automorphic spectrum and Fermi's golden rule*: *Festschrift on the occasion of the 70th birthday of Shmuel Agmon*, J. Anal. Math. **59** (1992), 179–187, DOI 10.1007/BF02790224. MR1226958

[23] R. S. Phillips and P. Sarnak, *On cusp forms for co-finite subgroups of* $\mathrm{PSL}(2,\mathbf{R})$, Invent. Math. **80** (1985), no. 2, 339–364, DOI 10.1007/BF01388610. MR788414

[24] Nicole Raulf, *Traces of Hecke operators acting on three-dimensional hyperbolic space*, J. Reine Angew. Math. **591** (2006), 111–148, DOI 10.1515/CRELLE.2006.016. MR2212881

[25] Zeév Rudnick and Peter Sarnak, *The behaviour of eigenstates of arithmetic hyperbolic manifolds*, Comm. Math. Phys. **161** (1994), no. 1, 195–213. MR1266075

[26] P. Sarnak, *The arithmetic and geometry of some hyperbolic three-manifolds*, Acta Math. **151** (1983), no. 3-4, 253–295, DOI 10.1007/BF02393209. MR723012

[27] A. I. Šnirel′man, *Ergodic properties of eigenfunctions* (Russian), Uspehi Mat. Nauk **29** (1974), no. 6(180), 181–182. MR0402834

[28] Kannan Soundararajan, *Quantum unique ergodicity for* $\mathrm{SL}_2(\mathbb{Z})\backslash\mathbb{H}$, Ann. of Math. (2) **172** (2010), no. 2, 1529–1538. MR2680500

[29] E. C. Titchmarsh, *The theory of the Riemann zeta-function*, 2nd ed., The Clarendon Press, Oxford University Press, New York, 1986. Edited and with a preface by D. R. Heath-Brown. MR882550

[30] Jimi L. Truelsen, *Quantum unique ergodicity of Eisenstein series on the Hilbert modular group over a totally real field*, Forum Math. **23** (2011), no. 5, 891–931, DOI 10.1515/FORM.2011.031. MR2836373

[31] Matthew P. Young, *The quantum unique ergodicity conjecture for thin sets*, Adv. Math. **286** (2016), 958–1016, DOI 10.1016/j.aim.2015.09.013. MR3415701

[32] Steven Zelditch, *Uniform distribution of eigenfunctions on compact hyperbolic surfaces*, Duke Math. J. **55** (1987), no. 4, 919–941, DOI 10.1215/S0012-7094-87-05546-3. MR916129

McGill University, Department of Mathematics and Statistics, Burnside Hall, 805 Sherbrooke Street West, Montreal, Quebec, H3A 0B9 Canada

Email address: `n.laaksonen@ucl.ac.uk`

Contemporary Mathematics
Volume **739**, 2019
https://doi.org/10.1090/conm/739/14897

Observability and quantum limits for the Schrödinger equation on $\mathbb{S}^d$

Fabricio Macià and Gabriel Rivière

ABSTRACT. In this note, we describe our recent results on semiclassical measures for the Schrödinger evolution on Zoll manifolds. We focus on the particular case of eigenmodes of the Schrödinger operator on the sphere endowed with its canonical metric. We also recall the relation of this problem with the observability question from control theory. In particular, we exhibit examples of open sets and potentials on the 2-sphere for which observability fails for the evolution problem while it holds for the stationary one. Finally, we give some new results in the case where the Radon transform of the potential identically vanishes.

1. Introduction

Let $\mathbb{S}^d$ be the sphere of dimension $d \geq 2$ endowed with its canonical metric and let V be a smooth real valued function on $\mathbb{S}^d$. Our goal here is to understand the behavior of the Schrödinger eigenfunctions:

$$\left(-\frac{1}{2}\Delta + V(x)\right) u(x) = \lambda u(x), \quad ||u||_{L^2(\mathbb{S}^d)} = 1, \tag{1}$$

in the high-frequency limit $\lambda \to \infty$. Such functions can be identified with stationary solutions of the following Schrödinger equation:

$$i\partial_t v(t,x) = \left(-\frac{1}{2}\Delta + V(x)\right) v(t,x), \quad v|_{t=0} = v_0 \in L^2(\mathbb{S}^d). \tag{2}$$

Solutions of (2) encode the position probability density of a quantum particle confined on the surface of the sphere and propagating under the action of the potential V. In this note, we revisit and extend some of the the results in [**25**], and reinterpret them from the light of control theory for the Schrödinger equation.

1.1. Controllability and observability. Let us briefly recall the basics of controllabilty theory for this equation. Fix ω an open set in $\mathbb{S}^d$ and some final time $T > 0$. The controllability problem for (2) is the following. Given ψ_0 and ψ_1 in

FM takes part into the visiting faculty program of ICMAT and is partially supported by grants ERC Starting Grant 277778 and MTM2013-41780-P (MEC) and MTM2017-85934-C3-3-P (MINECO, Spain).

GR is partially supported by the Agence Nationale de la Recherche through the Labex CEMPI (ANR-11-LABX-0007-01) and the ANR project GeRaSic (ANR-13-BS01- 0007-01).

$L^2(\mathbb{S}^d)$, is it possible to find $f(t,x)$ in $L^2([0,T]\times\mathbb{S}^d)$ such that the solution $\psi(t,x)$ of

$$i\partial_t\psi(t,x)+\left(\frac{1}{2}\Delta-V(x)\right)\psi(t,x)=\mathbf{1}_\omega(x)f(t,x),\quad \psi|_{t=0}=\psi_0 \tag{3}$$

satisfies $\psi|_{t=T}=\psi_1$? In other words, can you drive any ψ_0 to any u_1 in time T through the Schrödinger evolution by acting only the set ω? If this is possible, we say that *the Schrödinger equation is controllable in time T on the open set ω.*

It turns out that the controllability property is equivalent to a stability-type estimate for the solutions to the homogeneous Schrödinger equation (2). The *Schrödinger equation is said to be observable on the set ω in time $T>0$* if there exists $C_{\omega,T}>0$ such that

$$\forall v_0\in L^2(\mathbb{S}^d),\ \|v_0\|^2_{L^2(\mathbb{S}^d)}\leq C_{\omega,T}\int_0^T\|v(t,x)\|^2_{L^2(\omega)}dt, \tag{4}$$

where $v(t,x)$ is the solution to the homogeneous Schrödinger equation (2) with initial data v_0. It turns out that the controllability property for (3) and the observability for (2) are equivalent notions. The simple proof of this fact is part of the so-called Hilbert Uniqueness Method [**22**]. Let us briefly recall it here for the sake of completeness. Since the equation is linear and reversible, the problem reduces to studying the particular case $\psi_1=0$. One then considers the operator Λ defined by:

$$\Lambda: L^2((0,T)\times\omega)\ni f\longmapsto\psi_f|_{t=0}\in L^2(\mathbb{S}^d),$$

where ψ_f is the solution to (3) with control f that satisfies $\psi_f|_{t=T}=0$. The fact that the equation is controllable in time T on the open set ω is equivalent to the fact that the linear bounded operator Λ is onto. This property, in turn, is equivalent to the unique solvability of the adjoint equation with an estimate:

$$\Lambda^*v_0=f\in\operatorname{Im}\Lambda^*,\quad ||v_0||^2_{L^2(\mathbb{S}^d)}\leq C||\Lambda^*v_0||^2_{L^2((0,T)\times\omega)},$$

by the closed graph theorem. It is straightforward to check that $\Lambda^*v_0=-i\mathbf{1}_\omega v$, where v is the solution to (2) with initial datum v_0 and therefore the result follows with $C_{\omega,T}=C$.

A remarkable result of Lebeau states that observability (and thus control of the Schrödinger equation) holds for any $T>0$ on the open set ω provided that the following *geometric control condition* is satisfied [**21**]:

$$K_\omega:=\left\{\gamma\text{ closed geodesic of }\mathbb{S}^d:\gamma\cap\omega=\emptyset\right\}=\emptyset. \tag{5}$$

Conversely, one can show that, if $K_{\overline{\omega}}\neq\emptyset$, then observability fails *for any choice of V in $\mathcal{C}^\infty(\mathbb{S}^d;\mathbb{R})$* – see for instance [**25**, Prop. 2.2]. The same result holds if $\mathbb{S}^d$ is replaced by a Riemannian manifold all whose geodesics are closed (these are called *Zoll manifolds*), see [**24, 25**]. Whereas Lebeau's result holds for any compact Riemannian manifold, the Geometric Control Condition is not necessary in general. For instance, observability holds under weaker hypotheses on ω on flat manifolds, see for instance [**1–3, 7, 9**], or on negatively curved manifolds [**4, 13, 28**].

1.2. Observability and Quantum Limits. When particularized to stationary solutions of (2), Lebeau's theorem shows that for every ω satisfying $K_\omega=\emptyset$, there exists $C_\omega>0$ such that, for every u solution of (1), one has

$$0<C_\omega\leq\int_\omega|u(x)|^2\text{vol}(dx), \tag{6}$$

where vol is the canonical volume measure on $\mathbb{S}^d$. Since the constant C_ω is independent of the frequency, estimate (6) provides a restriction on the regions in $\mathbb{S}^d$ on which the L^2-mass of high-frequency eigenfunctions can concentrate. We refer to [**8**, **26**] for more explicit relations between observability for eigenmodes (or quasimodes) and for the Schrödinger evolution.

In the case where $V \equiv 0$, eigenfunctions are merely spherical harmonics. Using their explicit expression one can prove that the observability estimate (6) fails as soon as $K_{\overline{\omega}} \neq \emptyset$. We refer the reader to the work by Jakobson and Zelditch [**19**] for the proof of a stronger result – see also [**5**, **23**] for alternative proofs that extend to other manifolds than the sphere.

Note that in spite of the fact that the observability estimate for eigenfunctions (6) is weaker than the corresponding estimate for time-dependent solutions (4), the conditions on ω under which these estimates hold are exactly the same when V vanishes identically on $\mathbb{S}^d$. In fact, the same phenomenon takes place on the planar disk under a weaker geometric condition: both estimates hold if ω intersects the boundary on an open set, and fail if $\overline{\omega}$ is strictly contained in the interior of the disk [**2**]. On the flat torus, both estimates hold for any open set ω, even in the presence of a non-zero potential [**1**, **3**, **7**, **9**].

It is therefore natural to ask whether or not estimates (4) and (6) are equivalent, *i.e.* on any compact manifold both estimates hold for the same class of open sets ω. In this note, we answer this question by the negative. Our examples are precisely Schrödinger operators on the sphere with non-constant potentials, or more generally, Laplacians on Zoll manifolds.

Before stating our results, let us mention that these questions are naturally related to certain problems arising in mathematical physics. In fact, consider the set $\mathcal{N}(\infty)$ of probability measures in $\mathbb{S}^d$ that are obtained as follows. A probability measure ν belongs to $\mathcal{N}(\infty)$ provided there exists a sequence of eigenfunctions (u_n) :

$$-\frac{1}{2}\Delta u_n + V u_n = \lambda_n u_n, \quad ||u_n||_{L^2(\mathbb{S}^d)} = 1,$$

with eigenvalues satisfying $\lambda_n \to +\infty$ such that

$$\lim_{n\to\infty} \int_{\mathbb{S}^d} a(x)|u_n|^2(x)\,\mathrm{vol}(dx) = \int_{\mathbb{S}^d} a(x)\nu(dx), \quad \text{for every } a \in \mathcal{C}^0(\mathbb{S}^d).$$

Measures in $\mathcal{N}(\infty)$ therefore describe the asymptotic mass distribution sequences of eigenfunctions (u_n) whose corresponding eigenvalues tend to infinity. If one integrates these objects against $a = \mathbf{1}_\omega$, then one recovers the quantity we were considering before. In quantum mechanics, they describe the probability of finding a particle in the quantum state u_n on the set ω. The problem of characterizing the probability measures in $\mathcal{N}(\infty)$ has attracted a lot of attention in the last forty years especially in the context of the so-called quantum ergodicity problem – see e.g. [**27**, **29**, **34**] for recent surveys on that topic. Elements in $\mathcal{N}(\infty)$ are often called *quantum limits*. In the case of $\mathbb{S}^d$, it is well known that $\mathcal{N}(\infty)$ is contained in $\mathcal{N}$ which is, by definition, the closed convex hull (with respect to the weak-$\star$ topology) of the set of probability measures δ_γ, where γ is a closed geodesic of $(\mathbb{S}^d, \mathrm{Can})$. Recall that

$$\int_{\mathbb{S}^d} a(x)\delta_\gamma(dx) = \frac{1}{2\pi}\int_0^{2\pi} a(\gamma(s))ds,$$

where the parametrization $\gamma(s)$ has unit speed. In the case where $V \equiv 0$, it was proved by Jakobson and Zelditch [**19**] that $\mathcal{N}(\infty) = \mathcal{N}$ – the same result holds on other manifolds with positive curvature [**5, 23**]. Again, it is natural to ask if this property remains true when V does not identically vanish. This is of course related to the above observability question and we shall again answer to this question by the negative provided V satisfies certain generic properties. In [**25**] we showed that the answer remains negative on certain Zoll manifolds, even when V vanishes.

We finally present a simple criterium relating asymptotic separation properties of the spectrum of the Schrödinger operator to the structure of the set $\mathcal{N}(\infty)$. We extend the proof given in [**25**] to the case of potentials with vanishing Radon transform.

2. Statement of the main results

In order to state our results, we need to define the Radon transform of the potential V. Denote by $G(\mathbb{S}^d) \simeq S^*\mathbb{S}^d/\mathbb{S}^1$ the space of closed geodesics on $\mathbb{S}^d$, which is a smooth symplectic manifold [**6**]. Then, one can define the Radon transform of V as follows:

$$\forall \gamma \in G(\mathbb{S}^d),\ \mathcal{I}(V)(\gamma) = \int_{\mathbb{S}^d} V(x)\delta_\gamma(dx).$$

This is a smooth function on $G(\mathbb{S}^d)$ which can also be identified with a smooth 0-homogeneous function on $T^*\mathbb{S}^d - \{0\}$. We denote by $\varphi^t_{\mathcal{I}(V)}$ the corresponding Hamiltonian flow on $T^*\mathbb{S}^d - \{0\}$ which can itself be identified with an Hamiltonian flow on the symplectic manifold $G(\mathbb{S}^d)$. We also define the second order average:

$$\mathcal{I}^{(2)}(V) := \mathcal{I}(V^2) - \frac{1}{2\pi}\int_0^{2\pi}\int_0^t \{V\circ\varphi^t, V\circ\varphi^s\}dsdt,$$

where φ^t denotes the geodesic flow on $S^*\mathbb{S}^d$ and $\{\cdot,\cdot\}$ stands for the Poisson bracket. This extends into a smooth 0-homogeneous function on $T^*\mathbb{S}^d - \{0\}$ that is invariant by the geodesic flow, and it can again be viewed as a function acting on $G(\mathbb{S}^d)$. We denote by $\varphi^t_{\mathcal{I}^{(2)}(V)}$ its Hamiltonian flow.

2.1. Observability of eigenfunctions.

Our first result is the following

THEOREM 2.1. *Let ω be an open set in $\mathbb{S}^d$. Suppose that one of the following conditions holds:*

(i) $K_{\omega,V} := \left\{\gamma \in G(\mathbb{S}^d) : \forall t \in \mathbb{R},\ \varphi^t_{\mathcal{I}(V)}(\gamma) \cap \omega = \emptyset\right\} = \emptyset.$

(ii) $\mathcal{I}(V)$ *is constant and*

$$K^{(2)}_{\omega,V} := \left\{\gamma \in G(\mathbb{S}^d) : \forall t \in \mathbb{R},\ \varphi^t_{\mathcal{I}^{(2)}(V)}(\gamma) \cap \omega = \emptyset\right\} = \emptyset.$$

Then, there exists $C_{\omega,V} > 0$ such that, for every u solution of (1), *one has*

$$0 < C_{\omega,V} \leq \|u\|^2_{L^2(\omega)} = \int_\omega |u(x)|^2 \operatorname{vol}(dx). \tag{7}$$

Note that $K_{\omega,V} \subseteq K_\omega$ and equality holds when $\mathcal{I}(V)$ is constant. Therefore, for non-constant $\mathcal{I}(V)$ it may happen that $K_{\omega,V} = \emptyset$ while $K_{\overline{\omega}}$ contains a nonempty open set of closed geodesics – see Remark 2.2 below. In particular, this statement shows that observability for eigenfunctions may hold even if $K_{\overline{\omega}} \neq \emptyset$ provided that we choose a good V. This contrasts with the case of observability for the

Schrödinger evolution on $\mathbb{S}^d$ where $K_{\overline{\omega}}$ implies the failure of the observability property (4). As in the classical argument of Lebeau, this Theorem follows from the unique continuation principle (for the case of low frequencies) and from the study of the microlocal lift of eigenfunctions (for the case of high frequencies).

REMARK 2.2. Let us explain how to construct ω and V such that $K_{\omega,V} = \emptyset$ while $K_{\overline{\omega}} \neq \emptyset$. Recall first that the space of geodesics $G(\mathbb{S}^2)$ can be identified with $\mathbb{S}^2$ [**6**, p. 54]. This can be easily seen as follows. Take an oriented closed geodesic γ. It belongs to an unique 2-plane in $\mathbb{R}^3$ which can be oriented via the orientation of the geodesic, and γ can be identified with the unit vector in $\mathbb{S}^2$ which is directly orthogonal to this oriented 2-plane. With that identification in mind, we also know from the works of Guillemin that $\mathcal{I} : V \in \mathcal{C}^\infty_{\text{even}}(\mathbb{S}^2) \mapsto \mathcal{I}(V) \in \mathcal{C}^\infty_{\text{even}}(\mathbb{S}^2)$ is an isomorphism [**15**]. We can now explain how to construct ω and V. Write $\mathbb{S}^2 := \{(x,y,z) : x^2+y^2+z^2 = 1\}$. Suppose first that the open set ω contains the north pole $(0,0,1)$ and that it does not intersect a small enough neighborhood of the equator $\Gamma = \{(x,y,0) : x^2+y^2 = 1\}$. For instance, one can take ω to be equal to $\{(x,y,z) : x^2+y^2+z^2 = 1 \text{ and } z > \epsilon\}$ with $\epsilon > 0$ small enough. In particular, there are infinitely many geodesics which belong to $K_{\overline{\omega}} \subset K_\omega$, i.e. the geometric control condition fails. In the space of geodesics $G(\mathbb{S}^2) \simeq \mathbb{S}^2$, the geodesics belonging to $K_{\overline{\omega}}$ correspond to a small neighborhood of the two poles $(0,0,-1)$ and $(0,0,1)$ of $\mathbb{S}^2$. Hence, if one chooses $V \in \mathcal{C}^\infty_{\text{even}}(\mathbb{S}^2)$ in such a way that $\mathcal{I}(V)$ has no critical points in a slightly bigger neighborhood[1], then one finds that $K_{\omega,V} = \emptyset$. Indeed, in that case, the Hamiltonian flow $\varphi^t_{\mathcal{I}(V)}$ has no critical points inside K_ω. Thus, it transports the uncontrolled geodesics of K_ω to geodesics which are geometrically controlled by the geodesic flow. Note that the condition of having no critical points inside K_ω is a priori nongeneric among smooth functions.

2.2. Description of $\mathcal{N}(\infty)$. Let us now turn to the related problem of characterizing the elements inside $\mathcal{N}(\infty)$. In this direction, we prove the following results:

THEOREM 2.3. *Let ν be a measure in $\mathcal{N}(\infty)$ and let $\gamma \in G(\mathbb{S}^d)$.*
(i) One then has

$$d_\gamma \mathcal{I}(V) \neq 0 \Longrightarrow \nu(\gamma) = 0.$$

(ii) If $\mathcal{I}(V)$ is identically constant then:

$$d_\gamma \mathcal{I}^{(2)}(V) \neq 0 \Longrightarrow \nu(\gamma) = 0.$$

In particular, whenever $\mathcal{I}(V)$ is non-constant or $\mathcal{I}(V)$ is constant but $\mathcal{I}^{(2)}(V)$ is not, one has

$$\mathcal{N} \neq \mathcal{N}(\infty).$$

THEOREM 2.4. *If $d = 2$, any ν in $\mathcal{N}(\infty)$ can be decomposed as follows:*

$$\nu = f \operatorname{vol} + \alpha \nu_{\text{sing}}$$

where $f \in L^1(\mathbb{S}^2)$, $\alpha \in [0,1]$ and ν_{sing} belongs to $\mathcal{N}_{\text{Crit}}(V)$ which is by definition the closed convex hull (with respect to the weak-$\star$ topology) of the set of probability measures δ_γ, where $d_\gamma \mathcal{I}(V) = 0$.
If $\mathcal{I}(V)$ is constant then ν_{sing} is supported on the set of critical points of $\mathcal{I}^{(2)}(V)$.

[1]This is possible thanks to Guillemin's result.

Concerning the conclusion of Theorem 2.4, we recall from Remark 2.2 that $\mathcal{I} : V \in \mathcal{C}^{\infty}_{\text{even}}(\mathbb{S}^2) \mapsto \mathcal{I}(V) \in \mathcal{C}^{\infty}_{\text{even}}(\mathbb{S}^2)$ is an isomorphism. In particular, $\mathcal{I}(V)$ can always be identified with a smooth function on the real projective plane $\mathbb{R}P^2$. Hence, for a generic choice of potential V, the set $\mathcal{N}_{\text{Crit}}(V)$ is the convex hull of finitely many measures carried by closed geodesics depending only on V – see for instance [**11**, Sect. 3.4] for discussions (and also examples) on critical points in this geometric framework.

In the proofs of Theorems 2.1, 2.3, and 2.4 we will make use of classical methods from microlocal analysis that were originally developed for the study of the eigenvalue distribution by Duistermaat-Guillemin [**12**], Weinstein [**31**] and Colin de Verdière [**10**]. Even if it sounds natural, it seems that the problem of characterizing $\mathcal{N}(\infty)$ in this geometric framework has not been explicitly considered in the literature before except when $V \equiv 0$ [**5, 18, 19, 23**]. We will show that these methods from microlocal analysis allow to obtain in a rather simple manner nontrivial results on the high frequency behaviour of Schrödinger eigenfunctions.

2.3. Relation with the study of eigenvalue distribution. Finally, observe that the following Theorem allows to establish a relation between the study of $\mathcal{N}(\infty)$ and the level spacings:

THEOREM 2.5. *Let $\lambda_1 < \lambda_2 < \lambda_3 < \dots$ be the sequence of distinct eigenvalues of $-\frac{\Delta}{2} + V$. Suppose that*

$$\lim_{j \to +\infty} \sqrt{\lambda_j}(\lambda_{j+1} - \lambda_j) = +\infty.$$

Then, for every γ in $G(\mathbb{S}^d)$, $\delta_\gamma \in \mathcal{N}(\infty)$.

Moreover, the same conclusion holds if $\mathcal{I}(V)$ is constant and if we suppose

$$\lim_{j \to +\infty} \lambda_j^{\frac{3}{2}}(\lambda_{j+1} - \lambda_j) = +\infty.$$

The first part of this Theorem was proved in [**25**, Sect. 6] in the slightly more general framework of Zoll manifolds. It follows the strategy presented in [**23**] which consists in computing quantum limits using coherent states for the non-stationary Schrödinger equation – see also [**18**] for a recent, different, proof of the result in [**23**]. When $\mathcal{I}(V)$ is constant, the proof from [**25**] can be adapted and we shall briefly explain in paragraph 3.6 which modifications should be made to get the second part. This result combined with Theorem 2.3 shows that, if $\mathcal{I}(V)$ is non constant, then we can find a subsequence of distinct eigenvalues $(\lambda_j)_{j \in S}$ such that $\lambda_{j+1} - \lambda_j = \mathcal{O}(\lambda_j^{-\frac{1}{2}})$ for $j \in S$ tending to $+\infty$. When $\mathcal{I}(V)$ is constant but $\mathcal{I}^{(2)}(V)$ is not, then we deduce that $\lambda_{j+1} - \lambda_j = \mathcal{O}(\lambda_j^{-\frac{3}{2}})$. In other words, this gives simple criteria under which you can prove the existence of distinct eigenvalues which are asymptotically very close. In the case where $d = 2$ and where $\mathcal{I}(V)$ is constant, a much stronger result on level spacings was recently proved in [**17**]. Yet, in higher dimension or in the case of vanishing averages, it is not clear that such a result could be directly deduced from the classical results on the distribution of eigenvalues from [**10, 16, 30–32**].

2.4. The case of nonstationary solutions and general Zoll manifolds. A natural extension of all the above problems is to consider the case of quasimodes and of nonstationary solutions when (M, g) is a more general Zoll manifold [**6**]. These issues were discussed in great details in [**25**]. In order to emphasize the main geometric ideas and to avoid the technical issues inherent to these generalizations, we only focus here on the simpler framework described above. We refer the interested reader to this reference for more precise results. Let us just mention that the results remain true for quasimodes,

$$\left(-\frac{1}{2}\Delta + V(x)\right) u_\lambda(x) = \lambda u_\lambda(x) + r(\lambda), \quad ||u_\lambda||_{L^2(\mathbb{S}^d)} = 1,$$

with $r(\lambda) = o(\lambda^{-\frac{3}{2}})$. In the case where $\mathcal{I}(V)$ is nonconstant, it is in fact sufficient to suppose $r(\lambda) = o(\lambda^{-\frac{1}{2}})$. Finally, note that our method combined with earlier results of Zelditch [**32, 33**] also allows to show that, when $V \equiv 0$, one has $\mathcal{N}(\infty) \neq \mathcal{N}$ for many Zoll metrics of revolution on $\mathbb{S}^2$ – see [**25**, Th. 1.4] for the precise statement.

3. Semiclassical measures and their invariance properties

In order to prove the above results, we will make use of the so-called semiclassical measures [**14**] – see also [**35**, Chap. 5] for an introduction on that topic. In particular, we introduce the semiclassical parameter $\hbar = \lambda^{-1/2}$ and we are interested in the solutions of the following problem:

$$\left(-\frac{\hbar^2\Delta}{2} + \hbar^2 V(x)\right) u_\hbar(x) = u_\hbar(x), \quad ||u_\hbar||_{L^2(\mathbb{S}^d)} = 1, \tag{8}$$

in the semiclassical limit $\hbar \to 0^+$.

3.1. Semiclassical measures. One can define the Wigner distribution of the quantum states $u_\hbar$:

$$\mu_\hbar : a \in \mathcal{C}_c^\infty(T^*\mathbb{S}^d) \mapsto \langle u_\hbar, \mathrm{Op}_\hbar(a) u_\hbar \rangle_{L^2(\mathbb{S}^d)},$$

where $\mathrm{Op}_\hbar(a)$ is a pseudodifferential operator in $\Psi^{-\infty}(\mathbb{S}^d)$ with principal symbol a – see [**35**, Ch. 4 and 14]. From the Calderón-Vaillancourt [**35**, Ch. 5], the sequence $(\mu_\hbar)_{\hbar\to 0^+}$ is bounded in $\mathcal{D}'(T^*\mathbb{S}^d)$. Thus, one can extract subsequences and we denote by $\mathcal{M}(\infty) \subset \mathcal{D}'(T^*\mathbb{S}^d)$ the set of all possible accumulation points (as $\hbar \to 0^+$) when $(u_\hbar)_{\hbar\to 0^+}$ varies among sequences satisfying (8). From the Gårding inequality [**35**, Ch. 4], one can in fact verify that any $\mu \in \mathcal{M}(\infty)$ is a finite positive measure on $T^*\mathbb{S}^d$. Hence, any such μ is called a *semiclassical measure*. Then, applying the composition rule for pseudodifferential operators [**35**, Ch. 4], one can show that any $\mu \in \mathcal{M}(\infty)$ is a probability measure supported on the unit cotangent bundle $S^*\mathbb{S}^d$ and that

$$\mathcal{N}(\infty) := \left\{ \int_{S_x^*\mathbb{S}^d} \mu(x, d\xi) : \ \mu \in \mathcal{M}(\infty) \right\}. \tag{9}$$

For more details on these facts, we refer the reader to [**35**, Ch. 5]. Finally, as a warm up, let us briefly remind how to prove that these measures are invariant by the geodesic flow φ^s, i.e. the Hamiltonian flow associated with the function $\frac{\|\xi\|_x^2}{2}$. For that purpose, we write, given $u_\hbar$ satisfying (8) and any a in $\mathcal{C}_c^\infty(T^*\mathbb{S}^d)$,

$$\left\langle u_\hbar, \left[-\frac{\hbar^2\Delta}{2} + \hbar^2 V, \mathrm{Op}_\hbar(a)\right] u_\hbar \right\rangle = 0. \tag{10}$$

We can now apply the commutation rule for pseudodifferential operators [**35**, Ch. 4] combined with the Calderón-Vaillancourt Theorem:

$$\frac{\hbar}{i}\left\langle u_\hbar, \mathrm{Op}_\hbar\left(\left\{\frac{\|\xi\|^2}{2}, a\right\}\right) u_\hbar\right\rangle = \mathcal{O}(\hbar^2),$$

where $\{,\}$ is the Poisson bracket. Dividing this equality by $\hbar$ and letting $\hbar$ goes to 0 in this equality, one finds (after a possible extraction) that $\mu(\{\|\xi\|^2, a\}) = 0$ for every a in $\mathcal{C}_c^\infty(T^*\mathbb{S}^d)$. From the properties of μ, it exactly shows that elements in $\mathcal{M}(\infty)$ are probability measures which are invariant by the geodesic flow φ^s acting on $S^*\mathbb{S}^d$.

3.2. Weinstein's averaging method. Note that all the arguments so far are valid on a general compact Riemannian manifold and we shall now see which extra properties can be derived in the case of $\mathbb{S}^d$ endowed with its canonical metric. For that purpose, we need to fix some conventions and to collect some well-known facts on the spectral properties of the Laplace-Beltrami operator on $\mathbb{S}^d$. First, given any a in $\mathcal{C}_c^\infty(T^*\mathbb{S}^d - \{0\})$, we introduce the Radon transform of a:

$$\mathcal{I}(a)(x,\xi) := \frac{1}{2\pi}\int_0^{2\pi} a \circ \varphi^{s\|\xi\|_x}(x,\xi)ds.$$

In the case of V, this definition can be identified with the Radon transform that was defined in the introduction. We will now define the equivalent of this operator at the quantum level following the seminal work of Weinstein [**31**] – see also [**10, 12**] for more general geometric frameworks. Recall that the eigenvalues of $-\Delta$ are of the form

$$E_k = \left(k + \frac{d-1}{2}\right)^2 - \frac{(d-1)^2}{4},$$

where k runs over the set of nonnegative integer. In particular, we can write

$$-\Delta = A^2 - \left(\frac{d-1}{2}\right)^2, \tag{11}$$

where A is a selfadjoint pseudodifferential operator of order 1 with principal symbol $\|\xi\|_x$ and satisfying

$$e^{2i\pi A} = e^{i\pi(d-1)}\mathrm{Id}. \tag{12}$$

Given a in $\mathcal{C}_c^\infty(T^*\mathbb{S}^d - \{0\})$, we then set, by analogy with the Radon transfom of a,

$$\mathcal{I}_{\mathrm{qu}}(\mathrm{Op}_\hbar(a)) := \frac{1}{2\pi}\int_0^{2\pi} e^{-isA}\,\mathrm{Op}_\hbar(a)e^{isA}ds.$$

An important observation which seems to be due to Weinstein [**31**] is that the following exact commutation relation holds:

$$[\mathcal{I}_{\mathrm{qu}}(\mathrm{Op}_\hbar(a)), A] = 0.$$

In particular, from (11), one has

$$[\mathcal{I}_{\mathrm{qu}}(\mathrm{Op}_\hbar(a)), \Delta] = 0. \tag{13}$$

Finally, the Egorov Theorem allows to relate the operator $\mathcal{I}_{\mathrm{qu}}(\mathrm{Op}_\hbar(a))$ to the classical Radon transform as follows:

$$\mathcal{I}_{\mathrm{qu}}(\mathrm{Op}_\hbar(a)) = \mathrm{Op}_\hbar(\mathcal{I}(a)) + \hbar R, \tag{14}$$

where R is a pseudodifferential operator in $\Psi^{-\infty}(\mathbb{S}^d)$

3.3. Extra invariance properties on $\mathbb{S}^d$. Let us now apply these properties to derive some invariance properties of the elements in $\mathcal{M}(\infty)$. We fix μ in $\mathcal{M}(\infty)$ which is generated by a sequence $(u_\hbar)_{\hbar\to 0^+}$ and a in $\mathcal{C}_c^\infty(T^*\mathbb{S}^d - \{0\})$. We rewrite (10) with $\mathcal{I}_{\text{qu}}(\text{Op}_\hbar(a))$ instead of $\text{Op}_\hbar(a)$. According to (13), this implies that

$$\langle u_\hbar, [V, \mathcal{I}_{\text{qu}}(\text{Op}_\hbar(a))]\, u_\hbar\rangle = 0.$$

Combining (14) with the commutation formula for pseudodifferential operators and the Calderón-Vaillancourt theorem, we find then

$$\frac{\hbar}{i}\langle u_\hbar, \text{Op}_\hbar\left(\{V, \mathcal{I}(a)\}\right) u_\hbar\rangle = \mathcal{O}(\hbar^2).$$

Hence, after letting $\hbar$ go to 0, one finds that

$$\mu(\{V, \mathcal{I}(a)\}) = 0.$$

Applying the invariance by the geodesic flow twice, one finally gets that

$$\mu(\{\mathcal{I}(V), a\}) = \mu(\{\mathcal{I}(V), \mathcal{I}(a)\}) = 0. \tag{15}$$

This is valid for any smooth test function a in $\mathcal{C}_c^\infty(T^*\mathbb{S}^d - \{0\})$. Thus, we have just proved that **any μ in $\mathcal{M}(\infty)$ is invariant by the Hamiltonian flow $\varphi^t_{\mathcal{I}(V)}$ of** $\mathcal{I}(V)$ which is well defined on $S^*\mathbb{S}^d \subset T^*\mathbb{S}^d - \{0\}$. In other words, any element in $\mathcal{M}(\infty)$ is an invariant measure for the system

$$F : T^*\mathbb{S}^d - \{0\} \ni (x,\xi) \mapsto \left(\frac{\|\xi\|_x^2}{2}, \mathcal{I}(V)(x,\xi)\right) \in \mathbb{R}^2.$$

Theorems 2.3 and 2.4 follow then from classical arguments on integrable systems – see e.g. paragraph 3.3 in [**25**] for part (a) of Theorem 2.3 and Corollary 4.4 of that reference for the first conclusion of Theorem 2.4.

3.4. The case of vanishing averages. One can easily observe that the results we have proved so far are empty if we suppose that $\mathcal{I}(V)$ is constant. This is due to the fact that identity (15) does not provide any non-trivial information on μ in that case. We would now like to explain how one can obtain a new invariance relation in that case – namely invariance by the Hamiltonian flow of the second order average $\mathcal{I}_V^{(2)}$. This is enough to prove part (b) of Theorem 2.3 and complete the proof of Theorem 2.4. This problem was not considered in [**25**] and we will briefly expose how some ideas of Guillemin and Uribe [**16, 30**] can be applied to treat this case.

Recall that $\mathcal{I}(V)$ is constant if and only if V is an odd function on $\mathbb{S}^d$ plus a constant. Since a constant to the potential does not change the eigenfunctions of the operator,we will suppose, this point on, that V is *an odd function* on $\mathbb{S}^d$. In particular, its Radon transform identically vanishes. Recall also from [**16**, Lemma 3.1] that its quantum counterpart also identically vanishes, i.e.

$$\mathcal{I}_{\text{qu}}(V) := \frac{1}{2\pi}\int_0^{2\pi} e^{-isA} V e^{isA} ds = 0. \tag{16}$$

whenever V is an odd function. Following [**30**] and given a bounded operator C on $L^2(\mathbb{S}^d)$, one can define

$$\mathcal{I}_{\text{qu}}(C) := \frac{1}{2\pi}\int_0^{2\pi} e^{-isA} C e^{isA} ds,$$

and

$$\sigma(C) := -\frac{1}{2\pi}\int_0^{2\pi} dt\left(\int_0^t e^{-isA}Ce^{isA}ds\right).$$

As was already observed, one has $[A,\mathcal{I}_{\mathrm{qu}}(C)]=0$. For $\sigma(C)$, the following holds:

$$[A,\sigma(C)] = i(C-\mathcal{I}_{\mathrm{qu}}(C)). \tag{17}$$

We now set

$$U_\hbar(t) := \exp\left(-it\hbar\sigma(Q_\hbar)\right),$$

where $Q_\hbar$ is an $\hbar$-pseudodifferential operator in $\Psi^{-1}(\mathbb{S}^d)$ that has to be determined. We also fix a in $\mathcal{C}_c^\infty(T^*\mathbb{S}^d-\{0\})$. In order to motivate the upcoming calculation, we now write (10) with $U_\hbar(-1)\mathcal{I}_{\mathrm{qu}}(\mathrm{Op}_\hbar(a))U_\hbar(1)$ instead of $\mathrm{Op}_\hbar(a)$:

$$\left\langle u_\hbar,\left[\frac{\hbar^2A^2}{2}+\hbar^2V, U_\hbar(-1)\mathcal{I}_{\mathrm{qu}}(\mathrm{Op}_\hbar(a))U_\hbar(1)\right]u_\hbar\right\rangle = 0.$$

Equivalently, this can be rewritten as

$$\left\langle u_\hbar, U_\hbar(-1)\left[U_\hbar(1)\left(\frac{\hbar^2A^2}{2}+\hbar^2V\right)U_\hbar(-1),\mathcal{I}_{\mathrm{qu}}(\mathrm{Op}_\hbar(a))\right]U_\hbar(1)u_\hbar\right\rangle = 0. \tag{18}$$

Using the fact that V is odd, we would now like to choose an appropriate $Q_\hbar$ such that

$$U_\hbar(1)\left(\frac{\hbar^2A^2}{2}+\hbar^2V\right)U_\hbar(-1) = \frac{\hbar^2A^2}{2}+\hbar^4Q_\hbar^1,$$

for some bounded pseudodifferential operator $Q_\hbar^1$ to be determined. This kind of normal form for the Schrödinger operator on $\mathbb{S}^d$ was for instance obtained by Guillemin in [**16**, Sect. 3] and by Uribe in [**30**, Sect. 4 and 6]. Let us recall their argument.

We first use (17) and the composition formula for pseudodifferential operators to write that

$$U_\hbar(1)\hbar AU_\hbar(-1) = \hbar A-\hbar^2(Q_\hbar-\mathcal{I}_{\mathrm{qu}}(Q_\hbar))+i\frac{\hbar^3}{2}[\sigma(Q_\hbar),Q_\hbar-\mathcal{I}_{\mathrm{qu}}(Q_\hbar)]+\mathcal{O}_{\Psi^{-1}(\mathbb{S}^d)}(\hbar^5).$$

Hence, if we square this expression, we find, using the composition formula one more time,

$$\begin{aligned}U_\hbar(1)\hbar^2A^2U_\hbar(-1) \;=\;& \hbar^2A^2-\hbar^2[\hbar A(Q_\hbar-\mathcal{I}_{\mathrm{qu}}(Q_\hbar))-(Q_\hbar-\mathcal{I}_{\mathrm{qu}}(Q_\hbar))\hbar A]\\ &+i\frac{\hbar^3}{2}\left(\hbar A[\sigma(Q_\hbar),Q_\hbar-\mathcal{I}_{\mathrm{qu}}(Q_\hbar)]+[\sigma(Q_\hbar),Q_\hbar-\mathcal{I}_{\mathrm{qu}}(Q_\hbar)]\hbar A\right)\\ &+\hbar^4(Q_\hbar-\mathcal{I}_{\mathrm{qu}}(Q_\hbar))^2+\mathcal{O}_{\Psi^0(\mathbb{S}^d)}(\hbar^5).\end{aligned}$$

Similarly, one has

$$U_\hbar(1)\hbar^2VU_\hbar(-1) = \hbar^2V-i\hbar^3[\sigma(Q_\hbar),V]+\mathcal{O}_{\Psi^0(\mathbb{S}^d)}(\hbar^5).$$

Thus, if we want to cancel the term $\hbar^2V$ in (18), we have to impose that $(\hbar A)(Q_\hbar-\mathcal{I}_{\mathrm{qu}}(Q_\hbar))+(Q_\hbar-\mathcal{I}_{\mathrm{qu}}(Q_\hbar))(\hbar A)$ is equal to $2V$ (at least at first order). Define, for every bounded operator B,

$$\mathcal{L}(B) = BA^{-1}+A^{-1}B,$$

and set finally

$$Q_\hbar = \frac{1}{\hbar}\mathcal{L}\left(2V-\frac{AVA^{-1}+A^{-1}VA}{2}\right),$$

which is in $\Psi^{-1}(\mathbb{S}^d)$ with a principal symbol equal to $q(x,\xi) = V(x)/\|\xi\|_x$:

$$Q_\hbar = \mathrm{Op}_\hbar(q) + \hbar R_\hbar, \qquad \text{with } R_\hbar \text{ bounded.}$$

Observe that, as V is odd, one can verify that $\mathcal{I}(q) \equiv 0$. In the following, we will denote by $\sigma(q)$ the principal symbol of the operator $\sigma(Q_\hbar)$. Thanks to (16), we also know that $\mathcal{I}_{\mathrm{qu}}(Q_\hbar) = 0$ from which we can infer

$$(\hbar A)(Q_\hbar - \mathcal{I}_{\mathrm{qu}}(Q_\hbar)) + (Q_\hbar - \mathcal{I}_{\mathrm{qu}}(Q_\hbar))(\hbar A) = (\hbar A)Q_\hbar + Q_\hbar(\hbar A).$$

In other words, it remains to compute the difference between $(\hbar A)Q_\hbar + Q_\hbar(\hbar A)$ and $2V$:

$$(\hbar A)Q_\hbar + Q_\hbar(\hbar A) - 2V = A^{-1}VA + AVA^{-1} - V - \frac{A^{-2}VA^2 + A^2VA^{-2}}{2}.$$

We now write that $ABA^{-1} = [A, BA^{-1}] + B$ and $A^{-1}BA = [A^{-1}B, A] + B$ which implies

$$(\hbar A)Q_\hbar + Q_\hbar(\hbar A) - 2V = -\frac{[A,[A,VA^{-1}]A^{-1}] + [A^{-1}[A^{-1}V,A],A]}{2}.$$

According to the composition rules for pseudodifferential operators, we find that $(\hbar A)Q_\hbar + Q_\hbar(\hbar A) - 2V$ belongs to $\hbar^2\Psi^0(\mathbb{S}^d)$ with principal symbol equal to

$$\begin{aligned}\hbar^2 r(x,\xi)& \\ &= -\hbar^2\frac{\{\|\xi\|_x,\{\|\xi\|_x,V(x)\|\xi\|_x^{-1}\}\|\xi\|_x^{-1}\} + \{\|\xi\|_x^{-1}\{V(x)\|\xi\|_x^{-1},\|\xi\|_x\},\|\xi\|_x\}}{2}.\end{aligned}$$

Note that, as V is odd, one can verify that $\mathcal{I}(r) \equiv 0$. Combining these equalities, we find that

$$U_\hbar(1)\left(\frac{\hbar^2A^2}{2} + \hbar^2 V\right)U_\hbar(-1) = \frac{\hbar^2A^2}{2} + \hbar^4\,\mathrm{Op}_\hbar(q_1) + \mathcal{O}_{\Psi^0(\mathbb{S}^d)}(\hbar^5),$$

where

$$q_1(x,\xi) := \frac{q(x,\xi)^2 + \|\xi\|_x\{\sigma(q),q\}(x,\xi) - 2\{\sigma(q),V\}(x,\xi) - r(x,\xi)}{2}.$$

Insert this identity in (18) and apply (13) and (14) to derive that

$$\hbar^5\,\langle u_\hbar, \mathrm{Op}_\hbar(\{q_1,\mathcal{I}(a)\})u_\hbar\rangle = \mathcal{O}(\hbar^6). \tag{19}$$

If we let $\hbar$ go to 0, we find that the corresponding semiclassical measure μ verifies $\mu(\{q_1,\mathcal{I}(a)\}) = 0$. From the invariance of μ by the geodesic flow and from the relation $\mathcal{I}(r) \equiv 0$, this implies that

$$\mu(\{q(x,\xi)^2 + \|\xi\|_x\{\sigma(q),q\}(x,\xi) - 2\{\sigma(q),V\}(x,\xi),\mathcal{I}(a)\}) = 0. \tag{20}$$

Then, use that $\{\|\xi\|_x,\mathcal{I}(a)\} = 0$ and that μ is supported in $S^*\mathbb{S}^d$ in order to show that this is equivalent to

$$\mu(\{V^2 - \{\sigma(V),V\},\mathcal{I}(a)\}) = 0.$$

Using the invariance of μ by the geodesic flow, we find that

$$\mu\left(\left\{\mathcal{I}(V^2) - \frac{1}{2\pi}\int_0^{2\pi}\int_0^t\{V\circ\varphi^t, V\circ\varphi^s\}dsdt, a\right\}\right) = 0, \tag{21}$$

which replaces (15) when V is an odd function.

3.5. Observability estimates. Let us now give the proof of Theorem 2.1. Suppose by contradiction that this result is not true. It means that there exists a sequence $(u_n)_{n\geq 1}$ of solutions of (1) such that $\|u_n\|_{L^2(\omega)}\rightarrow 0$. From the unique continuation principle (see e.g. [**20**]) and using the fact that ω is a non empty open set, one can verify that λ_n has to converge to infinity. Up to an extraction, we can suppose that $(u_n)_{n\geq 1}$ generates an unique semiclassical measure μ. Using the invariance by the geodesic flow, one knows that $\mu(S^*\omega)=\mu(\mathcal{I}(\mathbf{1}_\omega))$. Suppose now that $\mathcal{I}(V)$ is non-constant. Then, as μ is a postive measure, one knows that

$$\lim_{n\rightarrow+\infty}\|u_n\|_{L^2(\omega)}^2\geq\mu(S^*\omega)\geq\frac{1}{T}\inf_{\rho\in S^*\mathbb{S}^d}\int_0^T\mathcal{I}(\mathbf{1}_\omega)\circ\varphi^s_{\mathcal{I}(V)}(\rho)ds.$$

From the fact that $K_{\omega,V}=\emptyset$, one knows that, for $T>0$ large enough, this lower bound is positive which implies the expected contradiction as the upper bound vanishes by hypothesis. When $\mathcal{I}(V)$ is constant it suffices to reproduce this argument using the invariance of semiclassical measures by the Hamiltonian flow of $\mathcal{I}^{(2)}(V)$.

3.6. Relation to eigenvalue distribution. In this last paragraph, we briefly explain the main lines of the proof of Theorem 2.5. We only treat the second part of the Theorem which was not discussed in [**25**]. Therefore, as in paragraph 3.4, we suppose that V is odd.

First of all, fix a point (x_0,ξ_0) in $S^*\mathbb{S}^d$ and a normalized sequence $(u_\hbar^{x_0,\xi_0})_{\hbar\rightarrow 0^+}$ of coherent states whose semiclassical measure is δ_{x_0,ξ_0}. Recall from [**25**, Sect. 6.1] that, up to some spectral truncation, we can always suppose that

$$u_\hbar^{x_0,\xi_0}=\sum_{\{j:\frac{1}{4}\leq\lambda_j\hbar^2\leq 1\}}c_\hbar^{x_0,\xi_0}(j)\hat{v}_\hbar^{x_0,\xi}(j),\tag{22}$$

where, for every choice of parameters, $c_\hbar^{x_0,\xi_0}(j)\geq 0$ and $\hat{v}_\hbar^{x_0,\xi_0}(j)$ is normalized in $L^2(\mathbb{S}^d)$ and verifies

$$\left(-\frac{\Delta}{2}+V\right)\hat{v}_\hbar^{x_0,\xi_0}(j)=\lambda_j\hat{v}_\hbar^{x_0,\xi_0}(j).$$

We now let $(\tau_\hbar)_{\hbar\rightarrow 0^+}$ be a sequence of times such that

$$\lim_{\hbar\rightarrow 0^+}\tau_\hbar\min\left\{\lambda_{j+1}-\lambda_j:\ \frac{1}{4}\leq\hbar^2\lambda_j\leq 1\right\}=+\infty.\tag{23}$$

From our assumption on the level spacing, we can in fact suppose that $\tau_\hbar=o(\hbar^{-3})$. As in the previous sections, we fix a in $\mathcal{C}_c^\infty(T^*\mathbb{S}^d-\{0\})$ and we consider the time dependent Wigner distribution:

$$\langle\mu_\hbar^{x_0,\xi_0}(t),a\rangle:=\langle v_\hbar^{x_0,\xi_0}(t\tau_\hbar),U_\hbar(-1)\mathcal{I}_\mathcal{Q}(\operatorname{Op}_\hbar(a))U_\hbar(1)v_\hbar^{x_0,\xi_0}(t\tau_\hbar)\rangle,$$

where $v_\hbar^{x_0,\xi_0}(t\tau_\hbar)$ is the solution at time $t\tau_\hbar$ of (2) with initial condition $u_\hbar^{x_0,\xi_0}$. If we differentiate this expression with respect to time and if we argue as in paragraph 3.4, we find that

$$\frac{d}{dt}\langle\mu_\hbar^{x_0,\xi_0}(t),a\rangle=\mathcal{O}(\hbar^3\tau_\hbar).$$

Recall that, for a general V, the proof from [**25**] gave a remainder term of order $\mathcal{O}(\tau_\hbar\hbar)$ and that we did not introduce the Fourier integral operator $U_\hbar(1)$ in our argument. Integrating this expression between 0 and t and using our assumption that $\tau_\hbar=o(\hbar^{-3})$, we find

$$\langle\mu_\hbar^{x_0,\xi_0}(t),a\rangle=\mathcal{I}(a)(x_0,\xi_0)+o(1).$$

If we now fix θ in $\mathcal{S}(\mathbb{R})$ whose Fourier transform $\mathcal{F}(\theta)$ is compactly supported and verifies $\mathcal{F}(\theta)(0) = 1$, then we find that

$$\int_{\mathbb{R}} \theta(t)\langle \mu_\hbar^{x_0,\xi_0}(t), a\rangle dt = \mathcal{I}(a)(x_0,\xi_0) + o(1).$$

Using the spectral decomposition (22) and (23), we obtain the following averaging formula:

$$\sum_{\{j:\frac{1}{4}\leq\lambda_j\hbar^2\leq 1\}} c_\hbar^{x_0,\xi_0}(j)^2\langle \hat{v}_\hbar^{x_0,\xi_0}(j), U_\hbar(-1)\mathcal{I}_{\mathcal{Q}}(\mathrm{Op}_\hbar(a))U_\hbar(1)\hat{v}_\hbar^{x_0,\xi_0}(j)\rangle = \mathcal{I}(a)(x_0,\xi_0) + o(1),$$

which yields after simplification

$$\sum_{\{j:\frac{1}{4}\leq\lambda_j\hbar^2\leq 1\}} c_\hbar^{x_0,\xi_0}(j)^2\langle \hat{v}_\hbar^{x_0,\xi_0}(j), \mathrm{Op}_\hbar(a)\hat{v}_\hbar^{x_0,\xi_0}(j)\rangle = \mathcal{I}(a)(x_0,\xi_0) + o(1),$$

Recall that, as $u_\hbar^{x_0,\xi_0}$ was chosen to be normalized in $L^2(\mathbb{S}^d)$, one has $\sum_j c_\hbar^{x_0,\xi_0}(j)^2 = 1$. Arguing as in the proof of the Quantum Ergodicity Theorem – see [**25**, Sect. 6.3] for details, we can obtain the following variance estimate:

$$\sum_{\{j:\frac{1}{4}\leq\lambda_j\hbar^2\leq 1\}} c_\hbar^{x_0,\xi_0}(j)^2\left|\left\langle \hat{v}_\hbar^{x_0,\xi_0}(j), \mathrm{Op}_\hbar(a)\hat{v}_\hbar^{x_0,\xi_0}(j)\right\rangle - \mathcal{I}(a)(x_0,\xi_0)\right|^2 = o(1), \tag{24}$$

which is sufficient to conclude the proof of the Theorem thanks to the Bienaymé-Tchebychev Theorem – see [**25**, Sect. 6.4] for details.

Acknowledgments

The present note has been written for the proceedings of the workshop *Probabilistic Methods in Spectral Geometry and PDE* which were held in the CRM of Montréal at the end of the summer 2016. The authors thank the organizers of this meeting for the opportunity to expose their work [**25**] and some further developments of it in these proceedings. We also warmly thank the anonymous referee for his nice and careful report.

References

[1] Nalini Anantharaman, Clotilde Fermanian-Kammerer, and Fabricio Macià, *Semiclassical completely integrable systems: long-time dynamics and observability via two-microlocal Wigner measures*, Amer. J. Math. **137** (2015), no. 3, 577–638, DOI 10.1353/ajm.2015.0020. MR3357117

[2] N. Anantharaman, M. Léautaud, F. Macià *Wigner measures and observability for the Schrödinger equation on the disk*, Invent. Math. **206**(2) (2016), 485–599.

[3] Nalini Anantharaman and Fabricio Macià, *Semiclassical measures for the Schrödinger equation on the torus*, J. Eur. Math. Soc. (JEMS) **16** (2014), no. 6, 1253–1288, DOI 10.4171/JEMS/460. MR3226742

[4] Nalini Anantharaman and Gabriel Rivière, *Dispersion and controllability for the Schrödinger equation on negatively curved manifolds*, Anal. PDE **5** (2012), no. 2, 313–338, DOI 10.2140/apde.2012.5.313. MR2970709

[5] Daniel Azagra and Fabricio Macià, *Concentration of symmetric eigenfunctions*, Nonlinear Anal. **73** (2010), no. 3, 683–688, DOI 10.1016/j.na.2010.03.056. MR2653740

[6] Arthur L. Besse, *Manifolds all of whose geodesics are closed*, Ergebnisse der Mathematik und ihrer Grenzgebiete [Results in Mathematics and Related Areas], vol. 93, Springer-Verlag, Berlin-New York, 1978. With appendices by D. B. A. Epstein, J.-P. Bourguignon, L. Bérard-Bergery, M. Berger and J. L. Kazdan. MR496885

[7] Jean Bourgain, Nicolas Burq, and Maciej Zworski, *Control for Schrödinger operators on 2-tori: rough potentials*, J. Eur. Math. Soc. (JEMS) **15** (2013), no. 5, 1597–1628, DOI 10.4171/JEMS/399. MR3082239

[8] Nicolas Burq and Maciej Zworski, *Geometric control in the presence of a black box*, J. Amer. Math. Soc. **17** (2004), no. 2, 443–471, DOI 10.1090/S0894-0347-04-00452-7. MR2051618

[9] Nicolas Burq and Maciej Zworski, *Control for Schrödinger operators on tori*, Math. Res. Lett. **19** (2012), no. 2, 309–324, DOI 10.4310/MRL.2012.v19.n2.a4. MR2955763

[10] Yves Colin de Verdière, *Sur le spectre des opérateurs elliptiques à bicaractéristiques toutes périodiques* (French), Comment. Math. Helv. **54** (1979), no. 3, 508–522, DOI 10.1007/BF02566290. MR543346

[11] Yves Colin de Verdière and San Vũ Ngọc, *Singular Bohr-Sommerfeld rules for 2D integrable systems* (English, with English and French summaries), Ann. Sci. École Norm. Sup. (4) **36** (2003), no. 1, 1–55, DOI 10.1016/S0012-9593(03)00002-8. MR1987976

[12] J. J. Duistermaat and V. W. Guillemin, *The spectrum of positive elliptic operators and periodic bicharacteristics*, Invent. Math. **29** (1975), no. 1, 39–79, DOI 10.1007/BF01405172. MR0405514

[13] Semyon Dyatlov and Long Jin, *Semiclassical measures on hyperbolic surfaces have full support*, Acta Math. **220** (2018), no. 2, 297–339, DOI 10.4310/ACTA.2018.v220.n2.a3. MR3849286

[14] P. Gérard, *Mesures semi-classiques et ondes de Bloch* (French), Séminaire sur les Équations aux Dérivées Partielles, 1990–1991, École Polytech., Palaiseau, 1991, pp. Exp. No. XVI, 19. MR1131589

[15] Victor Guillemin, *The Radon transform on Zoll surfaces*, Advances in Math. **22** (1976), no. 1, 85–119, DOI 10.1016/0001-8708(76)90139-0. MR0426063

[16] Victor Guillemin, *Some spectral results for the Laplace operator with potential on the n-sphere*, Advances in Math. **27** (1978), no. 3, 273–286, DOI 10.1016/0001-8708(78)90102-0. MR0478245

[17] Michael A. Hall, Michael Hitrik, and Johannes Sjöstrand, *Spectra for semiclassical operators with periodic bicharacteristics in dimension two*, Int. Math. Res. Not. IMRN **20** (2015), 10243–10277, DOI 10.1093/imrn/rnu270. MR3455866

[18] E. Humbert, Y. Privat, E. Trélat *Observability properties of the homogeneous wave equation on a closed manifold*, preprint arXiv:1607.01535 (2016).

[19] Dmitry Jakobson and Steve Zelditch, *Classical limits of eigenfunctions for some completely integrable systems*, Emerging applications of number theory (Minneapolis, MN, 1996), IMA Vol. Math. Appl., vol. 109, Springer, New York, 1999, pp. 329–354, DOI 10.1007/978-1-4612-1544-8_13. MR1691539

[20] Jérôme Le Rousseau and Gilles Lebeau, *On Carleman estimates for elliptic and parabolic operators. Applications to unique continuation and control of parabolic equations*, ESAIM Control Optim. Calc. Var. **18** (2012), no. 3, 712–747, DOI 10.1051/cocv/2011168. MR3041662

[21] G. Lebeau, *Contrôle de l'équation de Schrödinger* (French, with English summary), J. Math. Pures Appl. (9) **71** (1992), no. 3, 267–291. MR1172452

[22] J.-L. Lions, *Exact controllability, stabilization and perturbations for distributed systems*, SIAM Rev. **30** (1988), no. 1, 1–68, DOI 10.1137/1030001. MR931277

[23] Fabricio Macià, *Some remarks on quantum limits on Zoll manifolds*, Comm. Partial Differential Equations **33** (2008), no. 4-6, 1137–1146, DOI 10.1080/03605300802038601. MR2424392

[24] Fabricio Macià, *The Schrödinger flow in a compact manifold: high-frequency dynamics and dispersion*, Modern aspects of the theory of partial differential equations, Oper. Theory Adv. Appl., vol. 216, Birkhäuser/Springer Basel AG, Basel, 2011, pp. 275–289, DOI 10.1007/978-3-0348-0069-3_16. MR2858875

[25] Fabricio Macià and Gabriel Rivière, *Concentration and non-concentration for the Schrödinger evolution on Zoll manifolds*, Comm. Math. Phys. **345** (2016), no. 3, 1019–1054, DOI 10.1007/s00220-015-2504-8. MR3519588

[26] Luc Miller, *Resolvent conditions for the control of unitary groups and their approximations*, J. Spectr. Theory **2** (2012), no. 1, 1–55, DOI 10.4171/JST/20. MR2879308

[27] Stéphane Nonnenmacher, *Anatomy of quantum chaotic eigenstates*, Chaos, Prog. Math. Phys., vol. 66, Birkhäuser/Springer, Basel, 2013, pp. 193–238, DOI 10.1007/978-3-0348-0697-8_6. MR3204186

[28] G. Rivière *Quelques problèmes de dynamique classique et quantique*, Habilitation thesis, Université de Lille (2017)

[29] P. Sarnak *Recent progress on QUE*, Bull. of the AMS **48** (2011), 211–228.

[30] Alejandro Uribe, *Band invariants and closed trajectories on S^n*, Adv. in Math. **58** (1985), no. 3, 285–299, DOI 10.1016/0001-8708(85)90120-3. MR815359

[31] Alan Weinstein, *Asymptotics of eigenvalue clusters for the Laplacian plus a potential*, Duke Math. J. **44** (1977), no. 4, 883–892. MR0482878

[32] Steven Zelditch, *Maximally degenerate Laplacians* (English, with English and French summaries), Ann. Inst. Fourier (Grenoble) **46** (1996), no. 2, 547–587. MR1393525

[33] Steve Zelditch, *Fine structure of Zoll spectra*, J. Funct. Anal. **143** (1997), no. 2, 415–460, DOI 10.1006/jfan.1996.2981. MR1428823

[34] Steve Zelditch, *Recent developments in mathematical quantum chaos*, Current developments in mathematics, 2009, Int. Press, Somerville, MA, 2010, pp. 115–204. MR2757360

[35] Maciej Zworski, *Semiclassical analysis*, Graduate Studies in Mathematics, vol. 138, American Mathematical Society, Providence, RI, 2012. MR2952218

M^{2}ASAI, Universidad Politécnica de Madrid. ETSI Navales, Avda. de la Memoria, 4. 28040 Madrid, Spain

Email address: Fabricio.Macia@upm.es

Laboratoire Paul Painlevé (U.M.R. CNRS 8524), U.F.R. de Mathématiques, Université Lille 1, 59655 Villeneuve d'Ascq Cedex, France

Email address: gabriel.riviere@math.univ-lille1.fr

Contemporary Mathematics
Volume **739**, 2019
https://doi.org/10.1090/conm/739/14898

Random nodal lengths and Wiener chaos

Maurizia Rossi

ABSTRACT. In this survey we collect some of the recent results on the "nodal geometry" of random eigenfunctions on Riemannian surfaces. We focus on the asymptotic behavior, for high energy levels, of the nodal length of Gaussian Laplace eigenfunctions on the torus (arithmetic random waves) and on the sphere (random spherical harmonics). We give some insight on both Berry's cancellation phenomenon and the nature of nodal length second order fluctuations (non-Gaussian on the torus and Gaussian on the sphere) in terms of chaotic components. Finally we consider the general case of monochromatic random waves, i.e. Gaussian random linear combination of eigenfunctions of the Laplacian on a compact Riemannian surface with frequencies from a short interval, whose scaling limit is Berry's Random Wave Model. For the latter we present some recent results on the asymptotic distribution of its nodal length in the high energy limit (equivalently, for growing domains).

1. Introduction

1.1. Background and motivations. At the end of the 18th century, the musician and physicist Chladni noticed that sounds of different pitch could be made by exciting a metal plate with the bow of a violin, depending on where the bow touched the plate. The latter was fixed only in the center, and when there was some sand on it, for each pitch a curious pattern appeared (see Figure 1). About 60 years later, Kirchhoff, inspired also by previous contributions of Germain, Lagrange and Poisson, showed that Chladni figures on a plate correspond to the zeros of eigenpairs (eigenvalues and corresponding eigenfunctions) of the biharmonic operator with free boundary conditions. Nowadays nodal patterns arise in several areas, from the musical instruments industry to the study of natural phemomena such as earthquakes. See [**Chl02, GK12, Wig12**] and the references therein for more details.

Now let $(\mathcal{M}, g)$ denote a compact smooth Riemannian manifold of dimension 2 and $f : \mathcal{M} \to \mathbb{R}$ a function on $\mathcal{M}$. The set $f^{-1}(0) := \{x \in \mathcal{M} : f(x) = 0\}$ is usually called *nodal set* of f. We are interested in the case where f is an eigenfunction

2010 *Mathematics Subject Classification.* Primary 60G60, 60B10, 60D05, 35P20, 58J50.
Key words and phrases. Nodal length, chaos expansion, random eigenfunctions, limit theorems.
The author was financially supported by the grant F1R-MTH-PUL-15STAR (STARS) at the University of Luxembourg and the ERC grant no. 277742 *Pascal.* Her research is currently supported by the Fondation Sciences Mathématiques de Paris and the project ANR-17-CE40-0008 *Unirandom.*

FIGURE 1. Chladni's figures

of the Laplace-Beltrami operator Δ on the manifold. There exists an orthonormal basis $\{f_j\}_{j=0}^{+\infty}$ of $L^2(\mathcal{M})$ consisting of eigenfunctions for $-\Delta$ whose corresponding sequence of eigenvalues $\{\lambda_j^2\}_{j=0}^{+\infty}$

$$-\Delta f_j = \lambda_j^2 f_j \tag{1.1}$$

is non-decreasing (in particular $0 = \lambda_0 < \lambda_1 \leq \lambda_2 \leq \dots$).

The nodal set of f_j is, generically, a smooth curve (called *nodal line*) whose components are homeomorphic to the circle. We are interested in the geometry of $f_j^{-1}(0)$ for high energy levels, i.e. as $\lambda_j \to +\infty$.

Yau's conjecture [**Yau82**, **Yau93**] concerns, in particular[1], the length of nodal lines $f_j^{-1}(0)$: there exist two positive constant $c_{\mathcal{M}}, C_{\mathcal{M}}$ such that

$$c_{\mathcal{M}}\lambda_j \leq \text{length}(f_j^{-1}(0)) \leq C_{\mathcal{M}}\lambda_j, \tag{1.2}$$

for every $j \geq 1$. Yau's conjecture was proved by Donnelly and Fefferman in [**DF88**] for real analytic manifolds, and the lower bound was established by Logunov and Malinnikova [**Log16a**, **Log16b**, **LM16**] for the general case.

In [**Ber77**] Berry conjectured that the local behavior of high energy eigenfunctions f_j in (1.1) for generic *chaotic* surfaces is universal. He meant that it is comparable to the behavior of the centred Gaussian field $B_{\lambda_j} = \{B_{\lambda_j}(x)\}_{x\in\mathbb{R}^2}$ on the Euclidean plane whose covariance kernel is, for $x, y \in \mathbb{R}^2$,

$$\text{Cov}\,(B_{\lambda_j}(x), B_{\lambda_j}(y)) = J_0(\lambda_j \|x-y\|), \tag{1.3}$$

J_0 being the Bessel function of order zero [**Sze75**, §1.71] and $\|\cdot\|$ denoting the Euclidean norm (the latter model is known as Berry's Random Wave Model). In

[1] Yau stated its conjecture in any dimension

particular, nodal lines of B_{λ_j} should model nodal lines of deterministic eigenfunctions f_j for large eigenvalues, allowing the investigation of associated local quantities like the (nodal) length. See [**Wig12**] and the references therein for more details.

1.2. Short plan of the survey. In the case of surfaces $(\mathcal{M}, g)$ with spectral degeneracies, like the two dimensional standard flat torus or the unit round sphere, it is possible to construct a random model by endowing each eigenspace with a Gaussian measure. The asymptotic behavior, in the high energy limit, of the corresponding nodal length is the subject of §2. In particular, two comments are in order: the asymptotic variance is smaller than expected in both cases (*Berry's cancellation phenomenon*), and the nature of second order fluctuations is Gaussian for the spherical case but not on the torus. In §4 we give some insight on both these phenomena in terms of *chaotic projections*, notion introduced in §3.

For a generic Riemannian metric on a smooth compact surface, the eigenvalues of the Laplace-Beltrami operator are simple and one cannot define a meaningful random model as for the toral or the spherical case, that is, a single frequency random model. In view of this, in §5 we work with the so-called *monochromatic random waves* i.e., Gaussian random linear combination of eigenfunctions with frequencies from a short interval. Their scaling limit is Berry's RWM (1.3), and for the latter we present some recent results on the asymptotic behavior for its nodal length, which turns out to be Gaussian as for the spherical case.

2. Random nodal lengths

2.1. The toral case.

2.1.1. *Arithmetic random waves.* The Laplacian eigenvalues for the two-dimensional standard flat torus $\mathbb{T}^2 := \mathbb{R}^2/\mathbb{Z}^2$ are of the form $4\pi^2 n$, where n is an integer expressible as a sum of two squares ($n \in S := \{n = a^2 + b^2 : a, b \in \mathbb{Z}\}$). For $n \in S$, we define Λ_n to be the set of frequencies

$$\Lambda_n := \{\xi \in \mathbb{Z}^2 : \|\xi\| = \sqrt{n}\} \tag{2.4}$$

and we denote by $\mathcal{N}_n$ its cardinality ($\mathcal{N}_n$ is the multiplicity of eigenvalue $4\pi^2 n$). The set Λ_n in (2.4) induces a probability measure μ_n on the unit circle $\mathbb{S}^1 \subset \mathbb{R}^2$

$$\mu_n := \frac{1}{\mathcal{N}_n} \sum_{\xi \in \Lambda_n} \delta_{\xi/\sqrt{n}},$$

δ_z denoting the Dirac mass at $z \in \mathbb{R}^2$. For more details see [**KKW13**].

For $n \in S$, the arithmetic random wave T_n (of order n) is the Gaussian random eigenfunction on the torus defined as follows [**RW08**]:

$$T_n(x) := \frac{1}{\sqrt{\mathcal{N}_n}} \sum_{\xi \in \Lambda_n} a_\xi \, \mathrm{e}^{i2\pi\langle \xi, x\rangle}, \qquad x \in \mathbb{T}^2, \tag{2.5}$$

where $\{a_\xi\}_{\xi \in \Lambda_n}$ is a family of i.d. standard complex Gaussian random variables[2], and independent except for the relation $\overline{a_\xi} = a_{-\xi}$ that ensures T_n to be real. Recall that $\{\mathrm{e}^{i2\pi\langle \xi, \cdot\rangle}\}_{\xi \in \Lambda_n}$ is an orthonormal basis for the eigenspace related to the

[2]defined on some probability space $(\Omega, \mathcal{F}, \mathbb{P})$

eigenvalue $4\pi^2 n$. Equivalently, it can be defined as the centered Gaussian field on $\mathbb{T}^2$ whose covariance kernel is, for $x, y \in \mathbb{T}^2$,

$$\operatorname{Cov}(T_n(x), T_n(y)) = \frac{1}{\mathcal{N}_n} \sum_{\xi \in \Lambda_n} e^{i2\pi\langle \xi, x-y\rangle}. \tag{2.6}$$

There exists a density one subsequence $\{n_j\}_j \subset S$ of energy levels such that, as $j \to +\infty$,

$$\mu_{n_j} \Rightarrow d\theta/2\pi,$$

$d\theta$ denoting the uniform measure on $\mathbb{S}^1$ (see [**FKW06**]). From (2.6) we have, for $x, y \in \mathbb{T}^2$, as $j \to +\infty$,

$$\operatorname{Cov}(T_{n_j}(x/2\pi\sqrt{n_j}), T_{n_j}(y/2\pi\sqrt{n_j})) = \int_{\mathbb{S}^1} e^{i\langle \theta, x-y\rangle}\, d\mu_{n_j}(\theta) \to J_0(\|x-y\|),$$

J_0 still denoting the Bessel function of order zero (cf. (1.3)). There exist other weak-$\star$ partial limits of the sequence $\{\mu_n\}_{n\in S}$, partially classified in [**KW17**].

2.1.2. *Nodal lengths: some recent results.* The nodal set $T_n^{-1}(0)$ is a smooth curve a.s.; let us set $\mathcal{L}_n := \operatorname{length}(T_n^{-1}(0))$. By means of Kac-Rice formulas, the expected nodal length was computed by Rudnick and Wigman in [**RW08**]

$$\mathbb{E}[\mathcal{L}_n] = \frac{1}{2\sqrt{2}}\sqrt{4\pi^2 n}, \tag{2.7}$$

and in [**KKW13**] the exact asymptotic variance was found: as $\mathcal{N}_n \to +\infty$,

$$\operatorname{Var}(\mathcal{L}_n) \sim \frac{1+\widehat{\mu_n}(4)^2}{512}\frac{4\pi^2 n}{\mathcal{N}_n^2}, \tag{2.8}$$

$\widehat{\mu_n}(4)$ denoting the fourth Fourier coefficients of μ_n. In order to have an asymptotic law for the variance, it suffices to choose a subsequence $\{n_j\}_j \subset S$ such that as $j \to +\infty$ we have (a) $\mathcal{N}_{n_j} \to +\infty$ and (b) $|\widehat{\mu_{n_j}}(4)| \to \eta$, for some $\eta \in [0,1]$. Note that for each $\eta \in [0,1]$, there exists a subsequence $\{n_j\}$ such that both (a) and (b) hold (see [**KKW13, KW17**]). The second order fluctuations of $\mathcal{L}_n$ were investigated and fully resolved in [**MPRW16**], as follows.

THEOREM 2.1. *For $\{n_j\}_j \subset S$ such that, as $j \to +\infty$, $\mathcal{N}_{n_j} \to +\infty$ and $|\widehat{\mu_{n_j}}(4)| \to \eta$ for some $\eta \in [0,1]$, we have*

$$\frac{\mathcal{L}_{n_j} - \mathbb{E}[\mathcal{L}_{n_j}]}{\sqrt{\operatorname{Var}(\mathcal{L}_{n_j})}} \xrightarrow{d} \frac{1}{2\sqrt{1+\eta^2}}(2-(1-\eta)Z_1^2-(1+\eta)Z_2^2), \tag{2.9}$$

where Z_1 and Z_2 are i.i.d. standard Gaussian random variables.

A quantitative version (in Wasserstein distance) of (2.9) is given in [**PR17**], where an alternative proof for (2.8) is given by means of chaotic expansions. In the recent paper [**BMW17**] the authors study the nodal length of arithmetic random waves restricted to decreasing domains (shrinking radius-s balls) all the way down to Planck scale (i.e., for $s > n^{-1/2+\varepsilon}$, $\varepsilon > 0$). Remarkably, they prove that the latter is asymptotically fully correlated with $\mathcal{L}_n$. In [**DNPR16**] the interesection number of nodal lines corresponding to two independent arithmetic random waves with the same eigenvalue is studied, whereas the papers [**RW16, RoW17**] investigate the intersection number of the nodal lines $T_n^{-1}(0)$ and a fixed reference curve with nowhere zero curvature. The case of a straight line segment is studied in [**Maf17a**] by Maffucci. There are results also in the three dimensional setting [**BM17, Cam17, Maf17b, RWY16**].

2.2. The spherical case.

2.2.1. *Random spherical harmonics.* The Laplacian eigenvalues on the two-dimensional unit round sphere $\mathbb{S}^2$ are of the form $\ell(\ell+1)$, where $\ell \in \mathbb{N}$, and the multiplicity of the ℓ-th eigenvalue is $2\ell+1$. The ℓ-th random spherical harmonic on $\mathbb{S}^2$ is the Gaussian random eigenfunction T_ℓ (abusing notation) defined as follows (see e.g. [**Wig10**]):

$$T_\ell(x) := \sqrt{\frac{4\pi}{2\ell+1}} \sum_{m=-\ell}^{\ell} a_{\ell,m} Y_{\ell,m}(x), \qquad x \in \mathbb{S}^2, \tag{2.10}$$

where $\{a_{\ell,m}\}_{m=-\ell,\dots,\ell}$ is a family of i.d. standard complex Gaussian random variables, and independent except for the relation $\overline{a_{\ell,m}} = (-1)^\ell a_{\ell,-m}$ that ensures T_ℓ to be real. The family $\{Y_{\ell,m}\}_{m=-\ell,\dots,\ell}$ of spherical harmonics [**MP11**, §3.4] is an orthonormal basis for the eigenspace related to the eigenvalue $\ell(\ell+1)$. Equivalently, it can be defined as the centered Gaussian field on the sphere whose covariance kernel is, for $x, y \in \mathbb{S}^2$,

$$\mathrm{Cov}\,(T_\ell(x)), T_\ell(y)) = P_\ell(\cos d(x,y)),$$

where P_ℓ denotes the ℓ-th Legendre polynomial [**MP11**, §13.1.2] and $d(x,y)$ the geodesic distance between the two points x and y. Hilb's asymptotic formula [**Sze75**, Theorem 8.21.12] states that, *uniformly* for $\theta \in [0, \pi - \varepsilon]$ $(\varepsilon > 0)$, as $\ell \to +\infty$,

$$P_\ell(\cos\theta) \sim \sqrt{\frac{\theta}{\sin\theta}} J_0((\ell+1/2)\theta), \tag{2.11}$$

cf. (1.3).

2.2.2. *Nodal lengths: some recent results.* The nodal set $T_\ell^{-1}(0) := \{x \in \mathbb{S}^2 : T_\ell(x) = 0\}$ is a smooth curve a.s. The mean of the nodal length (abusing notation) $\mathcal{L}_\ell := \mathrm{length}(T_\ell^{-1}(0))$ was computed in [**Ber85**]

$$\mathbb{E}[\mathcal{L}_\ell] = 4\pi \cdot \frac{1}{2\sqrt{2}} \sqrt{\ell(\ell+1)}; \tag{2.12}$$

note that the coefficient $1/2\sqrt{2}$ in (2.12) is universal (cf. (2.7)). The asymptotic behaviour of the variance was given in [**Wig10**]: as $\ell \to +\infty$,

$$\mathrm{Var}(\mathcal{L}_\ell) \sim \frac{1}{32} \log \ell. \tag{2.13}$$

The second order fluctuations of $\mathcal{L}_\ell$ have been investigated and fully resolved in [**MRW17**]:

THEOREM 2.2. *As $\ell \to +\infty$,*

$$\frac{\mathcal{L}_\ell - \mathbb{E}[\mathcal{L}_\ell]}{\sqrt{\mathrm{Var}(\mathcal{L}_\ell)}} \xrightarrow{d} Z,$$

where Z is a standard Gaussian random variable.

2.3. Some comments.

REMARK 2.3. (i) The mean of the nodal length in both the toral (2.7) and the spherical case (2.12) is of the form $1/2\sqrt{2}$ (a universal constant) times the square root of the corresponding eigenvalue (cf (1.2)) times the area of the surface.

(ii) The order of magnitude of the asymptotic variance of the nodal length in both the toral (2.8) and the spherical setting (2.13) is smaller than expected. Indeed, Kac-Rice formula [**AT07**, Ch. 12] suggests that its variance is asymptotically proportional to the corresponding eigenvalue times the second moment of the covariance kernel which is, up to a constant factor, the inverse of the eigenvalue's multiplicity ($n/\mathcal{N}_n$ on $\mathbb{T}^2$ and ℓ for $\mathbb{S}^2$). This fact is known as Berry's *cancellation phenomenon*, indeed it was predicted by Berry in [**Ber02**] and proved by Wigman in [**Wig10**] on the sphere and in [**KKW13**] on the torus.
(iii) Theorem 2.1 reveals in particular the non-Gaussian asymptotic nature of the toral nodal length, in contrast with the Gaussian fluctuations for the spherical case (Theorem 2.2).

We will give some insights for this remark in §4.1.

3. Chaotic expansions

Before briefly introducing the notion of Wiener chaos, let us recall an integral representation for the nodal length (here we state it for the spherical case but it is – essentially – valid in general, thanks to the coarea formula [**AT07**, p.169]).

The nodal length on the sphere can be (formally) written as

$$\mathcal{L}_\ell = \int_{\mathbb{S}^2} \delta_0(T_\ell(x)) \|\nabla T_\ell(x)\| \, dx \tag{3.14}$$

where δ_0 denotes the Dirac mass in 0, ∇T_ℓ the gradient field and $\|\cdot\|$ the Euclidean norm in $\mathbb{R}^2$. Indeed, let us consider the ε-approximating random variable

$$\mathcal{L}_\ell^\varepsilon := \frac{1}{2\varepsilon} \int_{\mathbb{S}^2} 1_{[-\varepsilon,\varepsilon]}(T_\ell(x)) \, \|\nabla T_\ell(x)\| \, dx,$$

where $1_{[-\varepsilon,\varepsilon]}$ denotes the indicator function of the interval $[-\varepsilon, \varepsilon]$. It is possible to prove that

$$\lim_{\varepsilon \to 0} \mathcal{L}_\ell^\varepsilon = \operatorname{length}(T_\ell^{-1}(0)),$$

both a.s. and in $L^2(\mathbb{P})$, see [**MRW17**], thus justifying (3.14). In particular, $\mathcal{L}_\ell$ is a square integrable functional of a Gaussian field; this is the key point for the theory of chaotic expansions to apply.

Analogously, for the nodal length of arithmetic random waves we have (with obvious notation)

$$\mathcal{L}_n = \int_{\mathbb{T}^2} \delta_0(T_n(x)) \|\nabla T_n(x)\| \, dx. \tag{3.15}$$

See [**MPRW16**] for details.

3.1. Wiener chaos. In this part we introduce the notion of Wiener chaos restricting ourselves to our specific spherical setting (the toral case is analogous). For a complete discussion see [**NP12**, §2.2] and the references therein.

Denote by $\{H_k\}_{k\ge 0}$ the sequence of Hermite polynomials on $\mathbb{R}$; these polynomials are defined recursively as follows: $H_0 \equiv 1$ and

$$H_k(t) = tH_{k-1}(t) - H'_{k-1}(t), \qquad k \ge 1.$$

Recall that $\mathbb{H} := \{(k!)^{-1/2} H_k, k \ge 0\}$ constitutes a complete orthonormal system in the space of square integrable real functions $L^2(\gamma)$ w.r.t. the standard Gaussian density γ on the real line.

Random spherical harmonics (2.10) are a by-product of a family of complex-valued Gaussian random variables $\{a_{\ell,m} : \ell = 0, 1, 2, \dots, m = -\ell, \dots, \ell\}$ such that **(a)** every $a_{\ell,m}$ has the form $x_{\ell,m} + iy_{\ell,m}$, where $x_{\ell,m}$ and $y_{\ell,m}$ are two independent real-valued Gaussian random variables with mean zero and variance 1/2; **(b)** $a_{\ell,m}$ and $a_{\ell',m'}$ are independent whenever $\ell \neq \ell'$ or $m' \notin \{m, -m\}$, and **(c)** $\overline{a_{\ell,m}} = (-1)^\ell a_{\ell,-m}$. Define the space $\mathbf{A}$ to be the closure in $L^2(\mathbb{P})$ of all real finite linear combinations of random variables ξ of the form

$$\xi = z\, a_{\ell,m} + \overline{z}\, (-1)^\ell a_{\ell,-m}, \qquad z \in \mathbb{C},$$

thus $\mathbf{A}$ is a real centered Gaussian Hilbert subspace of $L^2(\mathbb{P})$.

Let us fix now an integer $q \geq 0$; the q-th Wiener chaos C_q associated with $\mathbf{A}$ is defined as the closure in $L^2(\mathbb{P})$ of all real finite linear combinations of random variables of the type

$$H_{p_1}(\xi_1) \cdot H_{p_2}(\xi_2) \cdots H_{p_k}(\xi_k)$$

for $k \geq 1$, where the integers $p_1, ..., p_k \geq 0$ satisfy $p_1 + \cdots + p_k = q$, and $(\xi_1, ..., \xi_k)$ is a standard real Gaussian vector extracted from $\mathbf{A}$ (in particular, $C_0 = \mathbb{R}$).

Taking into account the orthonormality and completeness of $\mathbb{H}$ in $L^2(\gamma)$, together with a standard monotone class argument (see e.g. [**NP12**, Theorem 2.2.4]), it is possible to prove that $C_q \perp C_m$ in $L^2(\mathbb{P})$ for every $q \neq m$, and moreover

$$L^2(\Omega, \sigma(\mathbf{A}), \mathbb{P}) = \bigoplus_{q=0}^{\infty} C_q,$$

that is, every real-valued functional F of $\mathbf{A}$ can be (uniquely) represented as a series, converging in L^2, of the form

$$F = \sum_{q=0}^{\infty} F[q], \tag{3.16}$$

where $F[q] := \mathrm{proj}(F \,|\, C_q)$ stands for the the projection of F onto C_q ($F[0] = \mathrm{proj}(F \,|\, C_0) = \mathbb{E}[F]$).

3.1.1. *Random nodal lengths.* Consider now the integral representation (3.14) for the nodal length of random spherical harmonics. It can be equivalently written as

$$\mathcal{L}_\ell = \int_{\mathbb{S}^2} \delta_0(T_\ell(x)) \|\nabla T_\ell(x)\| \, dx = \sqrt{\frac{\ell(\ell+1)}{2}} \int_{\mathbb{S}^2} \delta_0(T_\ell(x)) \|\widetilde{\nabla} T_\ell(x)\| \, dx, \tag{3.17}$$

where $\widetilde{\nabla}$ is the normalized gradient, i.e. $\widetilde{\nabla} := \nabla / \sqrt{\frac{2}{\ell(\ell+1)}}$ (see §3.2.1 in [**MRW17**] for details).

We are going to recall the chaotic expansion (3.16) for $\mathcal{L}_\ell$

$$\mathcal{L}_\ell = \sum_{q=0}^{+\infty} \mathcal{L}_\ell[2q], \tag{3.18}$$

where $\mathcal{L}_\ell[2q]$ denotes the orthogonal projection of $\mathcal{L}_\ell$ onto C_{2q}. Note that projections on odd chaoses vanish since the integrand functions in (3.17) are both even.

In [**MRW17**, §2] the terms of the series on the r.h.s. of (3.18) are explicitly given (see also [**Ros15, MPRW16**]). As it will be clear later in this section, it suffices to deal with the first three terms corresponding to $q = 0$ (the mean of the random variable) and $q = 1, 2$. See [**MRW17**, §2] for more details.

Let us introduce now two sequences of real numbers $\{\beta_{2k}\}_{k=0}^{+\infty}$ and $\{\alpha_{2n,2m}\}_{n,m=0}^{+\infty}$ corresponding to the chaotic coefficients of the Dirac mass at 0 and the Euclidean norm respectively: for $k = 0, 1, 2, \dots$

$$\beta_{2k} := \frac{1}{\sqrt{2\pi}} H_{2k}(0),$$

while for $n, m = 0, 1, 2, \dots$

$$\alpha_{2n,2m} := \sqrt{\frac{\pi}{2}} \frac{(2n)!(2m)!}{n!m!} \frac{1}{2^{n+m}} p_{n+m}\left(\frac{1}{4}\right),$$

where p_N is the swinging factorial coefficient

$$p_N(x) := \sum_{j=0}^{N} (-1)^j (-1)^N \binom{N}{j} \frac{(2j+1)!}{(j!)^2} x^j.$$

The first few terms are

$$\begin{aligned} \beta_0 &= \frac{1}{\sqrt{2\pi}}, \quad \beta_2 = -\frac{1}{\sqrt{2\pi}}, \quad \beta_4 = \frac{3}{\sqrt{2\pi}}, \\ \alpha_{00} &= \sqrt{\frac{\pi}{2}}, \quad \alpha_{02} = \frac{1}{2}\sqrt{\frac{\pi}{2}}, \quad \alpha_{04} = -\frac{3}{8}\sqrt{\frac{\pi}{2}}. \end{aligned} \tag{3.19}$$

The chaotic expansion of the nodal length is

$$\begin{aligned} \mathcal{L}_\ell = \sum_{q=0}^{+\infty} \mathcal{L}_\ell[2q] = & \sqrt{\frac{\ell(\ell+1)}{2}} \sum_{q=0}^{\infty} \sum_{u=0}^{q} \sum_{k=0}^{u} \frac{\alpha_{2k,2u-2k}\beta_{2q-2u}}{(2k)!(2u-2k)!(2q-2u)!} \times \\ & \times \int_{\mathbb{S}^2} H_{2q-2u}(T_\ell(x)) H_{2k}(\widetilde{\partial}_{1;x} T_\ell(x)) H_{2u-2k}(\widetilde{\partial}_{2;x} T_\ell(x))\, dx, \end{aligned} \tag{3.20}$$

where we use spherical coordinates (colatitude θ, longitude φ) and for $x = (\theta_x, \varphi_x)$ we are using the notation

$$\widetilde{\partial}_{1;x} = (\ell(\ell+1)/2)^{-1/2} \cdot \left.\frac{\partial}{\partial \theta}\right|_{\theta=\theta_x}, \quad \widetilde{\partial}_{2;x} = (\ell(\ell+1)/2)^{-1/2} \cdot \left.\frac{1}{\sin\theta} \frac{\partial}{\partial \varphi}\right|_{\theta=\theta_x, \varphi=\varphi_x}.$$

It is obvious that analogous formulas as (3.20) hold for the chaotic components of the nodal length $\mathcal{L}_n$ of arithmetic random waves. We are now in a position to give a sketch of the proof of Theorem 2.2 (see [**MRW17**, §3] for details).

4. On the proof of Theorem 2.2

PROOF. Substituting (3.19) into (3.20) we have

$$\mathcal{L}_\ell[0] = 4\pi \cdot \frac{1}{2\sqrt{2}} \sqrt{\ell(\ell+1)} = \mathbb{E}[\mathcal{L}_\ell]. \tag{4.21}$$

For the second chaotic projection, recalling that $H_2(t) = t^2 - 1$ and noting that $\alpha_{20} = \alpha_{02}$, we can write

$$\begin{aligned} \mathcal{L}_\ell[2] = \sqrt{\frac{\ell(\ell+1)}{2}} \Big(& \frac{\beta_2\alpha_{00}}{2!} \int_{\mathbb{S}^2} (T_\ell(x)^2 - 1)\, dx \\ & + \frac{\beta_0\alpha_{20}}{2!} \int_{\mathbb{S}^2} \Big(\langle \widetilde{\nabla} T_\ell(x)), \widetilde{\nabla} T_\ell(x) \rangle - 2 \Big)\, dx \Big). \end{aligned} \tag{4.22}$$

A standard application of Green's identity on manifolds[3] (see also the proof of [**Ros15**, Proposition 7.3.1.]) yields

(4.23)

$$\begin{aligned}\mathcal{L}_\ell[2] &= \sqrt{\frac{\ell(\ell+1)}{2}}\Big(\frac{\beta_2\alpha_{00}}{2!}\int_{\mathbb{S}^2}(T_\ell(x)^2-1)\,dx \\ &\qquad + \frac{\beta_0\alpha_{20}}{2!}\int_{\mathbb{S}^2}\Big(\langle\widetilde{\nabla}T_\ell(x)),\widetilde{\nabla}T_\ell(x)\rangle - 2\Big)\,dx\Big) \\ &= \sqrt{\frac{\ell(\ell+1)}{2}}\Big(\frac{\beta_2\alpha_{00}}{2!}\int_{\mathbb{S}^2}(T_\ell(x)^2-1)\,dx \\ &\qquad + \frac{\beta_0\alpha_{20}}{2!}\int_{\mathbb{S}^2}\Big(-\frac{2}{\ell(\ell+1)}T_\ell(x)\Delta T_\ell(x) - 2\Big)\,dx\Big) \\ &= \sqrt{\frac{\ell(\ell+1)}{2}}\Big(\frac{\beta_2\alpha_{00}}{2!}\int_{\mathbb{S}^2}(T_\ell(x)^2-1)\,dx + \frac{\beta_0\alpha_{20}}{2!}\int_{\mathbb{S}^2}\Big(2T_\ell(x)^2-2\Big)\,dx\Big),\end{aligned}$$

where the last equality we used the fact that T_ℓ is an eigenfunction of the Laplacian with eigenvalue $-\ell(\ell+1)$. From (4.23) we have

$$\mathcal{L}_\ell[2] = \sqrt{\frac{\ell(\ell+1)}{2}}\Big(\frac{\beta_2\alpha_{00}}{2!}+\beta_0\alpha_{20}\Big)\int_{\mathbb{S}^2}(T_\ell(x)^2-1)\,dx = 0, \tag{4.24}$$

where we used (3.19). Let us now investigate the fourth chaotic component. To this aim, consider the term

$$\mathcal{I}_\ell := -\frac{1}{4}\sqrt{\frac{\ell(\ell+1)}{2}}\frac{1}{4!}\int_{\mathbb{S}^2}H_4(T_\ell(x))\,dx \tag{4.25}$$

whose mean is zero, and whose variance [**MR15**, **MW14**] is

$$\mathrm{Var}(\mathcal{I}_\ell) = \frac{1}{16}\frac{\ell(\ell+1)}{2}\frac{1}{4!}2\cdot 4\pi\cdot 2\pi\int_0^{\pi/2}P_\ell(\cos\theta)^4\sin\theta\,d\theta.$$

A careful analysis of the fourth moment of Legendre polynomials (by means of Hilb's asymptotic formula (2.11)) give, as $\ell\to+\infty$,

$$\mathrm{Var}\,(\mathcal{I}_\ell) \sim \frac{1}{32}\log\ell \tag{4.26}$$

i.e., the variance of this single term is asymptotically equivalent to the total variance (2.13). Investigating asymptotic moments of product of powers of Legendre polynomials and their derivatives we prove that, as $\ell\to+\infty$,

$$\frac{\mathcal{L}_\ell[4]}{\sqrt{\mathrm{Var}(\mathcal{L}_\ell[4])}} \quad\text{and}\quad \frac{\mathcal{I}_\ell}{\sqrt{\mathrm{Var}(\mathcal{I}_\ell)}} \tag{4.27}$$

are fully correlated. By (3.18), the orthogonality of Wiener chaoses, (4.27), (4.26) and (2.13) we have

$$\frac{\mathcal{L}_\ell-\mathbb{E}[\mathcal{L}_\ell]}{\mathrm{Var}(\mathcal{L}_\ell)} = \frac{\mathcal{I}_\ell}{\sqrt{\mathrm{Var}(\mathcal{I}_\ell)}} + o_{\mathbb{P}}(1), \tag{4.28}$$

$o_{\mathbb{P}}(1)$ denoting a sequence converging to zero in probability, meaning that the fourth order chaotic component or more precisely the term $\mathcal{I}_\ell$ "dominates" the whole series. Thus in order to understand the asymptotic behavior of the total nodal length it

[3]The author thanks Domenico Marinucci for sharing with her this insight

suffices to study the second order fluctuations of $\mathcal{I}_\ell$. Proposition 3.4 in [**MW14**] states that, as $\ell \to +\infty$,

$$\frac{\mathcal{I}_\ell}{\sqrt{\mathrm{Var}(\mathcal{I}_\ell)}} \xrightarrow{d} Z,$$

where Z is a standard Gaussian random variable. This concludes the proof. □

Inspired by [**PR17**], it should be possible to show directly (4.28) by proving in addition that, as $\ell \to +\infty$,

$$\mathrm{Var}\left(\sum_{q=3}^{+\infty} \mathcal{L}_\ell[2q]\right) = O(1).$$

4.1. Some insight.

4.1.1. *Universality of the mean nodal length.* Consider point (i) in Remark 2.3. It is immediate to explain this phenomenon in terms of chaotic components, indeed behind (4.21) there is the following formula

$$\mathbb{E}[\mathcal{L}_\ell] = \mathcal{L}_\ell[0] = \sqrt{\frac{\ell(\ell+1)}{2}} \cdot \beta_0 \alpha_{0,0} \cdot \int_{\mathbb{S}^2} \underbrace{H_0(T_\ell(x))}_{=1} \, dx,$$

where $\ell(\ell+1)/2$ is the variance of the derivatives of T_ℓ. The same formula, accordingly modified, holds on the torus (and in more general settings – see also [**CM16**]). Of course, the same result can be obtained by Kac-Rice formula [**AT07**, Ch. 12].

4.1.2. *Berry's cancellation phenomenon.* As briefly explained in (ii) Remark 2.3, Berry's cancellation phenomenon concerns the order of magnitude of the asymptotic variance of the nodal length, which turns out to be smaller than expected. This fact can be expressed in terms of chaotic expansions as the vanishing of the second order chaotic component of the nodal length (see (4.24)). Indeed otherwise the variance of the nodal length would have been of order ℓ, the "natural" one. A careful inspection of (4.23) reveals that the steps of the proof of (4.24) are independent of the underlying manifold, in particular the same result holds for the toral case (i.e. the second chaotic component of the nodal length for arithmetic random waves vanishes [**Ros15,MPRW16**]). More generally, it is natural to guess that the same cancellation phenomenon appears for the nodal volume on so-called *isotropic* manifolds (compact two-point homogeneous spaces [**BM16**]) of any dimension (like the hyperspheres [**MR15**]), and multidimensional tori (see also [**Cam17**]). Moreover, this phenomenon has been observed also for other functionals of nodal sets of Gaussian random eigenfunctions, see [**CM16**] for a complete discussion.

4.1.3. *Gaussianity: torus vs. sphere.* In §4 we gave a sketch of the proof of Theorem 2.2: the second order fluctuations of the spherical nodal length $\mathcal{L}_\ell$ are Gaussian. From its proof it turns out that $\mathcal{L}_\ell$ and the term $\mathcal{I}_\ell$ in (4.25) have the same asymptotic behavior. As already said, the asymptotic Gaussianity of

$$\int_{\mathbb{S}^2} H_4(T_\ell(x))\, dx \tag{4.29}$$

and hence of $\mathcal{I}_\ell$ was proved in [**MW14**]. It is now natural to ask for the asymptotic behavior of the corresponding term on the torus, i.e.

$$\int_{\mathbb{T}^2} H_4(T_n(x))\, dx. \tag{4.30}$$

Indeed, also in the toral case the fourth chaotic component dominates the series of the total nodal length (see [**MPRW16**]). From Lemma 5.2 in [**MPRW16**] we have

$$\int_{\mathbb{T}^2} H_4(T_n(x))\,dx = \frac{6}{\mathcal{N}_n}\left(\frac{1}{\sqrt{\mathcal{N}_n/2}}\sum_{\xi\in\Lambda_n^+}(|a_\xi|^2-1)\right)^2 - \frac{3}{\mathcal{N}_n^2}\sum_{\xi\in\Lambda_n}|a_\xi|^4, \tag{4.31}$$

where $\Lambda_n^+ := \{\xi = (\xi_1,\xi_2)\in\Lambda_n : \xi_2 > 0\}$ if $\sqrt{n}$ is not an integer, otherwise $\Lambda_n^+ := \{\xi = (\xi_1,\xi_2)\in\Lambda_n : \xi_2 > 0\}\cup\{(\sqrt{n},0)\}$. From (4.31) the standard CLT applied to the following sum of i.i.d. random variables

$$\frac{1}{\sqrt{\mathcal{N}_n/2}}\sum_{\xi\in\Lambda_n^+}(|a_\xi|^2-1)$$

entails that (4.30) is not asymptotically Gaussian. This discussion gives moreover some insight on the reasons for the nature of the limiting distribution found in Theorem 2.1.

5. Further related work

5.1. Monochromatic random waves. Consider now the general setting as in §1.1. For a generic Riemannian metric on a smooth compact surface $\mathcal{M}$, the eigenvalues λ_j^2 of the Laplace-Beltrami operator (as in (1.1)) are simple and one cannot define a meaningful random model as for the toral or the spherical case, i.e. by endowing each eigenspace with a Gaussian measure. To overcome this, Zelditch introduced in [**Zel09**] the so-called *monochromatic random waves* as general approximate models of random Gaussian Laplace eigenfunctions defined on manifolds not necessarily having spectral multiplicities, see also [**CH16**] for more details. The monochromatic random wave on $(\mathcal{M},g)$ of parameter $\lambda>0$ is the random field

$$\phi_\lambda(x) := \frac{1}{\sqrt{\dim(H_\lambda)}}\sum_{\lambda_j\in[\lambda,\lambda+1]} a_j f_j(x), \quad x\in\mathcal{M}, \tag{5.32}$$

where the a_j are i.i.d. standard Gaussian random variables, and

$$H_\lambda := \bigoplus_{\lambda_j\in[\lambda,\lambda+1]} \operatorname{Ker}(\Delta_g + \lambda_j^2\,\mathrm{Id}),$$

Id being the identity operator. The field ϕ_λ is centered Gaussian and its covariance kernel is

$$K_\lambda(x,y) := \operatorname{Cov}(\phi_\lambda(x),\phi_\lambda(y)) = \frac{1}{\dim(H_\lambda)}\sum_{\lambda_j\in[\lambda,\lambda+1]} f_j(x)f_j(y), \quad x,y\in\mathcal{M}. \tag{5.33}$$

Now fix a point $x\in\mathcal{M}$, and consider the tangent plane $T_x\mathcal{M}$ to the manifold at x. The *pullback Riemannian random wave* (see [**CH16**]) associated with ϕ_λ is the Gaussian random field on $T_x\mathcal{M}$ defined as

$$\phi_\lambda^x(u) := \phi_\lambda\left(\exp_x\left(\frac{u}{\lambda}\right)\right), \qquad u\in T_x\mathcal{M},$$

where $\exp_x : T_x\mathcal{M}\to\mathcal{M}$ is the exponential map at x. The random field ϕ_λ^x is centered Gaussian and from (5.33) its covariance kernel is

$$K_\lambda^x(u,v) = K_\lambda\left(\exp_x\left(\frac{u}{\lambda}\right),\exp_x\left(\frac{v}{\lambda}\right)\right), \qquad u,v\in T_x\mathcal{M}.$$

DEFINITION 5.1 (See [**CH16**]). A point $x \in \mathcal{M}$ is of *isotropic scaling* if, for every positive function $\lambda \mapsto r(\lambda)$ such that $r(\lambda) = o(\lambda)$, as $\lambda \to \infty$, one has that

$$\sup_{u,v \in \mathbb{B}(r(\lambda))} \left| \partial^\alpha \partial^\beta [K_\lambda^x(u,v) - 2\pi \cdot J_0(\|u-v\|_{g_x})] \right| \to 0, \quad \lambda \to \infty, \tag{5.34}$$

where $\alpha, \beta \in \mathbb{N}^2$ are multi-indices labeling partial derivatives with respect to u and v, respectively, $\|\cdot\|_{g_x}$ is the norm on $T_x\mathcal{M}$ induced by g, and $\mathbb{B}(r(\lambda))$ is the corresponding ball of radius $r(\lambda)$ containing the origin. The manifold $\mathcal{M}$ is of isotropic scaling if every $x \in \mathcal{M}$ is a point of isotropic scaling, and the convergence in (5.34) is uniform on x for each $\alpha, \beta \in \mathbb{N}^2$.

It is difficult to prove directly that a point x is of isotropic scaling, except on flat tori, but sufficient conditions about the geodesics through x are known (see [**CH16**, §2.5] and the references therein). The condition that $\mathcal{M}$ is a manifold of isotropic scaling is *generic* in the space of Riemannian metrics on any smooth compact manifold, see [**CH16**, §2.5] for further details. Note that one can always choose coordinates around x to have $g_x = \mathrm{Id}$, so that the limiting kernel in (5.34) coincides with (1.3) up to a factor.

5.2. Nodal lengths: recent results. Let us now fix a C^1*-convex body* $\mathcal{D} \subset \mathbb{R}^2$ (that is: $\mathcal{D}$ is a compact convex set with C^1-boundary) such that $0 \in \mathring{\mathcal{D}}$ (i.e., the origin belongs to the interior of $\mathcal{D}$). For every $x \in \mathcal{M}$ we define

$$\mathcal{Z}_{\lambda,\mathcal{D}}^x := \mathrm{length}((\phi_\lambda^x)^{-1}(0) \cap \mathcal{D}).$$

THEOREM 5.2 (Special case of Theorem 1 in [**CH16**]). *Let x be a point of isotropic scaling, and assume that coordinates have been chosen around x in such a way that $g_x = \mathrm{Id}$. Then, as $\lambda \to \infty$,*

$$\mathcal{Z}_{\lambda,\mathcal{D}}^x \stackrel{d}{\to} \sqrt{2\pi} \cdot \mathrm{length}(B_1^{-1}(0) \cap \mathcal{D}),$$

where B_1 is Berry's RWM (1.3) with $\lambda_j = 1$.

Theorem 5.2 states that the local behavior of zeros of monochromatic random waves is universal (see also [**NPR17**] for more details). It is not an easy task to find the distribution of the limiting random variable $\mathrm{length}(B_1^{-1}(0) \cap \mathcal{D})$ in Theorem 5.2, so let us understand at least what happens when the domain $\mathcal{D}$ grows to $\mathbb{R}^2$. For $E > 0$ let us hence consider

$$\mathrm{length}(B_1^{-1}(0) \cap \sqrt{E} \cdot \mathcal{D}) = \sqrt{E} \cdot \mathrm{length}(B_{\sqrt{E}}^{-1}(0) \cap \mathcal{D}) =: \sqrt{E} \cdot \mathcal{L}_E, \tag{5.35}$$

where $B_{\sqrt{E}}$ is Berry's RWM (1.3) with $\lambda_j = \sqrt{E}$. We recall now the main results of [**NPR17**] concerning the behavior of $\mathcal{L}_E$, as $E \to +\infty$. Note that (5.36) and (5.37) confirm the results predicted by Berry in [**Ber02**].

THEOREM 5.3. *The expectation of the nodal length $\mathcal{L}_E$ is*

$$\mathbb{E}[\mathcal{L}_E] = \mathit{area}(\mathcal{D}) \frac{1}{2\sqrt{2}} \sqrt{E}, \tag{5.36}$$

the variance of $\mathcal{L}_E$ verifies the asymptotic relation

$$\mathrm{Var}(\mathcal{L}_E) \sim \mathit{area}(\mathcal{D}) \frac{1}{512\pi} \log E, \quad E \to \infty. \tag{5.37}$$

Moreover, as $E \to \infty$,

$$\frac{\mathcal{L}_E - \mathbb{E}[\mathcal{L}_E]}{\sqrt{\mathrm{Var}(\mathcal{L}_E)}} \stackrel{d}{\longrightarrow} Z,$$

where $Z \sim \mathcal{N}(0,1)$ is a standard Gaussian random variable.

Note that the mean (5.36) has the same form as in the toral or spherical case (see Remark 2.3), and the asymptotic variance (5.37) is of lower order than expected – as predicted by Berry in [**Ber02**]. Thanks to the symmetry of nodal lines on the two dimensional sphere [**Wig10**, §1.6.2], the result on the asymptotic variance for the nodal length on $\mathbb{S}^2$ (2.13) obtained by Wigman in [**Wig10**] is of course consistent to Berry's prediction ((5.37) or [**Ber02**]).

The proof of Theorem 5.3 relies on chaotic expansions as for the toral and spherical case (see §4). Here, the second order chaotic component does not vanishes identically but it is possible to prove, by means of Green's identity on manifold, that its contribution is negligible. In view of this, the next chaotic component to study is the fourth one which turns out to dominate the whole series, once proved that the contribution of higher order chaoses is negligible (for details see [**NPR17**]).

Acknowledgments

The author would like to thank the organizers of the conference *Probabilistic methods in spectral geometry and PDE* and the CRM Montréal for such a wonderful meeting. Some of the results presented in this short survey have been obtained in collaboration with Domenico Marinucci, Ivan Nourdin, Giovanni Peccati and Igor Wigman to whom the author is grateful; in particular, the author would like to thank Domenico Marinucci for the insight in §4 and Igor Wigman for fruitful discussions we had in Montréal. Moreover, the author thanks Valentina Cammarota for a number of useful discussions, and an anonymous referee for his/her valuable comments.

References

[AT07] Robert J. Adler and Jonathan E. Taylor, *Random fields and geometry*, Springer Monographs in Mathematics, Springer, New York, 2007. MR2319516

[BM16] V. S. Barbosa and V. A. Menegatto, *Strictly positive definite kernels on compact two-point homogeneous spaces*, Math. Inequal. Appl. **19** (2016), no. 2, 743–756, DOI 10.7153/mia-19-54. MR3458777

[BMW17] J. Benatar, D. Marinucci and I. Wigman (2017). *Planck-scale distribution of nodal length of arithmetic random waves.* Preprint arXiv:1710.06153.

[BM17] Jacques Benatar and Riccardo W. Maffucci, *Random Waves On $\mathbb{T}^3$: Nodal Area Variance and Lattice Point Correlations*, Int. Math. Res. Not. IMRN **10** (2019), 3032–3075, DOI 10.1093/imrn/rnx220. MR3952558

[Ber85] P. Bérard, *Volume des ensembles nodaux des fonctions propres du laplacien* (French), Bony-Sjöstrand-Meyer seminar, 1984–1985, École Polytech., Palaiseau, 1985, pp. Exp. No. 14 , 10. MR819780

[Ber77] M. V. Berry, *Regular and irregular semiclassical wavefunctions*, J. Phys. A **10** (1977), no. 12, 2083–2091. MR0489542

[Ber02] M. V. Berry, *Statistics of nodal lines and points in chaotic quantum billiards: perimeter corrections, fluctuations, curvature*, J. Phys. A **35** (2002), no. 13, 3025–3038, DOI 10.1088/0305-4470/35/13/301. MR1913853

[Cam17] V. Cammarota (2017). *Nodal area distribution for arithmetic random waves.* Preprint arXiv:1708.07679.

[CM16] Valentina Cammarota and Domenico Marinucci, *A quantitative central limit theorem for the Euler-Poincaré characteristic of random spherical eigenfunctions*, Ann. Probab. **46** (2018), no. 6, 3188–3228, DOI 10.1214/17-AOP1245. MR3857854

[CH16] Y. Canzani and B. Hanin (2016). *Local Universality for Zeros and Critical Points of Monochromatic Random Waves.* Preprint arXiv:1610.09438.

[Chl02] E.-F.-F. Chladni, (1802). *Die Akustik*, Leipzig. Available online from http://vlp.mpiwg-berlin.mpg.de/references?id=lit29494.

[DNPR16] F. Dalmao, I. Nourdin, G. Peccati and M. Rossi (2016). *Phase singularities in complex arithmetic random waves.* Preprint arXiv:1608.05631.

[DF88] Harold Donnelly and Charles Fefferman, *Nodal sets of eigenfunctions on Riemannian manifolds*, Invent. Math. **93** (1988), no. 1, 161–183, DOI 10.1007/BF01393691. MR943927

[FKW06] Laura Fainsilber, Pär Kurlberg, and Bernt Wennberg, *Lattice points on circles and discrete velocity models for the Boltzmann equation*, SIAM J. Math. Anal. **37** (2006), no. 6, 1903–1922, DOI 10.1137/040618916. MR2213399

[GK12] Martin J. Gander and Felix Kwok, *Chladni figures and the Tacoma bridge: motivating PDE eigenvalue problems via vibrating plates*, SIAM Rev. **54** (2012), no. 3, 573–596, DOI 10.1137/10081931X. MR2966726

[KKW13] Manjunath Krishnapur, Pär Kurlberg, and Igor Wigman, *Nodal length fluctuations for arithmetic random waves*, Ann. of Math. (2) **177** (2013), no. 2, 699–737, DOI 10.4007/annals.2013.177.2.8. MR3010810

[KW17] Pär Kurlberg and Igor Wigman, *On probability measures arising from lattice points on circles*, Math. Ann. **367** (2017), no. 3-4, 1057–1098, DOI 10.1007/s00208-016-1411-4. MR3623219

[Log16a] Alexander Logunov, *Nodal sets of Laplace eigenfunctions: polynomial upper estimates of the Hausdorff measure*, Ann. of Math. (2) **187** (2018), no. 1, 221–239, DOI 10.4007/annals.2018.187.1.4. MR3739231

[Log16b] Alexander Logunov, *Nodal sets of Laplace eigenfunctions: proof of Nadirashvili's conjecture and of the lower bound in Yau's conjecture*, Ann. of Math. (2) **187** (2018), no. 1, 241–262, DOI 10.4007/annals.2018.187.1.5. MR3739232

[LM16] A. Logunov and E. Malinnikova (2016).*Nodal sets of Laplace eigenfunctions: estimates of the Hausdorff measure in dimension two and three.* Preprint arXiv:1605.02595.

[Maf17a] Riccardo W. Maffucci, *Nodal intersections of random eigenfunctions against a segment on the 2-dimensional torus*, Monatsh. Math. **183** (2017), no. 2, 311–328, DOI 10.1007/s00605-016-1001-2. MR3641930

[Maf17b] Riccardo W. Maffucci, *Nodal intersections for random waves against a segment on the 3-dimensional torus*, J. Funct. Anal. **272** (2017), no. 12, 5218–5254, DOI 10.1016/j.jfa.2017.02.011. MR3639527

[MP11] Domenico Marinucci and Giovanni Peccati, *Random fields on the sphere*, London Mathematical Society Lecture Note Series, vol. 389, Cambridge University Press, Cambridge, 2011. Representation, limit theorems and cosmological applications. MR2840154

[MPRW16] Domenico Marinucci, Giovanni Peccati, Maurizia Rossi, and Igor Wigman, *Non-universality of nodal length distribution for arithmetic random waves*, Geom. Funct. Anal. **26** (2016), no. 3, 926–960, DOI 10.1007/s00039-016-0376-5. MR3540457

[MR15] Domenico Marinucci and Maurizia Rossi, *Stein-Malliavin approximations for nonlinear functionals of random eigenfunctions on $\mathbb{S}^d$*, J. Funct. Anal. **268** (2015), no. 8, 2379–2420, DOI 10.1016/j.jfa.2015.02.004. MR3318653

[MRW17] D. Marinucci, M. Rossi and I. Wigman (2017). *The asymptotic equivalence of the sample trispectrum and the nodal length for random spherical harmonics.* Preprint arXiv 1705.05747.

[MW14] Domenico Marinucci and Igor Wigman, *On nonlinear functionals of random spherical eigenfunctions*, Comm. Math. Phys. **327** (2014), no. 3, 849–872, DOI 10.1007/s00220-014-1939-7. MR3192051

[NP12] Ivan Nourdin and Giovanni Peccati, *Normal approximations with Malliavin calculus*, Cambridge Tracts in Mathematics, vol. 192, Cambridge University Press, Cambridge, 2012. From Stein's method to universality. MR2962301

[NPR17] I. Nourdin, G. Peccati and M. Rossi (2017). *Nodal statistics of planar random waves.* Preprint arXiv:arXiv:1708.02281.

[PR17] G. Peccati, M. Rossi (2017). *Quantitative limit theorems for local functionals of arithmetic random waves.* Preprint arXiv:1702.03765.

[Ros15] M. Rossi (2015). *The Geometry of spherical random fields.* PhD Thesis, University of Rome Tor Vergata. arXiv:1603.07575.

[RoW17] Maurizia Rossi and Igor Wigman, *Asymptotic distribution of nodal intersections for arithmetic random waves*, Nonlinearity **31** (2018), no. 10, 4472–4516, DOI 10.1088/1361-6544/aaced4. MR3846437

[RW08] Zeév Rudnick and Igor Wigman, *On the volume of nodal sets for eigenfunctions of the Laplacian on the torus*, Ann. Henri Poincaré **9** (2008), no. 1, 109–130, DOI 10.1007/s00023-007-0352-6. MR2389892

[RW16] Zeév Rudnick and Igor Wigman, *Nodal intersections for random eigenfunctions on the torus*, Amer. J. Math. **138** (2016), no. 6, 1605–1644, DOI 10.1353/ajm.2016.0048. MR3595496

[RWY16] Zeév Rudnick, Igor Wigman, and Nadav Yesha, *Nodal intersections for random waves on the 3-dimensional torus* (English, with English and French summaries), Ann. Inst. Fourier (Grenoble) **66** (2016), no. 6, 2455–2484. MR3580177

[Sze75] Gábor Szegő, *Orthogonal polynomials*, 4th ed., American Mathematical Society, Providence, R.I., 1975. American Mathematical Society, Colloquium Publications, Vol. XXIII. MR0372517

[Wig10] Igor Wigman, *Fluctuations of the nodal length of random spherical harmonics*, Comm. Math. Phys. **298** (2010), no. 3, 787–831, DOI 10.1007/s00220-010-1078-8. MR2670928

[Wig12] Igor Wigman, *On the nodal lines of random and deterministic Laplace eigenfunctions*, Spectral geometry, Proc. Sympos. Pure Math., vol. 84, Amer. Math. Soc., Providence, RI, 2012, pp. 285–297, DOI 10.1090/pspum/084/1362. MR2985322

[Yau82] Shing Tung Yau, *Survey on partial differential equations in differential geometry*, Seminar on Differential Geometry, Ann. of Math. Stud., vol. 102, Princeton Univ. Press, Princeton, N.J., 1982, pp. 3–71. MR645729

[Yau93] Shing-Tung Yau, *Open problems in geometry*, Differential geometry: partial differential equations on manifolds (Los Angeles, CA, 1990), Proc. Sympos. Pure Math., vol. 54, Amer. Math. Soc., Providence, RI, 1993, pp. 1–28. MR1216573

[Zel09] Steve Zelditch, *Real and complex zeros of Riemannian random waves*, Spectral analysis in geometry and number theory, Contemp. Math., vol. 484, Amer. Math. Soc., Providence, RI, 2009, pp. 321–342, DOI 10.1090/conm/484/09482. MR1500155

MAP5-UMR CNRS 8145, Université Paris Descartes, France
Email address: maurizia.rossi@parisdescartes.fr

Contemporary Mathematics
Volume **739**, 2019
https://doi.org/10.1090/conm/739/14899

Entropy bounds and quantum unique ergodicity for Hecke eigenfunctions on division algebras

Lior Silberman and Akshay Venkatesh

ABSTRACT. We prove the arithmetic quantum unique ergodicity (AQUE) conjecture for non-degenerate sequences of Hecke eigenfunctions on quotients $\Gamma\backslash G/K$, where $G \simeq \mathrm{PGL}_d(\mathbb{R})$, K is a maximal compact subgroup of G and $\Gamma < G$ is a lattice associated to a division algebra over $\mathbb{Q}$ of prime degree d.

More generally, we introduce a new method of proving positive entropy of quantum limits, which applies to higher-rank groups. The result on AQUE is obtained by combining this with a measure-rigidity theorem due to Einsiedler-Katok, following a strategy first pioneered by Lindenstrauss.

CONTENTS

1. Introduction

1.1. Result. In this paper, we shall show (a slightly sharper version of) the following statement. For precise definitions we refer to §6 especially §6.2.

THEOREM 1.1. *Let Γ be a lattice in $\mathrm{PGL}_d(\mathbb{R})$, with d prime, associated to a division algebra*[1]*, and ψ_i a non-degenerate sequence of Hecke-Maass eigenfunctions*

The second author was supported by a Clay Research Fellowship, and NSF Grant DMS–0245606.

Both authors were partly supported by NSF Grant DMS–0111298; they would also like to thank the Institute for Advanced Study for providing superb working conditions.

[1]This means that Γ is the image of $\mathcal{O}^\times$ in $\mathrm{PGL}_d(\mathbb{R})$, where $\mathcal{O}$ is an order in a $\mathbb{Q}$-division algebra so that $\mathcal{O}\otimes\mathbb{R} = M_d(\mathbb{R})$. We also impose a class number one condition, see §6.

on $Y := \Gamma\backslash \mathrm{PGL}_d(\mathbb{R})/\mathrm{PO}_d(\mathbb{R})$, *normalized to have* L^2*-norm* 1 *w.r.t. the Riemannian volume* dvol.

Then the measures $|\psi_i|^2$ dvol *converge weakly to the Haar measure, i.e. for any* $f \in C(Y)$,

$$\lim_{i\to\infty} \int_Y |\psi_i|^2 f d\mathrm{vol} = \int_Y f d\mathrm{vol}. \tag{1.1}$$

In words, Theorem 1.1 asserts that the eigenfunctions ψ_i become *equidistributed* – that they do not cluster too much on the manifold Y.

This theorem is a contribution to the study of the "Arithmetic Quantum Unique Ergodicity" problem. A detailed introduction to this problem may be found in our paper [**20**]. While it is hard to dispute that the spaces Y are far too special to provide a reasonable model for the physical problem of "quantum chaos," both the statement (1.1) and the techniques we use to prove it seem to the authors to be of interest because of the scarcity of results concerning analysis of higher rank automorphic forms. In particular, we believe our techniques will find applications to analytic problems beside QUE.

Our strategy follows that of Lindenstrauss in his proof of the arithmetic QUE conjecture for quotients of the hyperbolic plane (the case $d = 2$ of the Theorem above) and has three conceptually distinct steps. Let $A \subset \mathrm{PGL}_d(\mathbb{R})$ be the subgroup of diagonal matrices.

(1) *Microlocal lift:* notation as above, any weak limit (as $i \to \infty$) of $|\psi_i|^2$ dvol may be lifted to an A-invariant measure σ_∞ on $X := \Gamma\backslash \mathrm{PGL}_d(\mathbb{R})$, in a way compatible with the Hecke correspondence.
(2) *Mass of small tubes:* If σ_∞ is as in (1), then the σ_∞-mass of an ϵ-ball in $\Gamma\backslash \mathrm{PGL}_d(\mathbb{R})$ is $\ll \epsilon^{d-1+\delta}$, for some $\delta > 0$; note that the bound $\ll \epsilon^{d-1}$ is trivial from the A-invariance. [2]
(3) *Measure rigidity:* Any A-invariant measure satisfying the auxiliary condition prescribed by (2) must necessarily be a convex combination of algebraic measures. In our setting (Γ associated to a division algebra of prime degree) this means it must be Haar measure.

* * *

In the context of Lindenstrauss' proof, the analogues of steps 2 and 3 are due, respectively, to Bourgain–Lindenstrauss [**1**] and Lindenstrauss [**11**]. The analogue of step 1 is due to [**9**] (based on constructions of Schnirel'man [**21**], Zelditch [**26**,**27**], Colin de Verdiére [**4**] and Wolpert [**24**]).

We shall concern ourselves with the higher rank case ($d > 2$), where step 1 – under a nondegeneracy condition – has been established by the authors in [**20**], while step 3 was established by Einsiedler-Katok-Lindenstrauss in [**6**].

The contribution of the present paper is then the establishment of step 2.

1.2. Bounding mass of tubes – vague discussion. As was discussed in the previous Section, the main point of the present paper is to prove upper bounds for the mass of eigenfunctions in small tubes.

A *correspondence* on a manifold X, for our purposes, will be a subset $S \subset X \times X$ such that both projections are topological coverings. Such a correspondence induces

[2]This asserts, then, that σ_∞ has some "thickness" transverse to the A-direction; for instance, it immediately implies that the dimension of the support of σ_∞ is strictly larger than $d-1$.

an endomorphism of $L^2(X)$: pull back to S and push forward to X. We also think of a correspondence as a "multi-valued" or "set-valued" function h_S from X to X. In the latter view a correspondence induces a natural convolution action on functions on X, given by $(f * h_S)(x) = \sum_{y \in h_S(x)} f(y)$.

Two correspondences can be composed in a natural way and resulting algebra is, in general, non-commutative. However, the manifolds of interest to us ($X = \Gamma\backslash\mathbf{G}$ with Γ an arithmetic lattice in the semisimple Lie group G) come equipped with a large algebra of *commuting* correspondences, the Hecke algebra $\mathcal{H}$, which acts on $L^2(X)$ by normal operators. We will be interested in possible concentration of simultaneous eigenfunctions of the Hecke algebra.

As a concrete example, for $X = \mathrm{PGL}_d(\mathbb{Z})\backslash\mathrm{PGL}_d(\mathbb{R})$ the Hecke correspondences are induced by *left* multiplication with $\mathrm{PGL}_d(\mathbb{Q})$: given $\gamma \in \mathrm{PGL}_d(\mathbb{Q})$ and a coset $x \in X$, we consider the set of products γg as g varies over representatives in $\mathrm{PGL}_d(\mathbb{R})$ for x. It turns out that these products generate a finite set of cosets $h_\gamma(x) \subset X$. It is easy to check that the adjoint of h_γ is $h_{\gamma^{-1}}$, but the commutativity of the Hecke algebra is more subtle. An important feature of the Hecke correspondences on X is their equivariance w.r.t. the action of $G = \mathrm{PGL}_d(\mathbb{R})$ on X on the right.

Returning to the general $X := \Gamma\backslash G$, let $\mathcal{T}(\epsilon)$ be a small subset of G, with its size in certain directions on the order of ϵ (for us $\mathcal{T}(\epsilon)$ will be a tube of width ϵ around a compact piece of a Levi subgroup of G). Our goal will be to prove a statement of the following type, for some fixed $\eta > 0$ depending only on G:

$$\text{Each } \mathcal{H}\text{-eigenfunction } \psi \in L^2(X) \text{ satisfies } \mu_\psi(x\mathcal{T}(\epsilon)) \ll \epsilon^\eta. \tag{1.2}$$

Here $\mu_\psi := |\psi|^2\, d\mathrm{vol}$ is the product of the $\mathrm{PGL}_d(\mathbb{R})$-invariant measure and the function $|\psi|^2$, normalized to be a probability measure. (1.2) asserts that the eigenfunction ψ cannot concentrate too singularly on a small tube. This is proven, in the cases of interest for this paper, in Theorem 5.3.

We will sketch here our approach to the proof. A basic form of the idea appeared in the paper [**17**] of Rudnick and Sarnak. If ψ is an eigenfunction of a correspondence $h \in \mathcal{H}$, and ψ were large at some point x, it also tends to be quite large at points belonging to the orbit $h.x$. We can thereby "disperse" the local question of bounding the mass of a small tube to a global question about the size of ψ throughout the manifold.

Say that the image of the point x under $h \in \mathcal{H}$ is the collection of N points $h.x = \{x_i\}$. Equivariance implies that the image of the tube $x\mathcal{T}(\epsilon)$ under h is the collection of tubes $\{x_i\mathcal{T}(\epsilon)\}$. For any $t \in \mathcal{T}(\epsilon)$, we have, then

$$\lambda_h\psi(xt) = \sum_{i=1}^{N} \psi(x_i t),$$

where λ_h is so that $h.\psi = \lambda_h\psi$.

Squaring, applying Cauchy–Schwarz and integrating over $t \in \mathcal{T}(\epsilon)$ gives:

$$\begin{aligned} \mu_\psi(x\mathcal{T}(\epsilon)) &\leq \frac{N}{|\lambda_h|^2}\sum_{i=1}^{N}\mu_\psi(x_i\mathcal{T}(\epsilon)) \\ &\leq \frac{N}{|\lambda_h|^2}\max_i \#\left\{j | x_i\mathcal{T}(\epsilon)\cap x_j\mathcal{T}(\epsilon) \neq \emptyset\right\}. \end{aligned} \tag{1.3}$$

If the tubes $x_i\mathcal{T}(\epsilon)$ are *disjoint*[3] and, furthermore, $|\lambda_h|$ is "large" (w.r.t. N), (1.3) yields a good upper bound for $\mu_\psi(x\mathcal{T}(\epsilon))$.

The issue of choosing h so that λ_h is "large" turns out to be relatively minor. The solution is given in the appendix.

The more serious problem is that the tubes $x_i\mathcal{T}(\epsilon)$ might not be disjoint. It must be emphasized that this issue is not a technical artifact of the proof but related fundamentally to the analytic properties of eigenfunctions on arithmetic locally symmetric spaces: "returns" of the Hecke correspondence influence the sizes of eigenfunctions. For an instance of this phenomenon see the Rudnick–Sarnak example of a sequence of eigenfunctions on a hyperbolic 3-manifold with large L^∞-norms and the more recent work of Milicevic [**15, 16**].

Our approach to this difficulty is as follows: we prove a variant of (1.3) where the "worst-case" intersection number (i.e. $\max_j$) is replaced by an *average* intersection number (an average over j). This variant is presented in Lemma 3.4. The rest of the paper is then devoted to giving upper bounds for this average intersection number, which turns out to be much easier than bounding the worst-case intersection number.

REMARK 1.2. Let us contrast this approach to prior work. An alternate idea would be to choose a subset of translates $\{x_i\}$ for which we can prove an analogue of (1.3) and such that the tubes $x_i\mathcal{T}(\epsilon)$ are disjoint. Versions of this were used in the prior work of Rudnick–Sarnak and Bourgain–Lindenstrauss, with the quantitative version of Bourgain–Lindenstrauss requiring a sieving argument to find non-intersecting correspondences. Our original proof was based on a further refinement of this technique, which avoided sieves entirely by using some geometry of buildings. A presentation of that proof may be found in the PhD thesis of the first author [**18**]. The technique of this paper seems to us to be yet more streamlined.

REMARK 1.3. In order to "disperse" the eigenfunction we require the use of Hecke operators at many primes. The recent work [**2**] shows that in the case of hyperbolic surfaces, it suffices to use the Hecke operators at a single place. Generalizing that result to higher rank would be an interesting problem.

1.3. Spectrum of quotients. Significance of division algebras. More generally, the technique of the present paper can be interpreted as an implementation of the following philosophy, related to the work of Burger and Sarnak:

The analytic behavior of Hecke eigenfunctions on $\Gamma\backslash G$ *along orbits of a subgroup* $H \subset G$ *is controlled by the spectrum of quotients* $L^2(G_p/H_p)$.

Here G_p is the p-adic group corresponding to G, and $H_p \subset G_p$ is a p-adic Lie subgroup "corresponding" to H in a suitable sense. In the main situation of this paper, $G_p = \mathrm{PGL}_d(\mathbb{Q}_p)$ for almost all p, H will be a real Levi subgroup, and H_p will be a torus.

In this context, the possibilities for the subgroup H_p that can occur are closely related to the $\mathbb{Q}$-structure of the group underlying G. In general, the fewer $\mathbb{Q}$-subgroups $\mathbf{G}$ has, the fewer the possibilities for H_p. For this reason we can only reach Theorem 1.1 for quotients $\Gamma\backslash G$ arising from division algebras of prime rank: the corresponding $\mathbb{Q}$-groups have very few subgroups. As one passes to general

[3]It suffices for the number of tubes intersecting a given one to be uniformly bounded independenly of ϵ.

$\Gamma\backslash G$, the possibilities for H_p become wilder, and eventually the methods of this paper do not seem to give much information.

1.4. Organization of this paper. In Section 2 we describe our setup in the general setting of algebraic groups. Further notation regarding our special case of division algebras is discussed in Section 4.4.

Section 3 contains the derivation of our first technical result, Lemma 3.4, giving a bound for the integral on a small set of the squared modulus of an eigenfunction of an equivariant correspondence. This is a version of (1.3) which can be used for non-disjoint translates. The bound we obtain depends on the average multiplicity of intersection among the translates of the tube as well as on covering properties of the tubes (easily understood in natural applications).

In Section 4 we define the kind of tubes we shall be interested in and study the intersection patterns of their translates by elements of the Hecke algebra. We give two treatments of the analysis, one that is applicable to general $\mathbb{R}$-split groups, and another, more concrete, specific to division algebras of prime degree. In both cases we use a diophantine argument to show that under suitable hypotheses the intersection pattern is controlled by a torus in the underlying $\mathbb{Q}$-algebraic group.

Section 5 then obtains the desired power-law decay of the mass of small tubes. The considerations of this section are again fairly general.

Finally in Section 6 we recall our previous result ("step 1" of the strategy) and prove our main Theorems.

2. Notation

We shall specify here the "general" notation to be used throughout the paper. Later sections (after Section 3) will use, in addition to these notations, certain further setup about division algebras. This will be explained in §4.4.

Let $\mathbf{G}$ be a semisimple group over $\mathbb{Q}$. We choose an embedding $\rho : \mathbf{G} \to \mathrm{SL}_n$. Let $G = \mathbf{G}(\mathbb{R})$, $G = NAK$ a Cartan decomposition. At each prime p let $G_p = \mathbf{G}(\mathbb{Q}_p)$ and set $\mathbf{G}(\mathbb{Z}_p) = \rho^{-1}(\mathrm{SL}_n(\mathbb{Z}_p))$, an open compact subgroup of G_p. We write $\mathbb{A}_\mathrm{f} = \prod'_{p<\infty} \mathbb{Q}_p$ for the ring of finite adeles and set $\mathbf{G}(\mathbb{A}_\mathrm{f}) = \rho^{-1}(\mathrm{SL}_n(\mathbb{A}_\mathrm{f})) = \prod'_{p<\infty} G_p$. Finally let $K_\mathrm{f} \subset \mathbf{G}(\mathbb{A}_\mathrm{f})$ be an open compact subgroup contained in $\prod_p \mathbf{G}(\mathbb{Z}_p)$. We set $X = \mathbf{G}(\mathbb{Q})\backslash\mathbf{G}(\mathbb{A})/K_\mathrm{f}, Y = X/K$. We let $d\mathrm{vol}$ be the natural probability measures on either X or Y: in both cases, the projection of the $\mathbf{G}(\mathbb{A})$-invariant probability measure on $\mathbf{G}(\mathbb{Q})\backslash\mathbf{G}(\mathbb{A})$. We shall sometimes denote $d\mathrm{vol}_X$ by dx and $d\mathrm{vol}_Y$ by dy. We also normalize the Haar measure on K to be a probability measure.

For any subset $B \subset G$ or $B \subset \mathbf{G}(\mathbb{A})$, we denote by $\overline{B}$ the image of B in X under the natural projections $G \to X$ and $G(\mathbb{A}) \to X$.

We say that a prime p is *good* if $\mathbf{G}(\mathbb{Q}_p)$ is unramified and $K_p = \mathbf{G}(\mathbb{Z}_p) \subset \mathbf{G}(\mathbb{Q}_p)$ is a good maximal compact subgroup which is contained in K_f (via the natural embedding $\mathbf{G}(\mathbb{Q}_p) \hookrightarrow \mathbf{G}(\mathbb{A}_\mathrm{f})$). Then all but finitely many p are *good.* If p is not good, we say it is *bad.*

We denote by $\mathcal{H}$ the Hecke algebra of $\prod_{p\,\mathrm{good}} K_p$ bi-invariant functions on $\prod_{p\,good} \mathbf{G}(\mathbb{Q}_p)$. It forms a commutative algebra under convolution. It acts in a natural way on functions on X. We can identify an element of $\mathcal{H}$ with a K_f-invariant function on $\mathbf{G}(\mathbb{A}_\mathrm{f})/K_\mathrm{f}$ which is, moreover, supported on $K_\mathrm{f} \cdot \prod_{p\,good} \mathbf{G}(\mathbb{Q}_p)$. We shall abbreviate the latter condition to: *supported at good primes.*

If $\mathbf{H} \subset \mathbf{G}$ is a semisimple $\mathbb{Q}$-subgroup, we say a prime p is $\mathbf{H}$-good if it is good for $\mathbf{G}$ and moreover $\mathbf{H}(\mathbb{Q}_p) \cap K_p$ is a maximal compact subgroup of $\mathbf{H}(\mathbb{Q}_p)$.

We fix compact subsets $\Omega_\infty \subset G, \Omega \subset \mathbf{G}(\mathbb{A})$, and let $X_1 \subset X$ denote the (compact) projection of Ω to X.

DEFINITION 2.1. We call $\psi \in L^2(X)$ a *Hecke eigenfunction* if it is a joint eigenfunction of the Hecke algebra $\mathcal{H}$. We set $\mu_\psi = c|\psi|^2 d\text{vol}$, where the constant c is chosen so that μ_ψ is a probability measure on X.

In §3 we will deal with an abstract $\mathbf{G}$ as above. In §4 — §6 we sometimes specialize to the case of $\mathbf{G}$ arising from a division algebra of prime rank, D. The extra notation necessary for this specialization will be set up in §4.4.

Notational convention: We shall allow the implicit constants in the notation $\ll$ and $O(\cdot)$ to depend on $\mathbf{G}$ and the data $\rho, K_{\mathrm{f}}, \Omega, \Omega_\infty$ *without explicit indication.* In other words, the notation $A \ll B$ means that there exists a constant c, which may depend on all the data specified in this section, so that $A \leq cB$. Later we introduce further data concerning division algebras, and, as we specify there, we shall allow implicit constants after that point to depend on these extra data also.

3. Bounds on the mass of tubes

3.1. A covering argument. Let $B_0 \subset \Omega_\infty$ be an open set containing the identity. We set

$$B := B_0 \cdot B_0^{-1}, B_2 = B \cdot B, B_3 = B \cdot B \cdot B, \dots \tag{3.1}$$

In this section, we shall discuss estimating from above $\mu_\psi(\overline{xB})$ for $x \in G$. [4] We will repeatedly use the following fact. If $S \subset G$ is a compact subset and $x \in X_1$ is arbitrary, the cardinality of the fibers of the map $S \mapsto X$ defined by $s \mapsto x.s$ is bounded in a fashion that depends on S but not on x. This is an immediate consequence of the compactness of X_1: indeed, there exists an open neighbourhood U of the identity in G so that $u \in U \mapsto xu \in X$ is injective for any $x \in X_1$. In particular, we may apply this remark when $S = B, B_2, B_3$. The size of the relevant fibers will therefore be bounded by a constant depending on Ω.

LEMMA 3.1. *There exists a covering of X by translates $x_\alpha B$ so that any set zB, for $z \in X$, intersects at most $\frac{\text{vol} B_3}{\text{vol} B_0}$ of the $x_\alpha B$.*

PROOF. Choose a maximal subset $\{x_\alpha\} \subset X$ with the property that $x_\alpha B_0$ are disjoint.

Given $z \in X$, we must have $zB_0 \cap x_\alpha B_0 \neq \emptyset$ for some α. This means precisely that $X = \bigcup x_\alpha B$.

Next, fix $z \in X$. For any α so that $x_\alpha B \cap zB \neq \emptyset$, we may choose $\varpi_\alpha \in B \cdot B^{-1}$ so that $x_\alpha = z\varpi_\alpha$. Then the sets $z\varpi_\alpha B_0 \subset X$ are all disjoint.

A necessary condition for this is that the sets $\varpi_\alpha B_0$, considered as subsets of G, are disjoint. Since each $\varpi_\alpha B_0$ belongs to B_3, their number is bounded by $\frac{\text{vol} B_3}{\text{vol} B_0}$. □

[4] In fact, all our estimates really work for $\mu_\psi(xB)$ for $x \in X$, as the notation is meant to suggest. We prove our results only for $x \in G$ simply to keep notation to a minimum. If G acts with a single orbit on X, as we assume in our final applications anyway, there is no difference at all.

LEMMA 3.2. *Let ν be a probability measure on X and $y_1, \ldots, y_r \in X$. Then*

$$\sum_{i=1}^{r} \nu(y_i B)^{1/2} \ll \frac{\mathrm{vol} B_3}{\mathrm{vol} B_0} \left(\#\left\{(i,j) : y_i B_2 \cap y_j B_2 \neq \emptyset\right\}\right)^{1/2}$$

PROOF. Choose a collection x_α as in Lemma 3.1. Each set $y_i B$ is covered by at most $\frac{\mathrm{vol} B_3}{\mathrm{vol} B_0}$ sets $x_\alpha B$. Clearly $\nu(y_i B)^{1/2} \leq \sum_{\alpha : x_\alpha B \cap y_i B \neq \emptyset} \nu(x_\alpha B)^{1/2}$, and so

$$\text{(3.2)} \qquad \left(\sum_i \nu(y_i B)^{1/2}\right)^2 \leq \left(\sum_\alpha \nu(x_\alpha B)^{1/2} \sum_{i : y_i B \cap x_\alpha B \neq \emptyset} 1\right)^2$$
$$\leq \sum_\alpha \nu(x_\alpha B) \sum_\alpha \left(\sum_{i : y_i B \cap x_\alpha B \neq \emptyset} 1\right)^2$$

Now, $\sum_\alpha \nu(x_\alpha B) \leq \frac{\mathrm{vol} B_3}{\mathrm{vol} B_0}$ because each $z \in X$ belongs to at most $\frac{\mathrm{vol} B_3}{\mathrm{vol} B_0}$ of the $x_\alpha B$.

Moreover,

$$\sum_\alpha \left(\sum_{i : y_i B \cap x_\alpha B \neq \emptyset} 1\right)^2 = \#\{i, j, \alpha : y_i B \cap x_\alpha B \neq \emptyset, y_j B \cap x_\alpha B \neq \emptyset\}$$

For given i, j, the pertinent set of α is nonzero only if $y_i B_2 \cap y_j B_2 \neq \emptyset$. If it is nonempty, it has size at most $\mathrm{vol}(B_3)/\mathrm{vol}(B_0)$. □

3.2. General bound. Let ψ be a Hecke eigenfunction on X, μ_ψ the associated probability measure.

Let $h \in \mathcal{H}$, which we can think of as a function $s \mapsto h_s$ on $\mathbf{G}(\mathbb{A}_\mathrm{f})/K_\mathrm{f}$, supported on good primes. Let $S \subset \mathbf{G}(\mathbb{A}_\mathrm{f})/K_\mathrm{f}$ be the support of h.

Because ψ is a Hecke eigenfunction, there is $\Lambda_h \in \mathbb{C}$ so that:

$$\text{(3.3)} \qquad \Lambda_h \psi(x) = \sum_{s \in S} h_s \psi(x.s) \quad (x \in G)$$

Note that $x.s \in \mathbf{G}(\mathbb{A})/K_\mathrm{f}$ and therefore $\psi(x.s)$ makes sense. The following Lemma bounds $\mu_\psi(\overline{xB})$ in terms of the average mass of certain Hecke translates of $\overline{xB_2}$.

LEMMA 3.3. *Let $x \in X_1$. Then*

$$\mu_\psi(\overline{xB}) \ll \frac{\left(\sum_{s \in S} |h_s| \mu_\psi(\overline{xB}.s)^{1/2}\right)^2}{|\Lambda_h|^2}.$$

PROOF. This follows by squaring out equation (3.3), integrating over B and applying Cauchy-Schwarz. We use the fact that $b \in B \mapsto x.s.b \in X$ has fibers whose cardinality is bounded in terms of Ω_∞ (see discussion in §3.1). □

The next Lemma clarifies that the only necessary input to bound $\mu_\psi(\overline{xB})$ is an estimate for the average intersection multiplicity of Hecke translates of $\overline{xB}$.

LEMMA 3.4. *Suppose that h is supported on $S \subset \mathbf{G}(\mathbb{A}_{\mathrm{f}})/K_{\mathrm{f}}$ and $|h| \leq 1$. Then for $x \in G$*

$$\mu_\psi(\overline{xB}) \ll \left(\frac{\mathrm{vol}B_3}{\mathrm{vol}B_0}\right)^2 |\Lambda_h|^{-2} \left(\#\{s, s' \in S : \overline{xB_2 s} \cap \overline{xB_2 s'} \neq \emptyset\}\right) \tag{3.4}$$

PROOF. This follows from the prior Lemma and Lemma 3.2 (applied to the set $\{y_i\} = \{xs : s \in S\} \subset X$). □

Lemma 3.4 is our key technical Lemma. Setting $N = \#S$ as in the introduction we may rewrite the right-hand-side as:

$$\left(\frac{\mathrm{vol}B_3}{\mathrm{vol}B_0}\right)^2 \frac{N}{|\Lambda_h|^2} \cdot \frac{1}{N} \sum_{s \in S} \#\left\{s' \in S : \overline{xB_2 s} \cap \overline{xB_2 s'} \neq \emptyset\right\}.$$

This expression should be compared with the right-hand-side of equation (1.3). The key difference is that the bound now only depends on an "average" intersection number of Hecke translates (the average over s of the number of s' such that the translates by s and s' intersect), whereas the bound in (1.3) depended on a "worst case" intersection number (the supremum over s of the number of such s'). Our antecedent [**1**] relied on controlling this latter quantity, which imposed greater restrictions on the use of Hecke operators.

4. Diophantine Lemmata

Recall that we have fixed a maximal $\mathbb{R}$-split torus $A \subset G$. Given a nontrivial $a \in A$, we fix a compact neighbourhood of the identity $C \subset Z_G(a)$ in the centralizer of a inside G (the choice of C is immaterial; we will later take C to be small enough). Now let $B = B(C, \varepsilon)$ be an ε-neighbourhood of C inside G ("a tube around a piece of a Levi subgroup"). We intend to bound the μ_ψ-mass of sets of the form $\overline{xB(C, \varepsilon)} \subset X$.

In the previous Section we saw that such bounds require control on the intersection pattern of translates of these sets. This Section is devoted to obtaining this control in two situations.

We first analyze the case where $\mathbf{G}$ is $\mathbb{R}$-split and $a \in A$ is *regular*, that is when $Z_G(a)$ is a maximal torus. It turns out that the intersection is controlled by a $\mathbb{Q}$-subtorus of $\mathbf{G}$. Roughly, we show that all the $\gamma \in \mathbf{G}(\mathbb{Q})$ that lie "very close" to an $\mathbb{R}$-torus (where "very close" really means "very close relative to the denominator of γ") must all lie on a single $\mathbb{Q}$-torus. Using this we will show that the intersection pattern of Hecke translates of tubes of this type are controlled by the torus. The form of this argument seems relevant for more general results.

Our second example is the simplest case: when $\mathbf{G}$ arises from a division algebra of prime degree. In that case $\mathbf{G}$ has very few $\mathbb{Q}$-subgroups and one may remove the regularity assumption: elements close to any Levi subgroup (a subgroup of the form $Z_G(a)$) must lie on a $\mathbb{Q}$-torus.

An argument of this type (in the case $\dim_{\mathbb{Q}} D = 4$) already appears in [**1**]. That original argument does not suffice for the higher rank case because [**1**] uses commutativity of the relevant Levi subgroups in an important way. Our generalization here similarly is limited: it would be awkward to extend it further even to algebras which are not division algebras. However, this *ad hoc* approach for the case of division algebras is more transparent than the arguments for the first example above.

REMARK 4.1. Here is a toy model of the type of reasoning we use: Let ℓ be a line segment in $\mathbb{R}^2$ of length 1. Suppose $P_i = (x_i, y_i)$, for $1 \leq i \leq 3$, are points in $\mathbb{R}^2$ with rational coordinates all of which lie within ε of ℓ, and let M be an upper bound for the denominators of all x_i, y_i. Then, if $\varepsilon < \frac{1}{10}M^{-6}$, the P_i are themselves co-linear. Indeed, the area of the triangle formed by P_1, P_2, P_3 is a rational number with denominator $\leq 2M^6$. On the other hand the area of this triangle is $\leq 2\varepsilon$, whence the conclusion.

REMARK 4.2. Recall that we have fixed compact subsets $\Omega \subset \mathbf{G}(\mathbb{A})$ and $\Omega_\infty \subset G$. We will later take C and ϵ small enough to ensure that $B(C, \varepsilon) \subset \Omega_\infty$. In particular, $C \subset \Omega_\infty$. Consequently, in view of the convention discussed in §2, we shall not explicitly indicate that implicit constants in $\ll$ or $O(\dots)$ depend on C.

4.1. Denominators on an adelic group. As in the toy example, in our diophantine anaylsis it is convenient to use a notion of *denominator* rather than a notion of height. We define this notion progressively for scalars, for elements of $\mathrm{SL}_n(\mathbb{A}_\mathrm{f})$, and then for elements of $\mathbf{G}(\mathbb{A}_\mathrm{f})$. We write $\mathbb{N}$ for the set of natural numbers, where our denominator function will be valued.

Given $x \in \mathbb{Q}_p$ we let $\mathrm{d}(x)$ denote the minimal non-negative power of p such that $\mathrm{d}(x) \cdot x \in \mathbb{Z}_p$. In other words, p-adic integers have no denominator while a p-adic number of the form u/p^e, $u \in \mathbb{Z}_p^\times$, has denominator p^e. It is easy to check that the function $\mathrm{d}\colon \mathbb{Q}_p \to \mathbb{N}$ is invariant under translation by $\mathbb{Z}_p$ and is hence uniformly continuous. Next, for $x = (x_p)_{p<\infty} \in \mathbb{A}_\mathrm{f}$ we set $\mathrm{d}(x) = \prod_{p<\infty} \mathrm{d}(x_p)$ (almost all the factors are equal to 1). This gives a locally constant function $\mathrm{d}\colon \mathbb{A}_\mathrm{f} \to \mathbb{N}$. The diagonal embedding of $\mathbb{Q}$ in $\mathbb{A}_\mathrm{f}$ allows us to restrict d to $\mathbb{Q}$ and it is easy to check that if $a, b \in \mathbb{Z}$ are relatively prime with $b > 0$ then $\mathrm{d}(\frac{a}{b}) = b$. In other words, the restriction is the usual notion of the denominator of a rational number.

For $g = (g_p) \in \mathrm{SL}_n(\mathbb{A}_\mathrm{f})$ set $\mathrm{d}(g)$ to be the least common multiple of the denominators of the matrix entries. For $k_p \in \mathrm{SL}_n(\mathbb{Z}_p)$, or more generally for $k \in \prod_p \mathrm{SL}_n(\mathbb{Z}_p)$ we then have $\mathrm{d}(k_p) = \mathrm{d}(k) = 1$, and in fact $\mathrm{d}\colon \mathrm{SL}_n(\mathbb{A}_\mathrm{f}) \to \mathbb{N}$ is left- and right-invariant by $\prod_p \mathrm{SL}_n(\mathbb{Z}_p)$.

Finally, for $g \in \mathbf{G}(\mathbb{A}_\mathrm{f})$, we set $\mathrm{d}(g) = \mathrm{d}(\rho(g))$. This function is then bi-K_f-invariant. By restriction this induces a notion of denominator for elements $\gamma \in \mathbf{G}(\mathbb{Q})$, the least common multiple of the denominators of the matrix elements of $\rho(\gamma) \in \mathrm{SL}_n(\mathbb{Q})$.

For the infinite place we fix a left-invariant Riemannian metric dist on $\mathrm{SL}_n(\mathbb{R})$. Pulling the metric tensor back via ρ induces a left-invariant Riemannian metric dist_G on G. Since it maps geodesics on G to curves of the same length in $\mathrm{SL}_n(\mathbb{R})$, ρ is a nonexpansive map with respect to these metrics.

The "ε neighbourhoods" $B(C, \varepsilon)$ are taken with respect to the metric dist_G.

LEMMA 4.3. *Let $g, g' \in \mathbf{G}(\mathbb{A}_\mathrm{f})$. Then $\mathrm{d}(gg') \leq \mathrm{d}(g)\mathrm{d}(g')$ and $\mathrm{d}(g^{-1}) \leq \mathrm{d}(g)^{n-1}$.*

PROOF. Replacing g, g' by their images under ρ we may assume $g, g' \in \mathrm{SL}_n(\mathbb{A}_\mathrm{f})$. For the first claim we use an alternate characterization of $\mathrm{d}(g)$: it is the smallest positive integer m for which $m \cdot g \in \mathrm{GL}_n(\prod_p \mathbb{Z}_p)$. This implies

$$\mathrm{d}(g)\mathrm{d}(g')gg' \in \mathrm{GL}_n(\prod_p \mathbb{Z}_p),$$

and the claim follows. For the second note that the matrix entries of g^{-1} are polynomials of degree $n-1$ and integer coefficients in the matrix entries of g. □

4.2. A Diophatine Lemma for $\mathbb{R}$-split groups. Assume now that $\mathbf{G}$ is $\mathbb{R}$-split, and let $a \in A$ be a *regular* element. In that case $Z_G(a) = MA$ is a maximal torus in $\mathbf{G}(\mathbb{R})$, and $\rho(MA)$ is an $\mathbb{R}$-split torus in $\mathrm{SL}_n(\mathbb{R})$.

The following is the basic diophantine result.

LEMMA 4.4. *For $c > 0$ sufficiently large (in fact, depending only on $\mathbf{G}, \rho$), and $c' > 0$ sufficiently small (depending on $\mathbf{G}, \rho, A, \Omega_\infty$), for any $g \in \Omega_\infty$ the set of $\gamma \in \mathbf{G}(\mathbb{Q})$ such that*

$$\inf\{\mathrm{dist}_G(\gamma, t) \mid t \in g(MA \cap \Omega_\infty\Omega_\infty^{-1})g^{-1}\} \leq \varepsilon, \ \mathrm{d}(\gamma) \leq M \tag{4.1}$$

is contained in a $\mathbb{Q}$-subtorus $\mathbf{T} \subset \mathbf{G}$, provided that ε is small enough so that

$$\varepsilon M^c \leq c' \tag{4.2}$$

Let S be the set of γ defined in equation (4.1). We establish two preparatory results before the Lemma itself. The first is based on the analysis of the case of division algebras (see Lemma 4.9) where the idea is more fully exploited.

LEMMA 4.5 (enough to check subsets of fixed size). *A set $S \subset \mathbf{G}(\mathbb{Q})$ is contained in a $\mathbb{Q}$-torus iff the same holds for every subset S' of S of size $n+1$.*

PROOF. For each $S' \subset S$ let $A_{S'} \subset M_n(\mathbb{C})$ be the $\mathbb{C}$-algebra generated in $M_n(\mathbb{C})$ by the image set $\rho(S')$, and choose $S_0' \subset S$ such that $A_0' = A_{S_0'}$ has maximal dimension (as a vector space over $\mathbb{C}$).

Since $\rho(S_0')$ is contained in a subtorus of SL_n, the subalgebra A' is conjugate to a subalgebra of the algebra of diagonal matrices. In particular, it has dimension at most n. In a proper containment of $\mathbb{C}$-algebras, the two algebras have different dimensions, so there exists a subset $S_0 \subset S_0'$ of size at most n such that $A_{S_0} = A_0'$.

Now for any $\gamma \in S$, $A_{S_0 \cup \{\rho(\gamma)\}}$ contains A_0' and has at most the same dimension. It follows that they are equal, and hence that $A_S = A_0'$. Let $\mathbf{T}$ be the intersection with $\mathbf{G}$ of the $\mathbb{Q}$-subalgebra of $M_n(\mathbb{Q})$ generated by $\rho(S)$. This is a $\mathbb{Q}$-subgroup of $\mathbf{G}$ containing S but contained in A_0'. That $\mathbf{T}$ is a torus follows from the fact that its C-points are contained in A_0', hence conjugate to a set of diagonal matrices. □

LEMMA 4.6 (checking can be done algebraically). *There exist (finite) sets of polynomials $P_{ij}, Q_{ik} \in \mathbb{Z}[x_1, \ldots, x_{(n+1)n^2}]$ (to be thought of functions of $n+1$ matrices of size n), such that $\{\gamma_l\}_{l=1}^{n+1} \subset \mathbf{G}(\mathbb{Q})$ all lie on a $\mathbb{Q}$-torus iff for some i, $P_{ij}(\rho(\vec{\gamma})) = 0$ and $Q_{ik}(\rho(\vec{\gamma})) \neq 0$ hold for all j, k.*

PROOF. The analysis of the previous Lemma shows that the desired statement about $\rho(\vec{\gamma})$ is equivalent to the following formula in the language of fields:

> There exist matrices $D, A \in \mathrm{GL}_n(\mathbb{C})$ such that D is diagonal with distinct entries and such that each $\rho(\gamma_k)$ commutes with ADA^{-1}.

By Chevalley's Theorem on elimination of quantifiers for the theory of algebraically closed field, this formula is equivalent (over $\mathbb{C}$) to a one without quantifiers, which we may assume to be in the normal form

$$\bigvee_i \left(\bigwedge_j (P_{ij}(\rho(\vec{\gamma})) = 0) \bigwedge_k (Q_{ij}(\rho(\vec{\gamma})) \neq 0) \right)$$

We reiterate the point that the γ_k generate a $\mathbb{Q}$-torus iff they generate a torus over $\mathbb{C}$, and that the language of algebraically closed fields has no names for field elements other than $0, 1$ so that atomic formulas in it are equalities and inequalities of rational polynomials, which without loss of generality may be assumed integral. $\square$

Proof of the Diophantine Lemma 4.4. We note first that the polynomials P_{ij}, Q_{ik} were only constructed from $\mathbf{G}$ and ρ. Taking $c' < 1$ we may assume that $\varepsilon < 1$, so that all the γ are drawn from the 1-neighbourhood of the compact set $\Omega_\infty(MA\cap\Omega_\infty\Omega_\infty^{-1})\Omega_\infty^{-1}$. Let L be a bound for the Lipschitz constant of the smooth functions $P_{ij}\circ\rho, Q_{ik}\circ\rho$ in this domain. Again by compactness, there exists $b > 0$ such that for each $g \in \Omega_\infty$ and each $\vec{t} \in (MA\cap\Omega_\infty)^{n+1}$, we have $\left|Q_{ik}(g\vec{t}g^{-1})\right| > b$ for all k and for those i such that our system of constraints holds for $g\vec{t}g^{-1}$,

Fixing $g \in \Omega_\infty$, now let $\vec{\gamma}$ from S^{n+1}, and let $\vec{t} \in (MA\cap\Omega_\infty)^{n+1}$ so that γ_l is ε-close to gt_lg^{-1}. By the analysis above, there exists i such that $P_{ij}(\rho(\vec{t})) = 0$ while $Q_{ik}(\rho(\vec{t}))$ are at least b in magnitude. It follows that the magnitude $Q_{ik}(\rho(\vec{\gamma}))$ is at least $b - L\varepsilon$, which is positive as long as we ensure $c' < b/L$. It also follows that the magnitude of $P_{ij}(\rho(\vec{\gamma}))$ is at most $L\varepsilon$. Let c bound the total degree each P_{ij}. Then the denominator of each rational number $P_{ij}(\rho(\vec{\gamma}))$ is at most M^c. If $c' < 1/L$ this ensures that these rational numbers vanish. $\square$

4.3. The intersection pattern of translates of tubes around Levi subgroups. For the rest of Section 4 constants written c, c_i are suitably small, constants written κ or κ_i are suitably large.

PROPOSITION 4.7. *Let* $\mathbf{G}$ *be* $\mathbb{R}$*-split. Fix a relatively compact open neighbourhood of the identity* $C \subset A$ *and assume* $C \subset \Omega_\infty$*. There are* $c_2, \kappa > 0$*, depending only on the isomorphism class of* $\mathbf{G}$*, and* $c_1, c_3 = O_C(1)$*, so that*

For any $x = (x_\infty, x_{\mathrm{f}}) \in \Omega$ *and any* $0 < \varepsilon < c_1$ *there exists a* $\mathbb{Q}$*-subtorus* $\mathbf{T} \subset \mathbf{G}$ *so that:*

(1) *If* $s, s' \in \mathbf{G}(\mathbb{A}_{\mathrm{f}})$ *both have denominator* $\leq c_3\varepsilon^{-c_2}$ *and are so that*

$$\overline{xB(C,\varepsilon)s} \cap \overline{xB(C,\varepsilon)s'} \neq \emptyset \text{ in } \mathbf{G}(\mathbb{Q})\backslash\mathbf{G}(\mathbb{A})/K_{\mathrm{f}}$$

then there exists $\gamma \in \mathbf{T}(\mathbb{Q}) \subset \mathbf{G}(\mathbb{Q})$ *with*

$$\gamma xB(C,\varepsilon)s \cap xB(C,\varepsilon)s' \neq \emptyset \text{ in } \mathbf{G}(\mathbb{A})/K_{\mathrm{f}}. \tag{4.3}$$

(2) *There at most* $O(1-\kappa\log\varepsilon)$ *primes which are* $\mathbf{T}$*-bad.*

PROOF. Let $s, s' \in \mathbf{G}(\mathbb{A}_{\mathrm{f}})$ satisfy the intersection condition of item (1). By assumption there is – after replacing s, s' by suitable elements of sK_{f} and $s'K_{\mathrm{f}}$ respectively – an element $\gamma \in \mathbf{G}(\mathbb{Q})$ so that $\gamma x_{\mathrm{f}}s = x_{\mathrm{f}}s'$ (equality in $\mathbf{G}(\mathbb{A}_{\mathrm{f}})$) and, moreover, $\gamma \in x_\infty B(C,\varepsilon)B(C,\varepsilon)^{-1}x_\infty^{-1}$ (equality in $\mathbf{G}(\mathbb{R})$).

The former equality implies, in particular, that $\mathrm{d}(\gamma) \ll c_1'\varepsilon^{-c_2'}$, where each c_i' depends on c_i and $c_i' \to 0$ as $c_i \to 0$ for $i = 1, 2$.

Since conjugation by elements of a compact set is a map of bounded Lipchitz constant w.r.t. the metric dist_G, the latter inclusion shows that γ lies within $L\epsilon$ of $x_\infty(MA\cap\Omega_\infty\Omega_\infty{}^-1)x_\infty^{-1}$, where the Lischitz constant L depends only on $\mathbf{G}$ and Ω_∞.

Let R be the set of such γ, let $\mathbf{T} \subset \mathbf{G}$ be the closed subgroup they generate, and let E be the subalgebra of $M_n(\mathbb{Q})$ generated by $\rho(R)$. If c_1', c_2' are sufficiently small – this occurs, in particular, if c_1, c_2 are sufficiently small – then Lemma 4.4

shows that $\mathbf{T}$ is a torus, and E is its linear span, a semisimple abelian subalgebra of $M_n(\mathbb{Q})$. Analyzing the reasoning shows that the exponent c_2 may be taken to depend only on the isomorphism class of $\mathbf{G}$, whereas $c_1 = O(1)$. This proves (4.3).

For a $\mathbf{G}$-good prime p to be $\mathbf{T}$-good, it suffices that $E_p \cap M_n(\mathbb{Z}_p)$ is a maximal compact subring of $E_p = E \otimes \mathbb{Q}_p \subset M_n(\mathbb{Q}_p)$. For this it suffices to have generators $\{\gamma_i'\}_{i=1}^n \subset M_n(\mathbb{Z})$ for E as a $\mathbb{Q}$-algebra such that $\mathbb{Z}_p[\gamma_1', \dots, \gamma_d'] = E_p \cap M_n(\mathbb{Z}_p)$. That will happen as long as p does not divide the discriminant of the characteristic polynomial of each γ_i'.

By the proof of Lemma 4.5, there exist $\{\gamma_i\}_{i=1}^n \subset R$ which generate E, and let $\gamma_i' = \mathfrak{c}(\gamma_i) \cdot \gamma_i$. Then $\gamma_i' \in M_n(\mathbb{Z})$, still generate E as a $\mathbb{Q}$-algebra. Next, since $\gamma_i \in \Omega_\infty \Omega_\infty \Omega_\infty^{-1} \Omega_\infty^{-1}$ and as ρ is continuous, the matrix entries of γ_i' are $O(\epsilon^{-c_2'})$. Finally, the discriminant of γ_i' is a polynomial in the coefficients of its characteristic polynomial, themselves polynomials in the matrix entries of γ_i'.

It follows that the set of $\mathbf{G}$-good but $\mathbf{T}$-bad primes is contained in the set of prime divisors of an integer bounded by $O(\epsilon^{-O(1)})$. □

Generalizations. Proposition 4.7 is a statement of the following type:

> Let $H \subset G$ be a closed subgroup. Then given $x \in \Omega_\infty$ and a tubular neighbourhood $B(C, \varepsilon)$ of a piece $C \subset H$, there exists a $\mathbb{Q}$-subgroup $\mathbf{T} \subset \mathbf{G}$ such that intersection of Hecke translates of small denominator of $xB(C, \varepsilon)$ are controlled by $\mathbf{T}$, in the sense that if xBs and xBs' intersect in X, there exists $\gamma \in \mathbf{T}(\mathbb{Q})$ such that $\gamma x B s K_{\mathrm{f}} = xBs'K_{\mathrm{f}}$ holds in $\mathbf{G}(\mathbb{A})/K_{\mathrm{f}}$.

We have established this for $\mathbf{G}$ which is $\mathbb{Q}$-anisotropic and $\mathbb{R}$-split and H a maximal $\mathbb{R}$-split torus. Specializing further to the case of $\mathbf{G}$ associated to a division algebra of prime degree, we establish a result of this type for any Levi subgroup $H \subset G$ (that is, for a subgroup of the form $H = Z_G(a)$, $a \in A$). It turns out that the subgroup $\mathbf{T}$ remains a torus.

In general one would expect that points lying near pieces of orbits of Levi subgroups defined over $\mathbb{R}$ to lie on something like an orbit of a Levi subgroup defined over $\mathbb{Q}$. This is not quite correct, but precise versions of this intuition can be proved; see the work [**14**, §4].

4.4. Extra notations for the case of division algebras. Let D be a division algebra over $\mathbb{Q}$ of prime degree d, and fix a lattice $D_{\mathbb{Z}} \subset D$ (i.e. a free $\mathbb{Z}$-submodule of maximal rank) and a Euclidean norm $\|\cdot\|$ on $D \otimes_{\mathbb{Q}} \mathbb{R}$. In other words, we have chosen a norm on $D \otimes \mathbb{Q}_v$ for all v, finite or infinite. Since the only central division algebras over $\mathbb{R}$ are $\mathbb{R}$ itself and Hamilton's quaternions, assuming $d \geq 3$ ensures that $D \otimes \mathbb{R}$ is the full matrix algebra.

We will consider the case where $\mathbf{G}$ is the projectivized group of units (=invertible elements) of D, so (for $d \geq 3$) $G = \mathbf{G}(\mathbb{R})$ is isomorphic to $\mathrm{PGL}_d(\mathbb{R})$. The Lie algebra of G is identified with a quotient of $D \otimes \mathbb{R}$; as such, the norm on $D \otimes \mathbb{R}$ gives rise to a norm on the Lie algebra of G and thus to a left-invariant Riemannian metric on G.

We fix extra data $(\rho, K_{\mathrm{f}}, \Omega, \Omega_\infty)$ for the group $\mathbf{G}$, as discussed in §2. In the rest of this paper, when discussing this case the implicit constants in the notations $\ll$ and $O(\cdot)$ will be allowed to depend on $D, D_{\mathbb{Z}}$, the norm $\|\cdot\|$ and this extra data, without explicitly indicating this.

It should be noted that we do not assume that $D_{\mathbb{Z}}.D_{\mathbb{Z}} \subset D_{\mathbb{Z}}$; on the other hand, clearly there is an integer $K = O(1)$ so that $D_{\mathbb{Z}}.D_{\mathbb{Z}} \subset K^{-1}D_{\mathbb{Z}}$.

4.5. A diophantine lemma for $\mathbb{Q}$-algebras. In this section only we shall use an additional notion of denominator, special to the case of $\mathbb{Q}$-algebras. We fix a central simple $\mathbb{Q}$-algebra D of dimension d^2 and a $\mathbb{Z}$-lattice $D_{\mathbb{Z}} \subset \mathbb{Q}$. Given $x \in D$, we set

$$\widetilde{\mathrm{d}}(x) := \inf\{m \in \mathbb{N} : \; mx \in D_{\mathbb{Z}}\}. \tag{4.4}$$

Recall that for $\gamma \in \mathbf{G}(\mathbb{Q}) = D^\times/\mathbb{Q}^\times$, we also have the denominator $\mathrm{d}(\gamma)$ defined in Section 4.1. We first clarify the relation between the two notions.

LEMMA 4.8. *Let $\gamma \in \mathbf{G}(\mathbb{Q}) = D^\times/\mathbb{Q}^\times$ satisfy $\mathrm{d}(\gamma) \leq M$ and belong to a compact subset $E \subset \mathbf{G}(\mathbb{R})$. Then there exists $\alpha \in D^\times$ lifting γ so that the three quantities M^c, $\|\alpha^{-1}\|, \|\alpha\|$ are similarly bounded in the form*

$$\widetilde{\mathrm{d}}(\alpha) \ll_E M^c, \|\alpha^{-1}\|, \|\alpha\| \ll_E 1\,,$$

where c is a constant depending only on the isomorphism class of $\mathbf{G}$.

PROOF. In fact, let $\tilde{\mathbf{G}}$ be the algebraic group corresponding to $D^\times$, i.e. $\tilde{\mathbf{G}}(R) = (D \otimes_{\mathbb{Q}} R)^\times$ if R is a ring containing $\mathbb{Q}$.

Then $\tilde{\mathbf{G}}$ and $\mathbf{G}$ are affine algebraic groups. We first show that the map $\tilde{\mathbf{G}} \to \mathbf{G}$ *admits an algebraic section over a Zariski-open set $U \subset \mathbf{G}$.*

Let $\mathbf{G}^{(1)}$ denote the group of elements of norm 1 in $D^\times$. It is a (geometrically) irreducible variety, because SL_d is an irreducible variety. The map $\mathbf{G}^{(1)} \to \tilde{\mathbf{G}}$ is a covering map (i.e. étale) and its kernel is the group of dth roots of unity. Let E be the function field of $\mathbf{G}$, considered as $\mathbb{Q}$-variety. The generic point in $\eta \in \mathbf{G}(E)$ does not lift to a point of $\mathbf{G}^{(1)}(E)$, but it does at least to lift to a point of $\tilde{\eta} \in \mathbf{G}^{(1)}(\tilde{E})$ for some finite extension $\tilde{E}/E$, which we may assume to be Galois and to contain the dth roots of unity. Then $\sigma \mapsto \tilde{\eta}^\sigma/\tilde{\eta}$ defines a 1-cocycle of $\mathrm{Gal}(\tilde{E}/E)$ valued in the group of dth roots of unity. By Hilbert's theorem 90, there exists $\tilde{e} \in \tilde{E}$ so that this cocycle is $\sigma \mapsto \tilde{e}^\sigma/\tilde{e}$. Adjusting $\tilde{\eta}$ by $\tilde{e}$ gives a $\tilde{E}$-valued point of $\tilde{\mathbf{G}}$, which is invariant under $\mathrm{Gal}(\tilde{E}/E)$ and therefore is indeed an E-valued point of $\tilde{\mathbf{G}}$. This gives the desired section.

One may, by translating U, find a finite collection of open sets $U_1, \ldots, U_h$ which cover $\mathbf{G}$, and so that $\tilde{\mathbf{G}} \to \mathbf{G}$ admits a section $\theta_j : U_j \to \tilde{\mathbf{G}}$ over each U_j.

It follows from this that there exists $\alpha \in D^\times$ lifting γ so that

$$\widetilde{\mathrm{d}}(\alpha), \widetilde{\mathrm{d}}(\alpha^{-1}) \ll M^c$$

where c is a constant depending only on the choice of sets U_j and the sections, i.e. only the isomorphism class of $\mathbf{G}$.

From this bound, it follows in particular that $M^{-c} \ll_E \|\alpha\| \ll_E M^c$. The lower bound is clear; for the upper bound, we use the fact that α projects to the compact subset $E \subset \mathbf{G}(\mathbb{R})$.

Let p/q be a rational number satisfying $\|\alpha\| < p/q < 2\|\alpha\|$. We may choose p, q so that $\max(p, q) \ll M^c$. Replacing α by $q\alpha/p$, we obtain a representative α for γ that satisfies:

$$\widetilde{\mathrm{d}}(\alpha) \ll_E M^{2c}, \quad \|\alpha\| \asymp_E 1$$

We increase c as necessary.

Finally, the bound for $\|\alpha^{-1}\|$ follows from the bound for $\|\alpha\|$ together with the fact that α projects to the compact set $E \subset \mathbf{G}(\mathbb{R})$. □

The following should be compared with Lemma 4.4.

LEMMA 4.9. *Let $S \subset D \otimes \mathbb{R}$ be a proper $\mathbb{R}$-subalgebra.*

For $\kappa > 0$ (depending only on d) and $c > 0$ (depending only on $D, D_{\mathbb{Z}}, \|\cdot\|$), the set of $x \in D$ satisfying

$$\|x\| \leq R, \quad \inf_{s \in S} \|x - s\| \leq \varepsilon, \quad \widetilde{\mathrm{d}}(x) \leq M \tag{4.5}$$

is contained in a proper subalgebra $F \subset D$ as long as

$$\varepsilon R^{\kappa} M^{\kappa} < c \tag{4.6}$$

In other words: points of D near a proper subalgebra of $D \otimes \mathbb{R}$ lie on a proper $\mathbb{Q}$-subalgebra of D. This proof will not use the fact that D is a *division* algebra, nor the fact that it is of prime rank.

PROOF. We use here $f_1(d), f_2(d), \ldots$ to denote positive quantities that depend on the rank d alone.

Let $s = \dim(S) + 1$. Then there is a polynomial function $G : D^s \to \mathbb{Q}$, with integral coefficients with respect to $D_{\mathbb{Z}}$, so that $G(\alpha_1, \ldots, \alpha_s) = 0$ exactly when $\alpha_1, \ldots, \alpha_s$ span a linear space of dimension $\leq s - 1$. For example one may use the sum of the squares of the minors of a suitable matrix. The degree of G is $f_1(d)$ and the size of its coefficients is $O(1)$.

Take $x_1, \ldots, x_s$ belonging to the set defined by (4.5). There are $y_1, \ldots, y_s \in S$ so that $\|x_i - y_i\| \leq \varepsilon$. Then $G(x_1, \ldots, x_s) \ll R^{f_2(d)}\varepsilon$. On the other hand, if $G(x_1, \ldots, x_s) \neq 0$ then, because $\mathrm{d}(x_i) \leq M$, we must have $G(x_1, \ldots, x_s) \gg M^{-f_3(d)}$. It follows that, if a condition of the type (4.6) holds for suitable κ, c as stated, then $x_1, \ldots, x_s$ span a $\mathbb{Q}$-linear space of dimension $s - 1$.

Now let X be the $\mathbb{Q}$-algebra spanned by those x satisfying (4.5). It is clear that X is, in fact, spanned by monomials in such x of length at most $\dim_{\mathbb{Q}} D$. Each such monomial y satisfies $\|y\| \ll R^{f_4(d)}$, as well as $\inf_{s \in S} \|y - s\| \ll R^{f_5(d)}\varepsilon$ and $\mathrm{d}(y) \gg M^{f_6(d)}$. It follows that – increasing κ and decreasing c in (4.6) as necessary – the $\mathbb{Q}$-subalgebra generated by all solutions to (4.5) has dimension $\leq s - 1$ and, in particular, is a proper subalgebra of D. □

PROPOSITION 4.10. *Let $\mathbf{G}$ be the projectivized group of units of a division algebra $D/\mathbb{Q}$ of prime degree d. There are $c_2, c_3 > 0$, depending only on the isomorphism class of $\mathbf{G}$ and $c_1, \kappa = O_C(1)$, so that*

For any $x = (x_\infty, x_{\mathrm{f}}) \in \Omega$ and any $0 < \varepsilon < 1/2$ there exists a subfield $F \subset D$ so that:

(1) *If $s, s' \in \mathbf{G}(\mathbb{A}_{\mathrm{f}})$ both have denominator $\leq c_1\varepsilon^{-c_2}$ and are so that*

$$\overline{xB(C,\varepsilon)s} \cap \overline{xB(C,\varepsilon)s'} \neq \emptyset \text{ in } \mathbf{G}(\mathbb{Q})\backslash\mathbf{G}(\mathbb{A})/K_{\mathrm{f}}$$

then there exists $\gamma \in F^\times/\mathbb{Q}^\times \subset \mathbf{G}(\mathbb{Q})$ with

$$\gamma x B(C,\varepsilon)s \cap xB(C,\varepsilon)s' \neq \emptyset \text{ in } \mathbf{G}(\mathbb{A})/K_{\mathrm{f}}. \tag{4.7}$$

(2) *F is generated by $\alpha \in D^\times$, so that $\alpha D_{\mathbb{Z}} + D_{\mathbb{Z}}\alpha \subset D_{\mathbb{Z}}$, and with $\|\alpha\| \leq \kappa\varepsilon^{-c_3}$.*

PROOF. Let $s, s' \in \mathbf{G}(\mathbb{A}_f)$ satisfy the intersection condition of item (1). As in the general case before we find – after replacing s, s' by suitable elements of sK_f and $s'K_f$ respectively – an element $\gamma \in \mathbf{G}(\mathbb{Q})$ so that $\gamma x_f s = x_f s'$ (equality in $\mathbf{G}(\mathbb{A}_f)$) and, moreover, $\gamma \in x_\infty B(C,\varepsilon)B(C,\varepsilon)^{-1}x_\infty^{-1}$ (equality in $\mathbf{G}(\mathbb{R})$).

Again, this implies $\mathrm{d}(\gamma) \ll c_1'\varepsilon^{-c_2'}$, where c_i' depends on c_i, and $c_i' \to 0$ as $c_i \to 0$ for $i = 1, 2$.

The latter inclusion shows that γ lies in a fixed compact subset of $\mathbf{G}(\mathbb{R})$ depending only on Ω, C.

The element γ belongs to $D^\times/\mathbb{Q}^\times$. We may choose a representative $\alpha \in D^\times$ for γ as in Lemma 4.8. In that case α lies in a fixed compact subset of $(D \otimes \mathbb{R})^\times$ – depending only on Ω, C – and $\breve{\mathrm{d}}(\alpha) \ll c_1''\varepsilon^{-c_2''}$, where (for $i = 1, 2$) c_i'' depends on c_i and $c_i'' \to 0$ as $c_i' \to 0$.

Let E be the subalgebra of $D \otimes \mathbb{R}$ that centralizes $x_\infty a x_\infty^{-1}$. The assertion that $\gamma \in x_\infty B(C,\varepsilon)B(C,\varepsilon)^{-1}x_\infty^{-1}$ shows that α is "close" to E; in fact, it is clear that

$$\inf_{e \in E} \|\alpha - e\| \ll \varepsilon$$

and moreover $\|\alpha\| \ll 1$ (because α lies in a fixed compact subset of $(D \otimes \mathbb{R})^\times$).

By Lemma 4.9 we see that, if c_1'', c_2'' are sufficiently small – this occurs, in particular, if c_1, c_2 are sufficiently small – then all such α necessarily belong to a proper $\mathbb{Q}$-subalgebra of D; because D has prime degree, this must be a field F. Analyzing this reasoning shows that c_2 may be taken to depend only on the isomorphism class of $\mathbf{G}$, whereas $c_1 = O(1)$. This proves (4.7).

Also, there exists $K \in \mathbb{N}$ so that $D_\mathbb{Z}.D_\mathbb{Z} \subset K^{-1}D_\mathbb{Z}$. Then $\alpha.D_\mathbb{Z} \subset K^{-1}.\mathrm{d}(\alpha)^{-1}.D_\mathbb{Z}$ and similarly for $D_\mathbb{Z}.\alpha$.

Replacing α with $\alpha' = K.\mathrm{d}(\alpha).\alpha$, we see that $\alpha' D_\mathbb{Z} + D_\mathbb{Z}\alpha' \subset D_\mathbb{Z}$ and $\|\alpha'\| \leq \kappa\varepsilon^{-c_3}$, where $\kappa = O_C(1)$ and c_3 depends only on the isomorphism class of $\mathbf{G}$. □

We also need to know that there are only a few bad primes.

LEMMA 4.11. *Let $F \subset D$ be a subfield, and let $\mathbf{T}_F \subset \mathbf{G}$ be the torus defined by F, i.e. the centralizer of F in $\mathbf{G}$. Suppose that F is generated, over $\mathbb{Q}$, by an element α satisfying $\alpha D_\mathbb{Z} + D_\mathbb{Z}\alpha \subset D_\mathbb{Z}$. Then the number of primes which fail to be $\mathbf{T}_F$-good is at most $O(1 + \log\|\alpha\|)$.*

PROOF. Let K_f' be the stabilizer[5] of $D_\mathbb{Z}$ inside $\mathbf{G}(\mathbb{A}_f)$.

Then $\mathbf{T}_F(\mathbb{Q}_p) \cap K_f'$ is maximal compact inside $\mathbf{T}_F(\mathbb{Q}_p)$ as long as the maximal compact subring of $(F \otimes \mathbb{Q}_p)$ preserves $D_\mathbb{Z} \otimes \mathbb{Z}_p$ under both left and right multiplication. This will be so, in particular, at any prime where the maximal compact subring of $(F \otimes \mathbb{Q}_p)$ equals $\mathbb{Z}_p[\alpha]$. This will always be the case if p does not divide the discriminant of the ring $\mathbb{Z}[\alpha]$.

From this, one deduces that there at most $O(1 + \log\|\alpha\|)$ primes for which $\mathbf{T}_F(\mathbb{Q}_p) \cap K_f'$ fails to be maximal compact in $\mathbf{T}_F(\mathbb{Q}_p)$. But, for all but $O(1)$ primes, the intersection $\mathbf{G}(\mathbb{Q}_p) \cap K_f'$ coincides with $\mathbf{G}(\mathbb{Q}_p) \cap K_f$. So there are at most $O(1 + \log\|\alpha\|)$ primes for which $\mathbf{T}_F(\mathbb{Q}_p) \cap K_f$ fails to be maximal compact in $\mathbf{T}_F(\mathbb{Q}_p)$. □

[5]Note that $\mathbf{G}(\mathbb{A}_f)$ acts naturally on lattices inside D, in a fashion derived from the conjugation action of $\mathbf{G}$ on D. Indeed, if V is a $\mathbb{Q}$-vector space, the group $\mathrm{GL}(V \otimes_\mathbb{Q} \mathbb{A}_f)$ acts naturally on lattices inside V.

5. Bounds on the mass of tubes, II

We define "tubes" $B_0 := B(C,\varepsilon)$ as in Section 4. When $\mathbf{G}$ is merely assumed $\mathbb{R}$-split we take $C \subset MA$. In the special case of division algebras of prime degree C may be taken to lie in any Levi subgroup of G.

Set $B = B_0 B_0^{-1} \supset B(C,\varepsilon)$. There exists a compact subset $C' \subset Z_G(a)$ and a constant $M = O(1)$ such that B, B_2 $(= B_1.B_1)$ and B_3 are subsets of $B(C', M\varepsilon)$. Notations here are as in (3.1). Also, $\frac{\mathrm{vol} B_3}{\mathrm{vol} B_0} \ll 1$.[6]

5.1. Sets S for which intersections are controlled by tori. Let Q be so that $Q/2$ is larger than any bad prime for $\mathbf{G}$, and let ℓ be fixed. (In practice, $Q \to \infty$ as B becomes small, whereas ℓ is fixed depending only on $\mathbf{G}$).

Set $S_p = \{g_p \in \mathbf{G}(\mathbb{Q}_p)/K_p : \mathrm{d}_p(g_p) \le p^\ell\}$. Initially we shall consider the set of translates given by $\bigcup_{p\in[Q/2,Q]} S_p$, where we identify $\mathbf{G}(\mathbb{Q}_p)/K_p$ with a subset of $\mathbf{G}(\mathbb{A}_\mathrm{f})/K_\mathrm{f}$ in the natural way.

Part (1) of the conclusions of Propositions 4.7 and 4.10 can now be rephrased[7] as establishing (for $Q^l \ll \varepsilon^{-c_2}$) the following condition. In words, it states that *intersections between Hecke translates of B_2 by $\cup S_p$ all arise from a $\mathbb{Q}$-torus:*

$\boxed{\star = \star_{B,Q,\ell}}$: For any $x \in \Omega_\infty$ there is a $\mathbb{Q}$-torus $\mathbf{T} \subset \mathbf{G}$ so that

$$\overline{xB_2 s} \cap \overline{xB_2 s'} \ne \emptyset \quad \text{in } X$$

with $s, s' \in \bigcup_{p\in[Q/2,Q]} S_p$ only if there is $\gamma \in \mathbf{T}(\mathbb{Q})$ so that for these s, s',

$$\gamma x B_2 s \cap x B_2 s' \ne \emptyset \text{ in } \mathbf{G}(\mathbb{A})/K_\mathrm{f}.$$

LEMMA 5.1. *Suppose that condition $\star_{B,Q,\ell}$ is satisfied. Take $x \in \Omega_\infty$ and let $\mathbf{T}$ be the torus specified by $\star_{B,Q,\ell}$. Assume that $S \subset \bigcup_p S_p \subset \mathbf{G}(\mathbb{A}_f)/K_\mathrm{f}$, with the union taken over $p \in [Q/2,Q]$ which are $\mathbf{T}$-good. Then:*

$$\#\{s, s' \in S : \overline{xB_2 s} \cap \overline{xB_2 s'} \ne \emptyset\} \ll_\ell Q^2 + |S|$$

PROOF. Consider any intersection $\overline{xB_2 s} \cap \overline{xB_2 s'} \ne \emptyset$ in X when $s, s' \in S$. This means that there is $\gamma \in \mathbf{T}(\mathbb{Q})$ so that $\gamma x B_2 s \cap x B_2 s' \ne \emptyset$ in $\mathbf{G}(\mathbb{A})/K_\mathrm{f}$. Then $s \in S_p, s' \in S_q$ when p, q are $\mathbf{T}$-good primes in the range $[Q/2,Q]$; we distinguish two cases according to whether $p = q$ or not.

(1) $q = p$. In this case,

$$\gamma \in (xB_2B_2^{-1}x^{-1}).\mathbf{T}(\mathbb{Q}_p).K_\mathrm{f}.$$

For any fixed p, the number of K_f-cosets contained in $\mathbf{T}(\mathbb{Q}_p).K_\mathrm{f}$ satisfying $\mathrm{d}_p \le \ell$ is $O_\ell(1)$: since p is $\mathbf{T}$-good, the quotient $\mathbf{T}(\mathbb{Q}_p)/\mathbf{T}(\mathbb{Q}_p) \cap K_\mathrm{f}$ is a free abelian group of rank $\le \dim(\mathbf{T})$. Pick generators $t_1, \dots, t_r$ for this quotient; they generate a discrete subgroup. We need to show that the number of $(e_1, \dots, e_r) \in \mathbb{Z}^r$ so that $\mathrm{d}_p(t_1^{e_1} \dots t_r^{e_r}) \le p^\ell$ is $O_\ell(1)$. To see this, pass to an extension of $\mathbb{Q}_p$ where $\rho(t_i)$ become diagonalizable.

Therefore, γ is an element of $\mathbf{G}(\mathbb{Q})$ so that:

---

[6]In this and in the statement $M = O(1)$, the implicit constant certainly depends on C; however, C was assumed to belong to Ω_∞, and we have permitted implicit constants to depend on Ω_∞ without explicit mention.

[7]Note that, in what follows, we are applying the Proposition with $B(C,\varepsilon)$ replaced by the larger set $B_2 \subset B(C', M.\varepsilon)$. It is easy to see that the stated result holds even though this larger set may not be contained in Ω_∞.

(a) considered as an element of $\mathbf{G}(\mathbb{R})$, γ belongs to

$$xB_2^{-1}B_2x^{-1},$$

which in turn is contained in a compact set depending only on Ω_∞;

(b) considered as an element of $\mathbf{G}(\mathbb{A}_{\mathrm{f}})$, γ belongs to $O_\ell(1)$ right K_{f}-cosets.

The number of possibilities for γ is therefore $O_\ell(1)$. Since $(sK_f, s'K_f)$ is determined by (sK_f, γ), it follows the number of possibilities for (s, s') in the case "$p = q$" is at most $O_\ell(|S|)$.

(2) $p \neq q$.

In this case, $s \in \mathbf{T}(\mathbb{Q}_p).K_p$ and $s' \in \mathbf{T}(\mathbb{Q}_q).K_q$. By an argument already given, the number of K_p-cosets contained in $\mathbf{T}(\mathbb{Q}_p).K_p$ and satisfying $\mathrm{d}_p \leq \ell$ is $O_\ell(1)$, and similarly with q replacing p. It follows that the number of possibilities for (s, s') is $O_\ell(1)$ for *given* p, q.

The total number of possibilities for (s, s') in the case "$p \neq q$" is therefore $O_\ell(Q^2)$.

□

5.2. Conclusion. We shall apply Lemma A.1 to our setting. Take ℓ as in that Lemma.

LEMMA 5.2. *Suppose that condition $\star_{B,Q,\ell}$ is satisfied. Take $x \in \Omega_\infty$ and let $\mathbf{T}$ be the torus specified by $\star_{B,Q,\ell}$, $\mathcal{P}$ the set of $\mathbf{T}$-good primes in $[Q/2, Q]$.*

For any Hecke eigenfunction ψ on X.

$$\mu_\psi(\overline{xB}) \ll \frac{Q^{1.01}}{|\mathcal{P}|^2}.$$

PROOF. For $p \in \mathcal{P}$ let h_p be the K_p-bi-invariant function on G_p furnished in Lemma A.1. Now, if we consider h_p as a function on G_p/K_p, we have upper and lower bounds $p \ll \#\mathrm{supp}(h_p) \ll p^{\ell'}$, where c depends only on $\mathbf{G}, \rho$.

Therefore, by a dyadic decomposition argument, there exists a subset $\mathcal{P}_1 \subset \mathcal{P}$ of size $\gg |\mathcal{P}|/\log Q$ and $A \gg Q$ so that $\#\mathrm{supp}(h_p) \in [A, 2A]$ $(p \in \mathcal{P}_1)$. Set $h = \sum_{p \in \mathcal{P}_1} h_p$. Then, by what we have proven in Lemma 3.4 and Lemma 5.1,

$$\begin{aligned}
\mu_\psi(\overline{xB}) &\ll \left(\frac{\mathrm{vol}B_3}{\mathrm{vol}B_0}\right)^2 \frac{Q^2 + \sum_p \#\mathrm{supp}(h_p)}{\left(\sum_p \#\mathrm{supp}(h_p)^{1/2}\right)^2} \\
&\ll \left(\frac{\mathrm{vol}B_3}{\mathrm{vol}B_0}\right)^2 \frac{Q^2 + |\mathcal{P}_1|A}{|\mathcal{P}_1|^2 A} \\
&\ll \left(\frac{\mathrm{vol}B_3}{\mathrm{vol}B_0}\right)^2 \left(\frac{Q\log^2 Q}{|\mathcal{P}|^2} + \frac{\log Q}{|\mathcal{P}|}\right) \\
&\ll \left(\frac{\mathrm{vol}B_3}{\mathrm{vol}B_0}\right)^2 \frac{Q^{1.01}}{|\mathcal{P}|^2}.
\end{aligned}$$

Finally, as observed at the start of this section, $\frac{\mathrm{vol}B_3}{\mathrm{vol}B_0} \ll 1$. □

We now combine this Lemma with the results of Section 4, which give conditions under which $\star_{B,Q,\ell}$ is true.

THEOREM 5.3. *Let* $\mathbf{G}$ *be a semisimple group defined over* $\mathbb{Q}$ *which splits over* $\mathbb{R}$. *Let* ψ *be a Hecke eigenfunction on* $X = \mathbf{G}(\mathbb{Q})\backslash\mathbf{G}(\mathbb{A})/K_{\mathrm{f}}$.

Let $\Omega_\infty \subset G = NAK$ *be compact. Further, let* $B(C,\varepsilon) \subset \Omega_\infty$ *be a tube as in Section 4 such that either*

(1) $C \subset MA$; *or,*
(2) $\mathbf{G}$ *is the projectivized group of units of a division algebra* D *of prime degree over* $\mathbb{Q}$.

Then there is $c > 0$, *depending only on the isomorphism class of* $\mathbf{G}$ *so that, uniformly over* $x \in \Omega_\infty$,

$$\mu_\psi(xB(C,\varepsilon)) \ll \varepsilon^c.$$

PROOF. Let ℓ be as in Lemma A.1.

Recall, as remarked at the start of the present section, that $B_3 \subset B(C', M.\varepsilon)$, for a suitable compact set C' and a suitable constant M.

Proposition 4.7 (for case (1)) or 4.10 (for case (2)), applied to $B(C', M.\varepsilon)^8$, shows that property $\star_{B,Q,\ell}$ holds so long as

$$Q^\ell \leq a\varepsilon^{-b}, \tag{5.1}$$

where a, b depend only the isomorphism class of $\mathbf{G}$.

If Q satisfies this constraint, the previously quoted Propositions, in combination with Lemma 4.11, show that the number of primes that are not $\mathbf{T}$-good is $O(\log\varepsilon)$. (Here $\mathbf{T}$ is the torus occurring in the definition of $\star_{B,Q,\ell}$.)

This shows, in the notation of Lemma 5.2, that $|\mathcal{P}| \gg \frac{Q}{\log Q}$. Thus $\mu_\psi(xB(C,\varepsilon)) \ll Q^{-0.98}$. Choosing Q as large as allowable under (5.1) yields the desired result. □

6. The AQUE problem and the application of the entropy bound.

We now return to the AQUE problem discussed in the Introduction, recall our previous work on this problem, and explain how our main theorem concerning AQUE is deduced.

6.1. Quantum unique ergodicity on locally symmetric spaces.

PROBLEM 6.1. (QUE on locally symmetric spaces; Sarnak) Let G be a connected semi-simple Lie group with finite center. Let K be a maximal compact subgroup of G, $\Gamma < G$ a lattice, $X = \Gamma\backslash G$, $Y = \Gamma\backslash G/K$. Let $\{\psi_n\}_{n=1}^\infty \subset L^2(Y)$ be a sequence of normalized eigenfunctions of the ring of G-invariant differential operators on G/K, with the eigenvalues w.r.t. the Casimir operator tending to ∞ in absolute value. Is it true that $\bar{\mu}_{\psi_n} := |\psi_n|^2 d\mathrm{vol}$ converge weak-* to the normalized projection of the Haar measure to Y?

In the paper [**20**] we obtained Theorem 6.2, recalled below, constructing the microlocal lift in this setting. We needed to impose a non-degeneracy condition on the sequence of eigenfunctions (the assumption essentially amounts to asking that all eigenvalues tend to infinity, at the same rate for operators of the same order.) For the precise definition of *non-degenerate*, we refer to [**20**, Section 3.3].

With K and G as in Problem 6.1, let A be as in the Iwasawa decomposition $G = NAK$, i.e. $A = \exp(\mathfrak{a})$ where $\mathfrak{a}$ is a maximal abelian subspace of $\mathfrak{p}$. For

[8]instead of $B(C,\epsilon)$; it is easy to see the proof works verbatim even though $B(C', M.\varepsilon)$ need not be contained in Ω_∞

$G = \mathrm{PGL}_d(\mathbb{R})$ and $K = \mathrm{PO}_d(\mathbb{R})$, one may take A to be the subgroup of diagonal matrices with positive entries. Let $\pi\colon X \to Y$ be the projection. We denote by dx the G-invariant probability measure on X, and by dy the projection of this measure to Y.

THEOREM 6.2. *Let $\psi_n \subset L^2(Y)$ be a non-degenerate sequence of normalized eigenfunctions, whose eigenvalues approach ∞. Then, after replacing ψ_n by an appropriate subsequence, there exist functions $\tilde{\psi}_n \in L^2(X)$ and distributions μ_n on X such that:*

(1) *(Lift) The projection of μ_n to Y coincides with $\bar{\mu}_{\psi_n}$, i.e. $\pi_*\mu_n = \bar{\mu}_{\psi_n}$.*
(2) *Let σ_n be the measure $|\tilde{\psi}_n(x)|^2 dx$ on X. Then, for every $f \in C_c^\infty(X)$, we have $\lim_{n\to\infty}(\sigma_n(f) - \mu_n(f)) = 0$.*
(3) *(Invariance) Every weak-* limit σ_∞ of the measures σ_n (necessarily a positive measure of mass ≤ 1) is A-invariant.*
(4) *(Equivariance). Let $E \subset \mathrm{End}_G(C^\infty(X))$ be a $\mathbb{C}$-subalgebra of bounded endomorphisms of $C^\infty(X)$, commuting with the G-action. Noting that each $e \in E$ induces an endomorphism of $C^\infty(Y)$, suppose that ψ_n is an eigenfunction for E (i.e. $E\psi_n \subset \mathbb{C}\psi_n$). Then we may choose $\tilde{\psi}_n$ so that $\tilde{\psi}_n$ is an eigenfunction for E with the same eigenvalues as ψ_n, i.e. for all $e \in E$ there exists $\lambda_e \in \mathbb{C}$ such that $e\psi_n = \lambda_e\psi_n, e\tilde{\psi}_n = \lambda_e\tilde{\psi}_n$.*

We first remark that the distributions μ_n (resp. the measures σ_n) generalize the constructions of Zelditch (resp. Wolpert). Although, in view of (2), they carry roughly equivalent information, it is convenient to work with both simultaneously: the distributions μ_n are canonically defined and easier to manipulate algebraically, whereas the measures σ_n are patently positive and are central to the arguments of the present paper.

The existence of the microlocal lift already places a restriction on the possible weak-* limits of the measures $\{\bar{\mu}_n\}$ on Y. For example, the A-invariance of μ_∞ shows that the support of any weak-* limit measure $\bar{\mu}_\infty$ must be a union of maximal flats. Following Lindenstrauss, we term the weak-* limits σ_∞ of the lifts σ_n *quantum limits*.

More importantly, Theorem 6.2 allows us to pose a new version of the problem:

PROBLEM 6.3. (QUE on homogeneous spaces) In the setting of Problem 6.1, is the G-invariant measure on X the unique non-degenerate quantum limit?

The main result of this paper is the resolution of Problem 6.3 for certain higher rank symmetric spaces, in the context of *arithmetic* quantum limits. We refer to [**20**, Section 1.4] for a further discussion of the significance of these spaces and how the introduction of arithmetic helps to eliminate degeneracy.

6.2. Results: Arithmetic QUE for division algebra quotients. For brevity, we state the result in the language of automorphic forms; in particular, $\mathbb{A}$ is the ring of adèles of $\mathbb{Q}$.

Let $\mathbf{G}$ be a semisimple group over $\mathbb{Q}$, and let $G = \mathbf{G}(\mathbb{R})$. Let K_f be an open compact subgroup of $\mathbf{G}(\mathbb{A}_\mathrm{f})$ such that $X = \mathbf{G}(\mathbb{Q})\backslash\mathbf{G}(\mathbb{A})/K_\mathrm{f}$ consists of a single G-orbit (this condition is mainly cosmetic: see Remark 6.3.1 in §6.3). Then there exists a discrete subgroup $\Gamma < G$ such that $X = \Gamma\backslash G$. Let $\mathcal{H}$ be the Hecke algebra, as defined in Section 2. It acts on $L^2(X)$. Set $Y = \Gamma\backslash G/K$ the associated locally symmetric space, where K is a maximal compact subgroup inside G. A will denote a maximal $\mathbb{R}$-split torus in G compatible with K.

In the special case, let $D/\mathbb{Q}$ be a division algebra of prime degree d, and let $\mathbf{G}$ be the associated projective general linear group, i.e. the quotient of the group of units in D by its center. Assume that $\mathbf{G}$ is $\mathbb{R}$-split, i.e. $G = \mathbf{G}(\mathbb{R}) \simeq \mathrm{PGL}_d(\mathbb{R})$ (when $d \geq 3$ this is always the case). For this group let K be the standard maximal compact subgroup, A the group of diagonal matrices with positive entries (up to scaling).

Theorem 5.3 implies:

THEOREM 6.4. *Let $\tilde{\psi}_n \in L^2(X)$ be a sequence of $\mathcal{H}$-eigenfunctions on X such that the associated probability measures $\sigma_n := |\tilde{\psi}_n(x)|^2 dx$ on X converge weak-* to an A-invariant probability measure σ_∞. Then every* regular *$a \in A$ acts on every A-ergodic component of σ_∞ with positive entropy. When $\mathbf{G}$ is associated to a division algebra, the same holds for any $a \in A \setminus \{1\}$.*

PROOF. This is essentially a rephrasing of Theorem 5.3, where the uniformity of the estimate means it carries over to weak-* limits.

For a proof that the bound on measures of tubular neighbourhood of $Z_G(a)$ implies that $a \in A$ acts with positive entropy see [**10**, Sec. 8]. While written for the case of quaternion algebras ($d = 2$), that discussion readily generalizes to our situation by modifying its "Step 2" to account for the action of a on the Lie algebra. □

Using results on measure-rigidity due to Einsiedler and Katok, this has the following implication for the QUE problem:

THEOREM 6.5. *Let $\mathbf{G}$ be the projectivized unit group of a division algebra of prime degree, and maintain the other notations as above. Let $\{\psi_n\}_{n=1}^{\infty} \subset L^2(Y)$ be a non-degenerate sequence of eigenfunctions for the ring of G-invariant differential operators on G/K (cf.* [**20**, Sec. 3.3]*) which are also eigenforms of the Hecke algebra $\mathcal{H}$ (cf. Section 2). Such ψ_n are also called Hecke-Maass forms.*

Then the associated probability measures $\bar{\mu}_n$ converge weak- to the normalized Haar measure on Y, and their lifts μ_n (see Theorem 6.2) converge weak-* to the normalized Haar measure dx on $X = \Gamma\backslash \mathrm{PGL}_d(\mathbb{R})$.*

PROOF. The case $d = 2$ is Lindenstrauss's theorem, and we will thus assume $d \geq 3$.

Passing to a subsequence, let $\psi_n \in L^2(Y)$ be a non-degenerate sequence of Hecke-Maass forms on Y such that $\bar{\mu}_n \to \bar{\mu}_\infty$ weakly. Passing to a subsequence, let $\tilde{\psi}_n$ and σ_n be as in Theorem 6.2 such that $\sigma_n \to \sigma_\infty$ weakly and σ_∞ lifts $\bar{\mu}_\infty$. Then σ_∞ is a non-degenerate arithmetic quantum limit on X. By Theorem 6.4, σ_∞ is an A-invariant probability measure on X such that every $a \in A \setminus \{1\}$ acts on almost every A-ergodic component of σ_∞ with positive entropy. Remark also that the measure σ_∞ is invariant under the finite group $Z_K(A)$, the centralizer of A in K, by construction (see [**20**, Remark 1.7, (5)]).

Then [**5**, Thm. 4.1(iv)] shows that σ_∞ has a unique ergodic component, μ_{Haar}. □

Our methods also apply to the case of a split central simple $\mathbb{Q}$-algebra, that is when $\mathbf{G}(\mathbb{Q}) = \mathrm{PGL}_d(\mathbb{Q})$. The result is somewhat weaker, however:

THEOREM 6.6. *Let $G = \mathrm{PGL}_d(\mathbb{R})$ (d prime), and let $\Gamma < G$ be a lattice of the form* [9] $\mathrm{PGL}_d(\mathbb{Z}) \cap \gamma \, \mathrm{PGL}_d(\mathbb{Z}) \gamma^{-1}$ *for some $\gamma \in \mathrm{PGL}_d(\mathbb{Q})$. Let $X = \Gamma \backslash G$, $Y = X/K$ Let $\{\psi_n\}_{n=1}^{\infty} \subset L^2(Y)$ be a non-degenerate sequence of Hecke-Maass forms.*

Then the associated probability measures $\bar{\mu}_n$ converge weak- to the a Haar measure on Y, and their lifts μ_n (see Theorem 6.2) converge weak-* to a Haar measure cdx on $X = \Gamma \backslash \mathrm{PGL}_d(\mathbb{R})$, where $c \in [0,1]$.*

PROOF. Let $\mathbf{G} = \mathrm{PGL}_d / \mathbb{Q}$. For any prime p let $\mathcal{O}_p$ be the ("Eichler") order $M_n(\mathbb{Z}_p) \cap \gamma M_n(\mathbb{Z}_p) \gamma^{-1}$ of $M_n(\mathbb{Q}_p)$. Let $K_p = \mathcal{O}_p^{\times}$, $K_{\mathrm{f}} = \prod_p K_p$. Then $\Gamma = \mathrm{PGL}_d(\mathbb{Q}) \cap K_{\mathrm{f}}$, so that $X = \mathbf{G}(\mathbb{Q}) \backslash \mathbf{G}(\mathbb{A}) / K_{\mathrm{f}}$.

Passing to a subsequence, let μ be a weak-* limit of a sequence of lifts. Theorem 6.4 shows that there exist $a \in A$ which act with positive entropy on almost every A-ergodic component of μ. The measure rigidity results of [**6**] together with the orbit classification results of [**12**] (we use here the fact that d is prime) show that μ is a Haar measure on X. Since X is not compact, this method does not control the total mass c of μ. □

6.3. Remarks on generalizations.

6.3.1. *Class number one.* The assumption imposed that G act with a single orbit in $\mathbf{G}(\mathbb{Q}) \backslash \mathbf{G}(\mathbb{A}) / K_{\mathrm{f}}$ is, as we remarked, cosmetic. In general, if we remove this assumption, one would still know – making analogous definitions – that quantum limits remain G-invariant. However, this would not quite be a complete answer since the space of G-invariant measures on $\mathbf{G}(\mathbb{Q}) \backslash \mathbf{G}(\mathbb{A}) / K_{\mathrm{f}}$ is now finite dimensional, and we would not know the relative measures of the different components.

6.3.2. *Nondegeneracy.* The second author has obtained a version of Theorem 6.2 without the non-degeneracy assumption, see [**19**]. In that case the lifts are asymptotically invariant under (non-trivial) subgroups of A. The bounds on the mass of tubes obtained in this paper are, at their foundation, purely statements about Hecke eigenfunctions, and thus carry over to degenerate limits. However, the interpretation of these bounds as lower bounds on the *entropy* only applies to tubes associated to elements $a \in A$ by which the measure is invariant.

In the degenerate case, a-priori one only has invariance by *singular* elements $a \in A$. Thus our methods only show that those elements act with positive entropy in the case of division algebras.

An analogue of Theorem 6.5 would accordingly follow from a result classifying measures on X which are only invariant by a (potentially one-dimensional) subgroup of the Cartan subgroup, assuming a Hecke recurrence condition a-la [**11**]. An advance in this direction is necessary in order to show that every sequence of Hecke-Maass forms is equidistributed; in the rest of the remarks we only consider the case of non-degenerate limits.

It is worth noting that the mass of tubes correpsonding to the flow of a regular element would also be small, but since the measure is not invariant by the regular element it is not clear how to incorporate this information into the measure-classification result.

6.3.3. *Escape of mass.* When the quotient X is not compact (for example, in the split case of Theorem 6.6), there is an additional potential obstruction to equidistribution: weak-* limits are not necessarily probability measures – they may even be the zero measure. This possibility is known as "escape-of-mass".

[9] Again, the result is best expressed for an "adelic" quotient

In the case of congruence lattices in $SL_2(\mathbb{Z})$, escape-of-mass was ruled out by Soundararajan [**22**]. This was generalized to the congruence lattices in $SL_2(\mathcal{O})$, $\mathcal{O}$ the ring of integers of a number field, in the M.Sc. Thesis [**25**]. The extension to higher-rank groups is the subject of current research.

In the particular case of congruence lattices in $GL_d(\mathbb{Z})$ the normalization of the measure is already controlled by the degenerate Eisenstein series. Hence a sub-convexity result for the Rankin–Selberg L-function would control the escape, just as in the better-known case of GL_2 (though, notably, Soudararajan's argument does not rely on an L-function bound). In fact, such a sub-convexity result would also prove that the limits have *positive entropy* (this is demonstrated in [**8**]).

In the rest of our remarks we ignore this issue as well; the reader may assume the group $\mathbf{G}$ to be anisotropic.

6.3.4. *The case when* $\mathbf{G}$ *is not associated to a division algebra.* We expect the techniques developed for the proof of Theorem 6.5 will generalize at least to some other locally symmetric spaces, the case of division algebras of prime degree being the simplest; but there are considerable obstacles to obtaining a theorem for *any* arithmetic locally symmetric space at present. A brief discussion of some of these difficulties follows.

First, the intersection patters of Hecke translates will be controlled by subgroups more complicated than tori. Except for the simplest case, in general the best one can hope for is that intersections be controlled by Levi subgroups defined over $\mathbb{Q}$. Lemma 4.9 already establishes this for unit groups of semisimple $\mathbb{Q}$-algebras. Such subgroups will have *exponential* volume growth (say in terms of their orbit on the building of $\mathbf{G}(\mathbb{Q}_p)$), compared with the polynomial behaviour of tori. Even the purely local question of whether an eigenfunction on a building can concentrate appreciably on the orbit of such a large subgroup is difficult. To see where such an issue can arise note that when G is not $\mathbb{R}$-split, even the centralizer of the maximal $\mathbb{R}$-split torus is not a torus.

Dealing with intersections created by larger $\mathbb{Q}$-subgroups is essential for a more fundamental reason. The best possible outcome of the type of measure classification results one would use here (for state of the art see [**7**]) is that the measure is, in some sense, *algebraic*: it is a linear combination of measures supported on orbits of subgroups. From this point of view, to show that the limit measure is the G-invariant measure should at least require showing that the mass of orbits of these subgroups is zero a-la Rudnick–Sarnak. In our terms, this means showing that the mass concentrated in an an ϵ-neighbourhood of the orbit goes to zero with ϵ, even if it is not necessary to achieve power-law decay of the mass (i.e. positive entropy). In fact, for this reason it is hard to imagine an application of the current techniques that would rule out these intermediate measures without also establishing that all elements of A act with positive entropy.

Appendix A. Proof of Lemma A.1: how to construct a higher rank amplifier

To readers familiar with the usage of "amplification" in analytic number theory (as represented, for instance, in the work of Duke, Friedlander and Iwaniec): the Lemma A.1 in effect represents a way to construct an amplifier in higher rank.

Let $\mathbf{G}$ be a semisimple algebraic group over $\mathbb{Q}$, $\rho\colon \mathbf{G} \to \mathrm{SL}_N$ an embedding. For each prime p let $G_p := \mathbf{G}(\mathbb{Q}_p)$. There is p_0 such that for $p > p_0$, $K_p = \rho^{-1}(\mathrm{SL}_N(\mathbb{Z}_p))$ is a special maximal compact subgroup of the unramified group G_p.

We will allow the implicit constant in the symbols $\gg, \ll$ to depend on N (and hence, also on dimension of $\mathbf{G}$), but not on anything else.

LEMMA A.1. *Possibly increasing p_0 there exist integers ℓ, ℓ' depending only on N such that for any $p > p_0$ and for any character Λ of the Hecke algebra of G_p with respect to K_p, there is a K_p-bi-invariant function h_p with the following properties:*

(1) *Its support satisfies $p^{\ell'} \gg \#\mathrm{supp}(h_p) \gg p$, where we think of h_p as a function on G_p/K_p;*
(2) *$|h_p| \in \{0, 1\}$;*
(3) *$\Lambda(h_p)$ is positive and $\Lambda(h_p) \gg \#\mathrm{supp}(h_p)^{1/2}$*
(4) *For any $g \in \mathrm{supp}(h_p)$, the denominator of $\rho(g)$ is $\ll_N p^{\ell}$.*

A.1. Notation on p-adic groups. We shall use certain standard properties of semisimple algbraic groups over p-adic fields. Standard references are [**23**] and [**3**].

Increasing p_0 we may suppose that G_p is quasi-split and unramified, and that K_p is hyperspecial. Let $\mathbf{A}$ be a maximal $\mathbb{Q}_p$-split torus in G_p so that the corresponding apartment in the building of G_p contains the point fixed by K_p, and let $A_p = \mathbf{A}_p(\mathbb{Q}_p)$ be the corresponding subgroup of G_p.

Let $X_* = \mathrm{Hom}(\mathbb{G}_m, \mathbf{A})$ and $X^* = \mathrm{Hom}(\mathbf{A}, \mathbf{G}_m)$. Let $\Phi \subset X^*$ be the set of roots for the action of $\mathbf{A}$ on the Lie algebra of $\mathbf{G}$. We note that, as p varies, there will be at most finitely many distinct root systems. In particular, our bounds may depend on Φ.

Fix a positive system of roots for A_p, let N_p be the subgroup corresponding to all the positive roots. We have Iwasawa decomposition $G_p = N_p.A_p.K_p$. Let $\delta : A_p \to \mathbb{R}^\times$ be the character corresponding to the half-sum of positive roots, composed with $\|\cdot\|_p$ on $\mathbb{Q}_p$.

Let $\mathfrak{a} := A_p/(A_p \cap K_p)$, a free abelian group of rank equal to the rank of $\mathbf{G}(\mathbb{Q}_p)$. Then $\mathfrak{a}$ is identified with X_*: for, given $a \in \mathfrak{a}$, there exists a unique homomorphism $\theta : \mathbb{G}_m \to \mathbf{A}_p$ so that $\theta(p)$ and a lie in the same $A_p \cap K_p$ coset.

Next, let $V = X_* \otimes_{\mathbb{Z}} \mathbb{R}$, $V^* = X^* \otimes_{\mathbb{Z}} \mathbb{R}$, and $V^*_{\mathbb{C}} = V^* \otimes_{\mathbb{R}} \mathbb{C}$. Then to any $\nu \in V^*_{\mathbb{C}}$, and any $\theta \in X_*$ with $a = \theta(p)$ we set $a^\nu = p^{\langle \theta, \nu \rangle}$ with the obvious pairing. In particular this gives an identification between the unramified characters of A_p and the torus $\mathfrak{a}^*_{\mathrm{temp}} = iV^*/(2\pi i \log p X^*)$.

Let W be the Weyl group of A_p. It acts on $\mathfrak{a}$. Moreover, we fix a W-invariant inner product on $\mathfrak{a} \otimes \mathbb{R}$ in the following way: the elements of Φ, considered as belonging to $\mathrm{Hom}(\mathfrak{a}, \mathbb{Z})$, define elements of a root system. We require the longest root for each simple factor of $\mathbf{G}$ to have length 1. This uniquely normalizes a W-invariant inner product on the dual to $\mathfrak{a} \otimes \mathbb{R}$, so also on $\mathfrak{a} \otimes \mathbb{R}$.

This normalization has the following property: if $\alpha \in \Phi$ is any root and $a \in \mathfrak{a}$, then "$|\alpha(a)|$" – we implicitly identify a with an element of X_* – is bounded above by $p^{\|a\|}$.

Finally, let $M_p = Z_{K_p}(A_p)$ so that $N_p A_p M_p$ is a Borel subgroup of G_p.

A.2. Plancherel formula. To any character $\nu : \mathfrak{a} \to \mathbb{C}^\times$ (parametrized by an element of $V^*/(2\pi i \log p X^*)$), we associate the spherical representation[10] $\pi(\nu)$ of G_p obtained by extending $\nu\delta$ to $N_p A_p M_p$ trivially on $N_p M_p$, inducing to G_p, and taking the unique spherical subquotient.

For any K_p-bi-invariant function k on G_p, let $\hat{k}(\nu)$ be the scalar by which k acts on the spherical vector in $\pi(\nu)$. There is a unique K_p-bi-invariant function Ξ_ν on G_p (the spherical function with parameter ν) so that:

$$\hat{k}(\nu) = \int_{g\in G_p} k(g)\Xi_\nu(g)dg, \quad k(g) = \int_{\nu\in\mathfrak{a}^*_{\mathrm{temp}}} \hat{k}(\nu)\Xi_\nu(g)d\mu(\nu). \tag{A.1}$$

where the first integral is taken w.r.t. the Haar measure that assigns mass 1 to K_p, and the second integral is taken w.r.t. the Plancherel measure μ on $\mathfrak{a}^*_{\mathrm{temp}}$. In our normalization, μ is a probability measure.

The map $k \mapsto \hat{k}(\nu)$ is an isomorphism between the space of compactly supported, K_p-bi-invariant functions on $\mathbf{G}(\mathbb{Q}_p)$, and the space of W-invariant "trigonometric polynomials" on $\mathfrak{a}$; here "trigonometric polynomial" means "finite linear combination of characters." Also $k \mapsto \hat{k}$ is an isometry:

$$\int_{\mathbf{G}(\mathbb{Q}_p)} |k(g)|^2 dg = \int_{\nu\in\mathfrak{a}^*_{\mathrm{temp}}} |\hat{k}(\nu)|^2 d\mu(\nu) \tag{A.2}$$

The explicit form of the Plancherel measure is known [**13**]. From it we extract the following fact: "$\mu_p = \mu_\infty + O(p^{-1/2})$." More precisely, there exists a measure μ_∞ on $V^*/2\pi X^*$ (whose Fourier transform is supported in $\{X \in \mathfrak{a} : \|X\| \le 3\}$) such that the difference $\mu_p - \mu_\infty$ is a signed measure represented by a function of supremum norm $O(p^{-1/2})$. Here to consider μ_∞ as a measure on $\mathfrak{a}^*_{\mathrm{temp}}$ we identify with our reference torus $V^*/2\pi X^*$ via rescaling by $\log p$.

A.3. The Paley-Wiener theorem. Recall our fixed W-invariant inner product on $\mathfrak{a}$. The Paley-Wiener theorem asserts that under the transform $k \mapsto \hat{k}$, the preimage of characters of $\mathfrak{a}^*_{\mathrm{temp}}$ supported in $\{X \in \mathfrak{a} : \|X\| \le R\}$ is contained in functions supported in $K_p.\{X \in \mathfrak{a} : \|X\| \le R\}.K_p$

Let us briefly sketch the proof. Let $\mathfrak{a}^+$ be the closed positive Weyl chamber within $\mathfrak{a}$. For $\alpha, \beta \in \mathfrak{a}^+$, so that $\nu \mapsto \nu(\beta)$ occurs with a nonzero coefficient in $\widehat{K_p\alpha K_p}$, then, necessarily, $\alpha - \beta$ belongs to the dual cone to $\mathfrak{a}^+$.

Now, choose $\alpha \in \mathfrak{a}^+$ so that $k(K_p\alpha K_p) \neq 0$ and $\|\alpha\|$ is maximal subject to that restriction. We claim that $\nu \mapsto \nu(\alpha)$ necessarily occurs in $\hat{k}$. For, in view of the remarks above, if this were not the case there must exist $K_p\beta K_p$ in the support of k, with $\beta \in \mathfrak{a}^+$ and $\|\alpha\| < \|\beta\|$. This is a contradiction.

A.4. Proof of the amplification lemma. Let $\nu_0 \in \mathfrak{a}^*$ be the parameter of the character Λ of the Hecke algebra specified in the Lemma. We do not, of course, assume that $\nu_0 \in \mathfrak{a}^*_{\mathrm{temp}}$).

Take any $a \in \mathfrak{a}$ with a not in the support of the Fourier transform of μ_∞, but $\|a\|$ reasonably small. For example, we could take a to be twice the coroot associated

[10] Recall a spherical (irreducible) representation of G_p is one that possesses a one dimensional space of K_p-invariants

to any root of maximal length; then $\|a\| = 4$. Take $R = |W|\|a\|$. Construct the function k_1 with spherical transform

$$\hat{k}_1(\nu) = \sum_{|j|\le|W|} \overline{\sum_{w\in W} w\nu_0(ja)} \sum_{w\in W} w\nu(ja) \tag{A.3}$$

Note that if $\alpha_1, \ldots, \alpha_m$ are any nonzero complex numbers, then by a simple compactness argument.

$$\max_{j\neq 0, |j|\le m} \left|\alpha_1^j + \cdots + \alpha_m^j\right| \ge c(m) > 0 \tag{A.4}$$

So $M := \sum_{|j|\le|W|} \left|\sum_{w\in W} w\nu_0(ja)\right|^2 \gg 1$. We have $\sup_{\nu\in\mathfrak{a}^*_{\text{temp}}} |\hat{k}_1(\nu)| \ll M^{1/2}$. Thus, by (A.2), $\|k_1\|_{L^2}^2 \ll M$; by definition, $\hat{k}_1(\nu_0) = M$; and from the explicit form of Plancherel measure and (A.1), $|k_1(1)| = O(M^{1/2}p^{-1/2})$.

Put $k = k_1 - k_1(1)1_{K_p}$. Then $k(1) = 0$ and – if we suppose that the residue field size, p, is sufficiently large, as we may – $|\hat{k}(\nu_0)| \gg \|k\|_{L^2}$.

By the Paley-Wiener theorem, k is supported in $\bigcup_{|a|\le R} K_p a K_p$. The number of K_p-double cosets in this set is equal to the number of $a \in \mathfrak{a}$ with $|a| \le R$, which is, in turn $O(1)$. (a is completely determined by the value of all the roots on it; but the number of roots is $O(1)$ and each value is an integer $\le R$). We conclude that there is $|a| \le R$ so that

$$|\widehat{1_{K_paK_p}}(\nu_0)|^2 \gg \int_{K_paK_p} dg. \tag{A.5}$$

On the other hand, it is known ([**3**, Section 3.5]) that

$$\int_{K_paK_p} dg \asymp \delta(a)^2 \tag{A.6}$$

the notation $\asymp$ meaning that the quotient is bounded above and below, at least for $p \ge p_0(\dim \mathbf{G})$. Since δ is the half-sum of positive roots, and each root has length ≤ 1, we see that $p \le \delta(a)^2 \le p^{\dim(\mathbf{G})R}$. (cf. the end of §A.1.)

We take h_p to be the multiple $1_{K_pa_iK_p}$ by a suitable complex number of absolute value 1, so that $\hat{h}_p(\nu_0)$ is positive. The first three assertions of the lemma follow from remarks already made.

We now turn to establishing the necessary bounds on the denominator of $\rho(g)$. Let $\Phi_\rho \subset X^*$ be the weights of the representation of ρ with reference to $\mathbf{A}$; for each $\alpha \in \Phi_\rho$, let $V(\alpha)$ be the weight space. We require two preliminary observations:

Firstly, $\sup_{\alpha\in\Phi_\rho} |\alpha| \le N$. To verify this we may suppose that the representation ρ is irreducible. Because $\mathbf{G}$ is semisimple, we may choose $w \in W$ so that $\|w\alpha - \alpha\| \ge \|\alpha\|$. Call two elements of X^* *connected* if they differ by an element of Φ. There is a sequence $\alpha = \alpha_0, \alpha_1, \ldots, \alpha_r = w\alpha$ where α_i, α_{i+1} are connected and $\alpha_i \in \Phi_\rho$ for all i. So $\dim(V) \ge r + 1 \ge \|w\alpha - \alpha\| \ge \|\alpha\|$.

Secondly, $\mathbb{Z}_p^N = \oplus_\alpha \mathbb{Z}_p^N \cap V(\alpha)$ so long as the restriction map from characters of X^* to $\mathrm{Hom}(A_p \cap K_p, \mathbb{Z}_p^\times)$ is injective. For, this being the case, the spaces $\mathbb{Z}_p^N \cap V(\alpha)$ are characterized as the "eigenspaces" of a prime-to-p finite group acting on $\mathbb{Z}_p^N$. This is so, in particular, if $p \ge p_0(N)$.

Combining these remarks, we see at once that the denominator of $\rho(a)$ is $\le p^{NR}$ when $a \in A_p$ projects to the ball of radius $\le R$ in $\mathfrak{a}$. The desired bound on denominators follows. □

Acknowledgments

This paper owes a tremendous debt both to Peter Sarnak and Elon Lindenstrauss. It was Sarnak's realization, developed throughout the 1990s, that the quantum unique ergodicity problem on arithmetic quotients was a question that had interesting structure and interesting links to the theory of L-functions: it was Lindenstrauss' paper [**11**] which introduced ergodic-theoretic methods in a decisive way. Peter and Elon have both given us many ideas and comments over the course of this work, and it is a pleasure to thank them.

References

[1] Jean Bourgain and Elon Lindenstrauss, *Entropy of quantum limits*, Comm. Math. Phys. **233** (2003), no. 1, 153–171, DOI 10.1007/s00220-002-0770-8. MR1957735

[2] Shimon Brooks and Elon Lindenstrauss, *Joint quasimodes, positive entropy, and quantum unique ergodicity*, Invent. Math. **198** (2014), no. 1, 219–259, DOI 10.1007/s00222-014-0502-7. MR3260861

[3] P. Cartier, *Representations of p-adic groups: a survey*, Automorphic forms, representations and L-functions (Proc. Sympos. Pure Math., Oregon State Univ., Corvallis, Ore., 1977), Proc. Sympos. Pure Math., XXXIII, Amer. Math. Soc., Providence, R.I., 1979, pp. 111–155. MR546593

[4] Y. Colin de Verdière, *Ergodicité et fonctions propres du laplacien* (French, with English summary), Comm. Math. Phys. **102** (1985), no. 3, 497–502. MR818831

[5] Manfred Einsiedler and Anatole Katok, *Invariant measures on G/Γ for split simple Lie groups G*, Comm. Pure Appl. Math. **56** (2003), no. 8, 1184–1221, DOI 10.1002/cpa.10092. Dedicated to the memory of Jürgen K. Moser. MR1989231

[6] Manfred Einsiedler, Anatole Katok, and Elon Lindenstrauss, *Invariant measures and the set of exceptions to Littlewood's conjecture*, Ann. of Math. (2) **164** (2006), no. 2, 513–560, DOI 10.4007/annals.2006.164.513. MR2247967

[7] Manfred Einsiedler and Elon Lindenstrauss, *On measures invariant under tori on quotients of semisimple groups*, Ann. of Math. (2) **181** (2015), no. 3, 993–1031, DOI 10.4007/annals.2015.181.3.3. MR3296819

[8] Manfred Einsiedler, Elon Lindenstrauss, Philippe Michel, and Akshay Venkatesh, *Distribution of periodic torus orbits and Duke's theorem for cubic fields*, Ann. of Math. (2) **173** (2011), no. 2, 815–885, DOI 10.4007/annals.2011.173.2.5. MR2776363

[9] Elon Lindenstrauss, *On quantum unique ergodicity for $\Gamma\backslash\mathbb{H}\times\mathbb{H}$*, Internat. Math. Res. Notices **17** (2001), 913–933, DOI 10.1155/S1073792801000459. MR1859345

[10] Elon Lindenstrauss, *Adelic dynamics and arithmetic quantum unique ergodicity*, Current developments in mathematics, 2004, Int. Press, Somerville, MA, 2006, pp. 111–139. MR2459293

[11] Elon Lindenstrauss, *Invariant measures and arithmetic quantum unique ergodicity*, Ann. of Math. (2) **163** (2006), no. 1, 165–219, DOI 10.4007/annals.2006.163.165. MR2195133

[12] Elon Lindenstrauss and Barak Weiss, *On sets invariant under the action of the diagonal group*, Ergodic Theory Dynam. Systems **21** (2001), no. 5, 1481–1500, DOI 10.1017/S0143385701001717. MR1855843

[13] I. G. Macdonald, *Spherical functions on a group of p-adic type*, Ramanujan Institute, Centre for Advanced Study in Mathematics,University of Madras, Madras, 1971. Publications of the Ramanujan Institute, No. 2. MR0435301

[14] Simon Marshall, *Sup norms of maass forms on semisimple groups*, preprint available at `arXiv:math.NT/1405.7033`, 2014.

[15] Djordje Milićević, *Large values of eigenfunctions on arithmetic hyperbolic surfaces*, Duke Math. J. **155** (2010), no. 2, 365–401, DOI 10.1215/00127094-2010-058. MR2736169

[16] Djordje Milićević, *Large values of eigenfunctions on arithmetic hyperbolic 3-manifolds*, Geom. Funct. Anal. **21** (2011), no. 6, 1375–1418, DOI 10.1007/s00039-011-0144-5. MR2860192

[17] Zeév Rudnick and Peter Sarnak, *The behaviour of eigenstates of arithmetic hyperbolic manifolds*, Comm. Math. Phys. **161** (1994), no. 1, 195–213. MR1266075

[18] Lior Silberman, *Arithemtic quantum chaos on locally symmetric spaces*, Ph.D. thesis, Princeton University, 2005, available at `http://www.math.ubc.ca/~lior/work/`.

[19] Lior Silberman, *Quantum unique ergodicity on locally symmetric spaces: the degenerate lift*, Canad. Math. Bull. **58** (2015), no. 3, 632–650, DOI 10.4153/CMB-2015-023-0. MR3372878

[20] Lior Silberman and Akshay Venkatesh, *On quantum unique ergodicity for locally symmetric spaces*, Geom. Funct. Anal. **17** (2007), no. 3, 960–998, DOI 10.1007/s00039-007-0611-1. MR2346281

[21] A. I. Šnirel′man, *Ergodic properties of eigenfunctions* (Russian), Uspehi Mat. Nauk **29** (1974), no. 6(180), 181–182. MR0402834

[22] Kannan Soundararajan, *Quantum unique ergodicity for* $\mathrm{SL}_2(\mathbb{Z})\backslash\mathbb{H}$, Ann. of Math. (2) **172** (2010), no. 2, 1529–1538. MR2680500

[23] J. Tits, *Reductive groups over local fields*, Automorphic forms, representations and L-functions (Proc. Sympos. Pure Math., Oregon State Univ., Corvallis, Ore., 1977), Proc. Sympos. Pure Math., XXXIII, Amer. Math. Soc., Providence, R.I., 1979, pp. 29–69. MR546588

[24] Scott A. Wolpert, *Semiclassical limits for the hyperbolic plane*, Duke Math. J. **108** (2001), no. 3, 449–509, DOI 10.1215/S0012-7094-01-10833-8. MR1838659

[25] Asif Zaman, *Escape of mass on hilbert modular varieties*, Master's thesis, The University of British Columbia, 2012, M.Sc. thesis.

[26] Steven Zelditch, *Uniform distribution of eigenfunctions on compact hyperbolic surfaces*, Duke Math. J. **55** (1987), no. 4, 919–941, DOI 10.1215/S0012-7094-87-05546-3. MR916129

[27] Steven Zelditch, *The averaging method and ergodic theory for pseudo-differential operators on compact hyperbolic surfaces*, J. Funct. Anal. **82** (1989), no. 1, 38–68, DOI 10.1016/0022-1236(89)90091-8. MR976312

Lior Silberman, Department of Mathematics, University of British Columbia, Vancouver BC V6T 1Z2, Canada.
Email address: `lior@math.ubc.edu`

Akshay Venkatesh, Department of Mathematics, Stanford University, Stanford, California, 94035
Email address: `akshay@math.stanford.edu`